Genetics
and
Biochemistry
of
Pseudomonas

Genetics
and
Biochemistry
of
Pseudomonas

Edited by

P. H. Clarke

Department of Biochemistry,
University College London

and

M. H. Richmond

Department of Bacteriology,
University of Bristol

A Wiley–Interscience Publication

JOHN WILEY & SONS

LONDON · NEW YORK · SYDNEY · TORONTO

Copyright © 1975, by John Wiley & Sons Ltd.

Library of Congress Cataloging in Publication Data:

Clarke, Patricia H.
Genetics and biochemistry of pseudomonas

"A Wiley-Interscience publication."
1. Pseudomonas. 2. Bacterial genetics. 3. Microbial
metabolites. I. Richmond, Mark H., joint author.
II. Title.

QR82.P78C55   589.9′5   73–18926

ISBN 0 471 15896 8

Printed in Great Britain by J. W. Arrowsmith Ltd.,
Winterstoke Road, Bristol.

Contributors

CLARKE, PATRICIA H.   *Department of Biochemistry, University College London, Gower Street, London WC1E 6BT. England*

HOLLOWAY, BRUCE W.   *Department of Genetics, Monash University, Clayton, Victoria 3168, Australia*

KRISHNAPILLAI, VIJI.   *Department of Genetics, Monash University, Clayton, Victoria 3168, Australia*

LOWBURY, EDWARD J. L.   *M.R.C. Industrial Injuries & Burns Unit, Birmingham Accident Hospital, Bath Row, Birmingham, B15 1NA. England*

MEADOW, PAULINE.   *Department of Biochemistry, University College London, Gower Street, London, WC1E 6BT. England*

ORNSTON, L. NICHOLAS.   *Department of Biology, Yale University, New Haven, Connecticut, 06520. U.S.A.*

PALLERONI, NORBERTO J.   *Hoffman-La Roche Research Division, Nutley, New Jersey, 07110. U.S.A.*

RICHMOND, MARK H.   *Department of Bacteriology, University of Bristol, University Walk, Bristol BS8 1TD. England*

STANISICH, VILMA A.   *Department of Bacteriology, University of Bristol, University Walk, Bristol, BS8 1TD. England*

Preface

Over the years microbial biochemists have been challenged by the diversity of the metabolic reactions catalysed by members of the genus *Pseudomonas* and consequently their investigation of members of this genus has played a vital part in building up our knowledge of bacterial intermediary metabolism, particularly with respect to possible degradative routes. Latterly, interest in the pseudomonads has become much wider. Not only has the relative eclipse of Gram-positive bacteria as human pathogens revealed the presence of the so-called opportunistic pathogens (of which *Pseudomonas aeruginosa* is a particularly vicious example), but also pseudomonads have assumed a steadily increasing importance in industrial microbiology. Much of our knowledge of modern molecular biology has come from studies on *Escherichia coli*, but this organism is far from universally representative. It is to try to define what is known of another group of bacterial species of potential practical importance that these contributions have been collected and published in this form.

This book is intended for advanced students and research workers in the fields of microbiology, biochemistry and molecular biology. *Pseudomonas aeruginosa*, the species that receives the most detailed treatment in this book, is a human pathogen of particular concern in burns. Indeed, observations reported by E. J. L. Lowbury of newly acquired carbenicillin resistance in *P. aeruginosa* strains isolated in the Birmingham Accident Hospital, led to the discovery of an important class of bacterial plasmids that can be freely transmitted across species and even generic boundaries. This development is a good example of the benefits that flow from collaboration between medical and molecular microbiologists. The clinical occurrence of *P. aeruginosa* is discussed by E. J. L. Lowbury in Chapter 2, while the more general implications of the class of drug resistance plasmids found in *Pseudomonas aeruginosa* is discussed with reference to the genetic structure of this species in later chapters.

It was fortunate that, at a time when the attention of almost every other microbial geneticist was directed towards *Escherichia coli* and its bacterio-phages, B. W. Holloway started investigating the exchange of genetic information in *P. aeruginosa*. In Chapter 4 he and V. Krishnapillai describe the bacteriophages which infect *P. aeruginosa*, and also the aeruginocins which are specific bacteriocins produced by strains of this species. Genetic material can be transferred by conjugation and also by bacteriophages. Both conjugation and transduction have been widely used for studies on gene linkage and chromosome mapping. In Chapter 5 B. W. Holloway discusses the organization of genetic material in *Pseudomonas* and compares the arrangement of genes on the chromosome with that found in *Escherichia coli*. Genetic analysis has also been carried out with *P. putida*, and in so far as comparisons have been made between the two species, the linkage structures of genes for biosynthetic pathways are similar to those of *P. aeruginosa*. The most remarkable finding for *P. putida* was that some of the genes for catabolic pathways may be carried on transmissible plasmids which may be of considerable size. The details and implications of gene transfer are discussed in Chapter 6 by M. H. Richmond and V. A. Stanisich.

Among the species included in the genus *Pseudomonas* are some plant pathogens, and a wide variety of strains that can be isolated from soil and water. In Chapter 1, N. J. Palleroni explains that methods traditionally used by bacteriologists for the Enterobacteriaceae are inadequate for classifying *Pseudomonas* strains and discusses comparative studies of metabolic pathways, and recent work on protein and nucleic acid homologies. In Chapter 3, P. M. Meadow discusses the membranes and cell wall structures of *P. aeruginosa* and compares them with other *Pseudomonas* species.

Microbiologists interested in biochemical pathways have exploited the enormous biochemical versatility of *Pseudomonas* for a very long time. The majority of this work has been carried out with *P. aeruginosa* and *P. putida* and this has made it possible to combine genetic and biochemical observations, but some of the most valuable contributions to an understanding of biochemical pathways have come from studies with a number of other *Pseudomonas* species. The work of R. Y. Stanier and his colleagues at Berkeley, and of I. C. Gunsalus and his team at Urbana, has been outstanding in this respect and has laid the foundations on which much of our current knowledge of the biochemistry of *Pseudomonas* now rests. In Chapters 7 and 8, P. H. Clarke and L. N. Ornston describe the biosynthetic and catabolic pathways of pseudomonads and discuss the evolutionary implications of the regulatory patterns which have developed in this genus.

M. H. Richmond and P. H. Clarke speculate on the future evolution of *Pseudomonas* and discuss the possibilities with respect to contributions which might be made to fundamental studies on molecular biology, and also to practical problems such as the biodegradation of man-made organic molecules that have been introduced into the natural environment.

In the last few years the literature on the genetics and biochemistry of *Pseudomonas* species has expanded steadily. We are now able to present a fairly detailed picture of many aspects of this extremely interesting group of bacteria; we feel that this is a suitable moment to introduce them to a wider audience.

The authors are grateful to many of their colleagues who have given advice and valuable criticisms of the manuscripts and have generously provided research results in advance of publication. They are also grateful to those bodies who have been generous in supporting their own research. These include the Australian Research Grants Committee (B.W.H., V.K., V.A.S.), the Medical Research Council (P.H.C., P.M.M., M.H.R.), the Science Research Council (P.H.C.), the National Science Foundation of the United States and National Institutes of Health, U.S.A. (L.N.O.).

P.H.C.
M.H.R.

Contents

CHAPTER 1

General Properties and Taxonomy of the Genus *Pseudomonas*

NORBERTO J. PALLERONI

1. INTRODUCTION

The aerobic pseudomonads are bacteria of considerable scientific and practical importance. They are among the most active participants in the process of mineralization of organic matter in nature, a role that can be easily inferred from their widespread occurrence in soil and water, and from the fact that many members of the group are endowed with the capacity of attacking a large variety of organic compounds. This catabolic versatility, combined with a fast rate of growth in simple media frequently has resulted in the isolation of pseudomonads from enrichment cultures with various

1

carbon sources for use in the investigation of numerous degradative pathways. The group includes etiological agents of important diseases of plants and animals. Some species have become widespread in clinical specimens due to their resistance to deleterious agents including many antibiotics; among them, *P. aeruginosa*, normally an occasional pathogen, is the cause of considerable concern in medical practice. The genus has also attracted the attention of geneticists, who are now tapping the interesting resources of some of the species.

A general definition of the aerobic pseudomonads was proposed in 1966 by Stanier, Palleroni and Doudoroff. The bacteria possess the following general phenotypic properties. Unicellular rods straight or curved but not helical, measuring 0·5 to 1 μm by 1·5 to 4 μm. Motile by means of one or more polar flagella. One member of the group is permanently immotile; others produce lateral flagella in addition to the polar. Gram negative. Do not form endospores, stalks or sheaths. The energy-yielding metabolism is respiratory, never fermentative or photosynthetic. Molecular oxygen is used as terminal oxidant, but some members of the group can live anaerobically in media containing nitrate. All members are chemoorganotrophs, but some are facultative chemolithotrophs capable of using hydrogen gas as an energy source. The guanine plus cytosine content of the DNA of these bacteria is considered to range from 58 to 69 moles %, these limits having been established by the analysis of a large number of representative strains.

This introductory chapter will cover species of the genus *Pseudomonas* including the species of the polarly flagellated hydrogen bacteria assigned until recently to the genus *Hydrogenomonas*. A picture of our present knowledge of the general properties of the species that have been thoroughly characterized through a given set of bacteriological methods will be presented here, and the emphasis will be inclined toward the important subgroup of fluorescent pseudomonads that includes the type species of the genus, *P. aeruginosa*. The group has been the focus of attention of several laboratories in recent years and progress in our knowledge of the genus has been substantial. However, in some areas the taxonomic structure is still highly unsatisfactory and the complexity of some of the subgroups constitutes a formidable challenge to the bacterial taxonomist.

The burst of interest in the taxonomy of the genus *Pseudomonas* in the last few years stems principally from the revival of the fundamental contributions of den Dooren de Jong, who worked about fifty years ago in the laboratory of Beijerinck, the great Dutch microbiologist. A brief account of his work, with comments on its ecological and taxonomic importance, has been presented elsewhere (Stanier *et al.*, 1966).* A rational study of the classification

---

* This paper by no means represents the first attempt at rescuing this important work from oblivion, since comments and extensive citations of it had appeared much earlier. One good example is M. Stephenson's pioneer book on bacterial metabolism first published in 1930.

of the genus has now become possible through the adoption of den Dooren de Jong's extensive nutritional characterization, supplemented with a number of modern techniques. One of the techniques that has become particularly important to the bacterial taxonomist in recent years is nucleic acid hybridization. This has played a particularly important role in the internal subdivision of the genus. The homology groups that can be defined by application of these techniques probably come very close to representing natural arrangements. However, phylogenetic conclusions seem perhaps unwarranted at the present moment. In a later section some approaches to the phylogeny of the genus will be discussed, but this will be done mainly with the intention of mentioning techniques by which we can record observations that we believe bear on phylogeny, although rather distantly. It is a well-known fact that in bacteriology we can gather data only from the present cross section of the living bacterial world, without the help of a fossil record that could relieve us in part from the burden of reconstructing past history from the study of its present consequences.

Relation of the Genus Pseudomonas *to Other Bacterial Genera*

The morphological and physiological properties collected in the definition given in the previous section apply to a number of species the majority of which are now assigned to the genus *Pseudomonas*. The morphological characters of our definition are shared by a number of groups which are excluded for possessing some physiological properties at variance with those already outlined. Such is the case for the fermentative species belonging to *Aeromonas, Zymomonas, Photobacterium* and *Vibrio*, and also for the physiological specialized groups included in the genera *Nitrosomonas, Thiobacillus, Ferrobacillus, Methylomonas (Methanomonas), Halobacterium, Gluconobacter* and *Xanthomonas*. In a later section of this book (Chapter 7) work on one-carbon compound utilizers will be discussed. The taxonomic situation of these organisms is quite confusing at the present moment and those with typical pseudomonad morphology will not be treated here. The species named *Pseudomonas methanica* by Dworkin and Foster (1956) is considered to be identical with *Methylomonas methanica* of Whittenbury, Phillips and Wilkinson (1970) and its nutritional properties and DNA base composition (52 % G + C) suggest an allocation different from the group to be discussed in this chapter. In fact, the assignment of this species to a genus other than *Pseudomonas* appears to be perfectly reasonable. This and other pseudomonads of similar physiological properties are discussed by Quayle (1972).

This radical process of elimination leaves us a group which, however, is far from uniform. As we shall see in a later section, some subgroups within *Pseudomonas* are as different from one another as they are from bacterial species of genera assumed to be by all available criteria very remote from the genus *Pseudomonas*. The taxonomic hierarchy of these clusters within the genus is far from clear at present, particularly because of the impossibility of providing a practical definition for their circumscription.

2. CHARACTERS USEFUL FOR THE DIFFERENTIATION OF *PSEUDOMONAS* SPECIES

2.a. Morphology

Morphological attributes are very important in taxonomy, but it is well known that bacteria do not offer a great wealth of morphological information, which is perhaps the main reason for the unsatisfactory state of bacterial taxonomy. The cells of *Pseudomonas* are typically straight rods, although in any population it is almost invariably possible to find a number of curved cells. In some strains, the length of the cells greatly exceeds the maximum 4 μm indicated in our definition; this has been found to be the case in strains of fluorescent plant pathogenic pseudomonads and in some strains of *P. putida*. The cells do not show particular features under the light microscope, except in those species which are able to accumulate substantial amounts of the polymer poly-β-hydroxybutyrate (PHB). The accumulation of this compound is enhanced by growth of the cells in media of low nitrogen content, and normally the granules can be seen clearly under the microscope with phase contrast equipment. Some species, however, accumulate very little PHB, and its presence can only be demonstrated beyond reasonable doubt by isolation and chemical characterization.

Yamamoto (1967) has reported the presence of the microtubular structures known as rhapidosomes in all the species included in his study (*P. aeruginosa, P. fluorescens, P. chlororaphis, P. putida, P. riboflavina, P. boreopolis, P. fragi*) which were examined under the electron microscope after negative contrast staining. The rhapidosomes appear to be in close association with the nucleoplasm. Interestingly, among the organisms that failed to reveal these structures was the hydrogen bacterium *Hydrogenomonas facilis*, now placed in the genus *Pseudomonas*. The biological meaning of the rhapidosomes is at present obscure; however, evidence favouring the identity of rhapidosomes with the polymerized sheath of bacteriocins in *P. fluorescens* has been presented by Amako, Yasunaka and Takeya (1970).

In thin sections of *P. aeruginosa* cells mesosome-like structures have been observed (Carrick & Berk, 1971; Hoffmann *et al.*, 1973) these structures are also present in *P. saccharophila* (Young *et al.*, 1972).

The cells of *Pseudomonas* show active swimming motility, especially when taken from young cultures. This motility is due to polar flagella (Plate 1.1). The number and wavelength of the flagella are taxonomically important characters. The number can be most conveniently expressed on a statistical basis (Lautrop & Jessen, 1964). The flagella are unsheathed in most species.

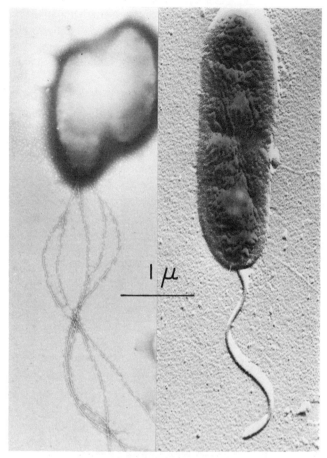

Plate 1.1. Left, *P. viridiflava*, negative staining with uranyl acetate; right, *P. stizolobii*, chromium shadowing. (Courtesy of Alicia Palleroni)

P. stizolobii (a plant pathogenic species) has sheathed flagella (Fuerst & Hayward, 1969b); the single polar flagellum of this species has a structure comparable to the sheathed flagella described in other genera (*Vibrio, Bdellovibrio, Beneckea, Photobacterium*).

In some strains polar and lateral flagella have been observed. The lateral flagella have a shorter wavelength than the polar and they are easily shed so that when the culture becomes older the cells are only polarly flagellated (Palleroni *et al.*, 1970). The insertion of the flagella in some species is not exactly polar but somewhat lateral or sub-polar. One species, *P. mallei*, has non-flagellated cells but it is included in the genus on the bais of a very clear relationship with other species having normal flagellation (Redfearn, Palleroni & Stanier, 1966). These cases of abnormal flagellation have reduced the taxonomic significance of one of the few morphological characters, perhaps the most striking, which until recently was considered to be absolutely typical of the pseudomonads.

Twitching motility on the surface of solid media has been reported for non-flagellated strains of *P. aeruginosa* (Lautrop, 1965). This phenomenon has also been observed in different species of flagellated non-fluorescent *Pseudomonas* (Heinrichsen, 1972) and can be distinguished from the flagellar motility. None of the fifteen strains of *P. mallei* examined by Heinrichsen showed twitching motility.

Fimbriae or pili have been observed in eight out of fifteen species of *Pseudomonas* examined under the electron microscope by Fuerst and Hayward (1969a). The appendages are polarly inserted in *P. aeruginosa, P. acidovorans, P. testosteroni, P. maltophilia, P. alcaligenes* and *P. solanacearum*; on the other hand, *P. cepacia* (*P. multivorans*) (Plate 1.2) and *P. fragi* have peritrichous fimbriae. No pili have been observed in *P. fluorescens, P.*

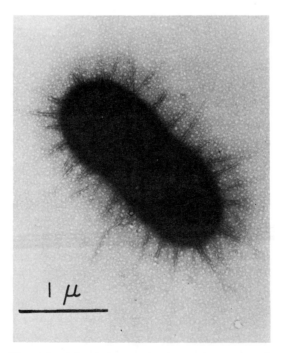

Plate 1.2. Cell of *P. cepacia* (*P. multivorans*) with peritrichous fimbriae. Reproduced with permission from J. A. Fuerst and A. C. Hayward (1969) *Journal of General Microbiology,* **58,** 227, 239.

chlororaphis (*P. fluorescens* biotype D), *P. aureofaciens* (*P. fluorescens* biotype E), *P. putida*, *P. putrefaciens* (a species now excluded from the genus), *P. oleovorans* and *P. stizolobii*. The polar pili of *P. aeruginosa* have been studied by Bradley (1972), who found that some of the appendages are receptors for RNA-containing phages, while others are probably receptors for filamentous phages (Plate 1.3). No pili functional in the transfer of the

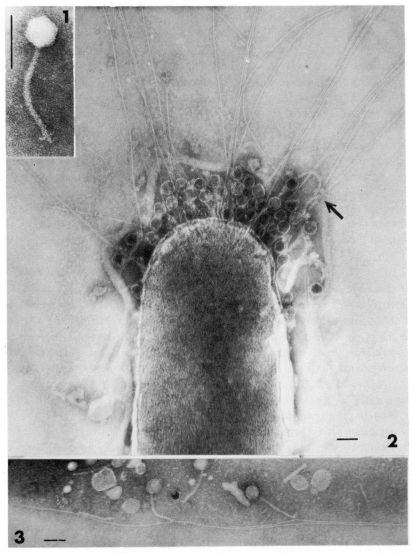

Plate 1.3. Polar pili of *P. aeruginosa*. Tailed pilus phage. Reproduced with permission from D. E. Bradley (1972) *Genetical Research*, **19**, 39.

P. aeruginosa sex factor FP2 have yet been observed. The pili of *P. echinoides* (a species of uncertain taxonomic position) are clearly involved in the formation of cell aggregates (Heumann & Marx, 1964).

2.b. Pigmentation

The pigmentation of some species of *Pseudomonas* is very characteristic. Usually special media have to be used in order to elicit or enhance the production of pigments. The purity of the chemicals and the agar used in the preparation of these media seems to be very important and often the highest purity does not guarantee the best results. This obviously points to our ignorance of the proper conditions necessary to elicit pigment production and, as a consequence, the character is one of the most erratic of all phenotypic properties. Colour production can be lost after repeated transfers in laboratory media and occasionally apigmented strains of normally pigmented species are isolated from nature.

Several types of pigments can be found to be produced by *Pseudomonas* species. One is the water-soluble fluorescent type, to which belong pigments that have been known to bacteriologists for a very long time. Recently, the chemical structures of four fluorescent pteridine derivatives found in the supernatants of *P. ovalis* (*P. putida*) were elucidated (Suzuki & Goto, 1971); three of these pigments are also produced by other microorganisms and the fourth ('putidolumazine') has never been reported for other species. Pigments of this type, characteristic of the so-called fluorescent pseudomonads are produced in many different media, particularly in those of low iron content (King, Ward & Raney, 1964; Garibaldi, 1967) and it has been suggested that they function as iron scavenging compounds. Another type of pigment belongs to the chemical family of phenazine compounds, whose structure is better known. At present, significant progress is being made in the elucidation of their biosynthesis. Pyocyanin, one of the phenazine pigments of *P. aeruginosa*, is synthesized via the aromatic amino acid pathway (Calhoun, Carson & Jensen, 1972). Some members of the fluorescent group and species of the *P. cepacia* group produce various other phenazine pigments. This last group of organisms may produce pigments that diffuse into the culture medium and resemble the fluorescent pigments when viewed under normal light. They can be differentiated under an ultraviolet source of short wavelength (*ca.* 254 nm), where only true fluorescent pigments are capable of fluorescence.

Carotenoid pigments can be produced by four of the well-characterized species of the genus, namely, *P. vesicularis*, *P. mendocina*, *P. flava* and *P. palleronii*. These pigments remain associated with the cells and are responsible for the yellow or orange colour of the colonies of these organisms. Various other pigments have been reported (Sneath, 1960; Palleroni & Doudoroff, 1972).

2.c. Physiological Characters

The dearth of morphological data in bacteria is compensated by a great deal of physiological information easily obtainable. Every bacteriologist must be constantly aware of physiological phenomena because in most cases bacteria can only be studied in populations that have grown, or are growing, under well-specified conditions. *Pseudomonas* are particularly suitable for physiological observations; their growth requirements are in most cases extremely simple, the generation time is relatively short even in mineral media, and the physiological diversity is very pronounced.

Various physiological properties used for the characterization of the species discussed in this chapter are listed below but will not be discussed further, since they have been adequately described elsewhere (Stanier *et al.*, 1966; Palleroni & Doudoroff, 1972). These include temperature relationships, oxidase reaction and differential cytochrome spectra; acid production from sugars and sugar alcohols (tested in a limited number of cases); production of slime (levan) from sucrose; denitrification; the arginine dihydrolase reaction; mode of cleavage of intermediates in the degradation of aromatic compounds; detection of extracellular enzymatic activities such as hydrolysis of gelatin, hydrolysis of starch, hydrolysis of poly-β-hydroxybutyrate, lecithinase action by the so-called egg yolk reaction, and lipase action as determined by the hydrolysis of 'Tween' 80. It must be remarked here that, considering the marked biochemical versatility of the pseudomonads, these organisms produce relatively few extracellular enzymes capable of degrading large molecules and in the nutritional analysis of the species the majority of the substrates utilized are soluble compounds of low molecular weight. *Pseudomonas* species are relatively intolerant to acid, but the diminuta group includes strains capable of oxidizing ethanol and tolerant to pH values lower than most other species. Several lines of evidence suggest that the diminuta can be separated from both *Pseudomonas* and *Acetomonas* in terms of important phenotypic properties (flagellar morphology, nutritional properties, acid tolerance, pigmentation and intracellular carbon reserve materials) (Ballard *et al.*, 1968).

A number of cultural and physiological characters considered of significance in systematic bacteriology have been left out for various reasons. Among these characters are the reduction of nitrate to nitrite, the production of H_2S, the urease test, the action on litmus milk, growth on potato slices, and various routine tests useful in the study of enteric bacteria. Some of these tests are designed to reveal properties that are either absent from all the *Pseudomonas* group, or appear erratically and uncorrelated with other important characters.

Outstanding among the physiological attributes of the species are the nutritional characters. Den Dooren de Jong (1927) was the first bacteriologist to show that species of *Pseudomonas* could utilize a large number of organic compounds for growth in an otherwise purely mineral solution containing an ammonium salt as nitrogen source. One of the species studied by den Dooren de Jong, *P. putida*, appeared to be exceptionally versatile in this respect, since out of 200 compounds tested, about 80 could be utilized as substrates. The compounds utilized by these strains belonged to various chemical classes, such as carbohydrates, alcohols, saturated and unsaturated fatty acids, amino acids, amines and amides. Aside from their physiological and ecological implications, these findings opened the road for taxonomic studies of the genus based on a thorough phenotypic characterization; in principle, several other genera should be equally amenable to such type of analysis.

In the investigations in which the author has been concerned, attention has been focused on the nutritional properties which can be considered as characteristic of the strains of a given species. The nutritional analysis can be performed in several different ways. The work involved in the screening of a large number of strains is considerable, and attempts at simplification are not always free of important disadvantages. One of the simplifications was the adoption of the replica plate technique, which permits the collection of a large mass of data on nutritional characters with relatively little effort. The imperfections of this technique for this particular purpose have been discussed in detail (Stanier *et al.*, 1966; Palleroni & Doudoroff, 1972). A more critical control of such factors as purity of chemicals in the media, and a degree of standardization of techniques is required than may be general in some routine bacteriological laboratories.

The data obtained from nutritional screenings are heterogeneous in nature. In some cases, the utilization of a given substrate provides information on the enzymatic constitution of a given strain, but this is generally not the rule, since alternative pathways may be possible for the degradation of organic compounds. Thus, strains of *P. fluorescens*, *P. acidovorans* and *P. cepacia* can utilize L-tryptophan as carbon source, but *P. fluorescens* and *P. cepacia* degrade tryptophan *via* the aromatic pathway, while *P. acidovorans* follows the quinoline pathway in which kynurenate is an intermediate. Strains of the three species may be able to grow on kynurenate, but while *P. fluorescens* does not seem able to involve this compound as intermediate in the oxidation of tryptophan, *P. cepacia* can do so under certain conditions (unpublished observations). In summary, the immediate applications of the nutritional analyses are the use of the information for taxonomic purposes and also the collection of observations that may be starting points for biochemical investigations. That the nutritional analysis has been an important source of suggestions for biochemical work is witnessed by a number of papers on various pathways and their regulation, which represent outgrowths of those initial observations.

3. DNA COMPOSITION OF *PSEUDOMONAS* SPECIES

The analysis of the base composition of DNA of a large number of strains of *Pseudomonas* of the Berkeley collection was performed in the laboratory of Dr. Manley Mandel and the information appears in a number of papers (Mandel, 1966; Ballard *et al.*, 1968; Ballard *et al.*, 1970; Davies *et al.*, 1970; Palleroni & Doudoroff, 1971; Palleroni *et al.*, 1970; Ralston, Palleroni & Doudoroff, 1973). The data are comprised in the approximate range 58 to 69 moles % of guanine + cytosine, a range evidently wide and suggestive of

Table 1.1. Classification of *Pseudomonas* species into RNA homology groups

rRNA homology group	GC in DNA (%)	Species[a]	Competition (%)[b] rRNA/DNA	DNA/DNA
1	67	*P. aeruginosa* (131)	87–101	4–87
	59–63	*P. fluorescens* (D-31)		
	61–62	*P. putida*		
	58–61	Fluorescent plant pathogens		
	61–66	*P. stutzeri*		
	63–64	*P. mendocina*		
	66	*P. alcaligenes*		
	63	*P. pseudoalcaligenes*		
2	67–68	*P. cepacia* (382)	87–101	23–79
	68	*P. marginata*		
	65	*P. caryophylli*		
	69	*P. pseudomallei* [c]		
	69	*P. mallei* [c]		
	64	*P. pickettii*		29–56
	66–67	*P. solanacearum*		
3	67	*P. acidovorans* (14)	79–92	33
	62	*P. testosteroni*		
	65–66	*P. delafieldii* (134)		42–100
	62–64	*P. facilis*		
	69	*P. saccharophila*		
4	66–67	*P. diminuta* (501)	96	27
	66	*P. vesicularis*		
5	67	*P. maltophilia* (67)	95	—
	66–68	*Xanthomonas* spp. (Xc-1)		

[a] Numbers in parentheses after the species name correspond to the strain used as reference in rRNA/DNA hybridization experiments.

[b] Calculated according to formula given by Ballard *et al.* (1970). Each vial contains 1 μg of labelled DNA or rDNA, 100 μg of unlabelled rRNA or 150 μg of unlabelled DNA (competitor) in a final volume of 0·25 ml of 0·3 M NaCl and 0·03 M Na citrate; one nitrocellulose filter per vial with about 20 μg of unsheared immobilized DNA.

[c] Not included in RNA/DNA experiments.

phylogenetic heterogeneity. However, it is impossible to subdivide the genus into natural subgroups of different G + C content without a great deal of overlapping, and some of the subgroups cover a G + C range almost as wide as that for the whole genus (see Table 1.1).

Recent work on a number of strains of marine origin (Baumann *et al.*, 1972) has revealed a group of bacteria sharing the structural and physiological properties of *Pseudomonas*, whose DNA has a G + C content ranging from about 30 to 67 moles %. These organisms could be subdivided into three clusters according to DNA base composition, and the authors have incorporated in the genus *Pseudomonas* the group having 56·4 or more moles % G + C. This decision is clearly arbitrary, since there are no good reasons at present to exclude several organisms of G + C content as low as 53 %. The cluster having a G + C content ranging from 43 to 48 % has been assigned to the new genus *Alteromonas*, and in the extensive characterization of the species of this genus, it has been impossible to find phenotypic properties other than DNA base composition that would permit a clear separation from the members of the genus *Pseudomonas* as currently defined. Very likely, the application of other approaches, such as nucleic acid hybridization, and in particular ribosomal RNA/DNA hybridization, may be able to reveal the relationships of these recently defined groups to other well-characterized species of *Pseudomonas*.

4. THE COMPARATIVE BIOCHEMISTRY OF *PSEUDOMONAS*

In this section will be presented a brief discussion of the results of several studies selected mainly on the basis of their significance in the comparative biochemistry of the genus, since these investigations contribute a great deal to our understanding of the relationships among different species and species groups of *Pseudomonas*.

4.a. Biochemical Pathways and Regulatory Mechanisms

The biochemical study of strains related by a number of phenotypic properties usually reveals identity in the pathways for the catabolism of various organic compounds. On the other hand, unrelated organisms may follow different biochemical pathways, these differences being the expression of fundamental dissimilarities in the makeup of the organisms being compared. In some cases, however, unrelated organisms share complicated pathways in all their biochemical intricacies. Since purely chemical considerations may impose severe constraints on the routes available for the solution of a given biochemical problem, the hypothesis of independent origin for identical pathways does not necessarily contradict the bases of biochemical evolution; two unrelated species may have adopted a given metabolic method among relatively few chemical possibilities in the course of evolution (Stanier, 1968).

Some inducible catabolic pathways that have been studied in members of the genus *Pseudomonas* are very rich in biochemical and taxonomic information. Examples often quoted in this respect are taken from the extensive investigations performed by Stanier and his collaborators on the degradation of aromatic compounds by species of *Pseudomonas* and other genera. The members of the fluorescent group (*P. aeruginosa*, *P. fluorescens*, *P. putida*) and related non-fluorescent bacteria (*P. stutzeri*, *P. mendocina*) use the β-ketoadipate pathway for the degradation of various aromatic compounds such as mandelate, benzoate, *p*-hydrozybenzoate, tryptophan. In contrast, the strains of the acidovorans group (*P. acidovorans*, *P. testosteroni*) use a different pathway for the metabolism of some of these compounds, involving α-keto acids as intermediates.

The β-ketoadipate pathway is also followed by pseudomonads unrelated to the fluorescent group (for instance, *P. cepacia*) and by bacteria belonging to genera other than *Pseudomonas* (for instance, *Acinetobacter*). In these cases, further investigation on the regulatory mechanisms of these pathways often can provide important clues at a level that is thought to be of a more conservative nature than the biochemical pathways (Cànovas, Ornston & Stanier, 1967); the regulation of the β-ketoadipate pathway is fundamentally different in the two genera (Stanier, 1968). It is fair to say, however, that not all workers in this field share these views about the phylogenetic meaning of the identity of pathways and regulatory mechanisms (Datta, 1969).

The apparent absence of a pathway in a given organism may in some cases be a reflexion of our ignorance about the proper inducer of the pathway. A good example is the presence of the 'meta cleavage' mechanism of catechol degradation that can be induced in *P. putida* by growth on phenol. Catechol is also produced from benzoate in this organism, but then cleavage is of the 'ortho' type (Feist & Hegeman, 1969). Until it was discovered that the β-ketoadipate pathway could be induced in *P. acidovorans* by growth on *cis,cis*-muconate (Robert-Gero, Poiret & Stanier, 1969), it was thought that this pathway did not occur in this species. In fact, members of the acidovorans group can also grow on β-carboxy-*cis,cis*-muconate which is metabolized through the β-ketoadipate pathway. The mode of regulation of the synthesis of the enzymes of this pathway is unique to the group (Ornston & Ornston, 1972).

Induction of L-tryptophan oxidation by one of the intermediates in the pathway, L-kynurenine, was the first demonstration of metabolite induction reported in the literature (Palleroni & Stanier, 1964). This mechanism was first described in a fluorescent strain of dubious taxonomic position, but is probably of widespread occurrence in the fluorescent group. It has subsequently been demonstrated in *P. acidovorans* (Rosenfeld & Feigelson, 1969) and in *P. cepacia* (unpublished observations). Metabolite induction has also been demonstrated in *P. aeruginosa* for the histidine degradative pathway (Lessie & Neidhardt, 1967). Urocanate, the product of the first enzyme, induces the formation of the histidine-degrading enzymes in a fashion similar to that found for *Aerobacter aerogenes* (Schlesinger, Scotto & Magasanik, 1965). This pattern is not common to the degradation of other amino acids; in *P. putida*, the degradation of hydroxyproline can be induced by either hydroxy-L-proline or allohydroxy-D-proline, the first and second compound of the sequence, respectively (Gryder & Adams, 1969). The regulation of catabolic pathways is discussed in more detail in Chapters 7 and 8.

In the field of biosynthetic pathways, Jensen and his collaborators have performed interesting comparative studies on the control of a branch-point enzyme, 3-deoxy-D-arabinoheptulosonate-7-phosphate (DAHP) synthetase in different organisms, and the studies have included a number of the strains of *Pseudomonas* from the Berkeley collection. In tests with crude cell extracts of *P. aeruginosa*, *P. putida*, *P. fluorescens*, *P. alcaligenes*, *P. pseudoalcaligenes*, *P. stutzeri*, *P. cepacia*, it was found that there was feedback inhibition of a single enzyme species by only one of the three final products of the diverging pathways, namely tyrosine. The DAHP synthetase of *P. acidovorans* and *P. testosteroni*, on the other hand, appeared to be also sensitive to phenylalanine inhibition (Jensen, Nasser & Nester, 1967).

In the biosynthetic pathway of amino acids of the aspartate family, *Pseudomonas* species have a single aspartokinase, which is sensitive to concerted feedback inhibition by lysine and threonine. High concentrations of either amino acid inhibit the enzyme in the organisms of the fluorescent group and in *P. cepacia*, whereas in the acidovorans group the sensitivity to single amino acids is markedly lower. Homoserine dehydrogenase in the acidovorans group is absolutely specific for NAD, and in all other pseudomonads examined, NADP is more effective than NAD (Cohen, Stanier & Le Bras, 1969). The characteristics of feedback regulation of branch-chain pathways in *Pseudomonas* is discussed further in Chapter 8.

Further evidence of the differences between the acidovorans group and the fluorescent group has been presented by Queener and Gunsalus (1970) in their studies on anthranilate synthetase. This enzyme can be resolved into two sub-units and hybrid molecules can be formed with sub-units of different species. The hybrid molecules made with sub-units from *P. aeruginosa* and *P. putida* are fully active, and the same is true of the hybrids between sub-units from *P. acidovorans* and *P. testosteroni*. When sub-units from different groups are hybridized, however, the resulting molecules are lower in activity than either native form of enzyme. Interestingly, *P. stutzeri* (presently incorporated to the fluorescent DNA complex) has an anthranilate synthetase resembling the putida–aeruginosa enzyme, while *P. cepacia* (*P. multivorans*) enzyme is similar to the acidovorans–testosteroni class.

4.b. Immunological Analysis of Proteins

Two enzymes of the catechol branch in the β-ketoadipate pathway have been compared in a number of *Pseudomonas* strains by measuring their immunological cross-reactivity with antibody obtained from the enzymes of a known strain of *P. putida* (Stanier *et al.*, 1970). Strong cross-reactions are observed with *P. putida* and *P. aeruginosa* extracts, although in this last case the homology is only partial. Other members of the *P. fluorescens* complex belonging to both the fluorescent and non-fluorescent types, also reacted to an appreciable extent. *P. cepacia* extracts, on the other hand, showed no cross-reactions with the *P. putida* antibody. *P. cepacia* shares with the fluorescent organisms both common biochemical pathways and control mechanisms in the metabolism of some aromatic compounds; the analogy between these two distantly related groups breaks down at the level of the structure of the enzymes involved, which do not show any immunological cross-reactivity (Figure 1.1). Similarly, enzymes of the β-ketoadipate pathway in *P. acidovorans* induced by growth on *cis,cis*-muconate, do not resemble immunologically those of the fluorescent organisms.

Benzoate metabolism

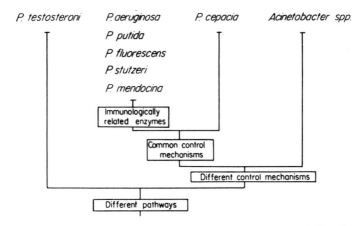

Figure 1.1. Comparative biochemistry of benzoate metabolism in various species of *Pseudomonas* and in *Acinetobacter*

Recently Clarke (1972) found that the aliphatic amidases of *P. aeruginosa* and *P. putida* strains gave strong cross-reactions with antibody prepared against the amidase from a known *P. aeruginosa* strain, but the *P. cepacia* strains gave only very faint and diffuse reactions. Surprisingly, *P. acidovorans* strains gave strong cross-reactions with the *P. aeruginosa* antibody although it is clear from the study of a number of properties that the two species are very different. In view of the fact that there are several reports in the literature of genetic transfer between different species of *Pseudomonas*, Clarke concluded that it is imperative 'to be cautious in interpreting similarities in protein properties of catabolic enzymes of pseudomonads to long-standing evolutionary relationships, since they may have exchanged the relevant genes in relatively recent times by interspecific genetic transfer'.

Unfortunately, the immunological evidence is only negative for distant groups and it is obvious that the absence of cross-reactivity can still hide a substantial degree of homology, since in many cases the immunological similarity may be one of the first features to be lost in the evolution of proteins. It is also important to bear in mind the fact that the rate of change of proteins endowed with catalytic activity (and therefore limited in the freedom with which they can change) may be different for different proteins (Stanier *et al.,* 1970).

In general, in spite of obvious limitations, the immunological approach is a powerful complement to the comparative studies on biochemical pathways and their mode of regulation, and as more enzymes become available in purified form, its use is likely to be extended.

4.c. Amino Acid Sequences of Proteins

The comparison of amino acid sequences of homologous proteins is one of the most direct approaches to the study of phylogeny. Sequence differences may be hard to evaluate (Dickerson, 1971), but the efforts have been very successful in the comparative studies involving groups of eukaryotes whose major evolutionary relationships are firmly established.

Due mainly to their small size, ease of purification and widespread occurrence, cytochromes *c* have been favourite subjects for sequence determination and comparisons over a very wide range of living creatures. The work has been extended to bacterial cytochromes, and the matrix shown below is a summary of the amino acid differences for cytochromes *c*-551 of four *Pseudomonas* species of the *P. fluorescens* DNA homology complex (Ambler & Wynn, 1973):

Difference per 100 amino acids

		1	2	3	4
1: *P. aeruginosa*	1	0	32	32	39
2: *P. fluorescens* biotype C	2		0	28	33
3: *P. stutzeri*	3			0	22
4: *P. mendocina*	4				0

The above data support the conclusions from DNA homology studies on these four species (Palleroni *et al.*, 1972). Recently, Ambler and Murray (1973) have reported data on the sequences of the cytochromes c_4 of denitrifying pseudomonads (*P. aeruginosa, P. stutzeri*) which show an unexpectedly high degree of homology with the cytochrome c_4 of the unrelated species *Azotobacter vinelandii*. Sequence comparison, however, offers the appropriate resolving power for the comparative study of the major bacterial groups, according to Crawford and Yanofsky (1971). The work on tryptophan synthetase is a good example. The enzyme of *P. putida* has several properties in common with that of *E. coli* (Enatsu & Crawford, 1971), but no interspecies association of the sub-units could be demonstrated. A segment of 50 residues in the α-chain was compared in both organisms by Crawford and Yanofsky (1971); 24 of the amino acids were different, and a minimum of 37 mutations would be required for the conversion of one segment into the other. Among enterobacteria, only 12 % variation is found in the same region of the molecule. Differences between the amino acid composition of the α-chain of the fluorescent species and of *E. coli* thus parallel differences in the regulation and genetic organization of the tryptophan biosynthetic machinery of the two groups (Enatsu & Crawford, 1971).

4.d. *In vitro* Nucleic Acid Hybridization

A direct approach to the phylogeny of a given group is that which can provide information at the genetic level. In *Pseudomonas*, the direct measure of genetic relationships is not yet feasible, because genetic transfer is not known to be of general occurrence and, moreover, it is perhaps unlikely that it will ever be of much comparative value due to the marked heterogeneity of this group of organisms. In the absence of a practical means for genetic studies involving many species of the genus, nucleic acid hybridization *in vitro* is obviously an appropriate substitute. The usefulness of this approach for the examination of taxonomic relationships has been proved in work on several different groups of organisms (McCarthy & Bolton, 1963; Hoyer, McCarthy & Bolton, 1964; Brenner, Martin & Hoyer, 1967; Kingsbury, 1967; Johnson & Ordal, 1968; Bendich & McCarthy, 1970). In *Pseudomonas*, investigations performed with both DNA–DNA and ribosomal RNA–DNA systems have yielded very interesting information which can be used as a basis for the construction of a natural taxonomic system. In the experiments in which the author has been concerned, the competition method modified from the procedure of Johnson and Ordal (1968) was adopted. The homology between pairs of strains is measured by the efficiency with which nucleic acid of one of the strains competes for the reassociation or binding of nucleic acid of another strain to its homologous DNA. The competition method is very convenient for surveying large number of strains. In many cases, direct binding techniques have also been followed for comparison. For the DNA–

DNA hybridization experiments, temperatures ranging from 70 to 72 °C (25 °C below the T_M of the immobilized DNA) were used for the annealing, and in many cases competition experiments were repeated at 80 °C. These stringent conditions of reassociation were an adaptation of the work performed in the enterobacteria and in *Neisseria* (Brenner & Cowie, 1968; Kingsbury *et al.*, 1969). Their main disadvantage is the substantial loss of DNA from the membrane filters but the results are of considerable interest; the values obtained at 80 °C are similar to those obtained at the 'normal' temperature only when closely related strains are compared. With the exception of this simplified procedure for the analysis of the thermal stability of the reassociation product, in the work with *Pseudomonas* the close examination of the hybrid molecules has not been attempted, and thus the data do not permit to decide whether a low level of DNA homology represents homology over a limited segment of the genome with very little mismatching or else, a less perfect resemblance of nucleotide sequence over larger portions of the bacterial chromosome.

DNA–DNA hybridization is effective in revealing relatively close relationships and the information can be correlated effectively with the general phenotypic differences of the strains to which, obviously, DNA composition is directly related. DNA–DNA experiments cease to be of much value when members of different sub-groups within the genus *Pseudomonas* are compared; the competition values obtained are negligible or zero. A number of 'DNA homology complexes' can thus be precisely defined (Palleroni & Doudoroff, 1972); members of a given complex are related directly or indirectly to one another at various levels of DNA homology and show no detectable homology with the members of other complexes.

Since ribosomal RNA is coded by a markedly conservative region of the genome, *in vitro* hybridization of ribosomal RNA with immobilized DNA is capable of revealing distant relationships among organisms (Bendich & McCarthy, 1970); this technique should therefore be useful in comparisons involving members of the different *Pseudomonas* DNA homology complexes. In spite of the fact that ribosomal RNA is coded by a small portion of the genome, the very basic importance of ribosomes as essential elements in the cellular organization may warrant phylogenetic conclusions from *r*RNA–DNA hybridization experiments which otherwise might be subjected to the same criticisms that can be applied to conclusions drawn from the comparative study of a few cellular proteins.

The results of experiments carried out with *r*RNA in *Pseudomonas* permit the grouping of the species of the genus into at least five sharply defined clusters to which the rather non-committal name of 'RNA homology groups' will be applied in this discussion (Palleroni *et al.*, 1973). The composition of these groups correlates very well with that of the DNA homology groups (DNA homology complexes) previously drawn on the basis of DNA–DNA competition experiments (Palleroni & Doudoroff, 1972). The results obtained with the two types of nucleic acids agree very well when they are compared in the proper direction, that is, there is no case in which a detectable level of DNA–DNA homology is not accompanied by high values in rRNA–DNA experiments. In addition, by virtue of revealing distant relationships, the rRNA experiments have shown groups whose existence was not suspected before on the basis of an extensive characterization of many other properties of the component species (Table 1.1). RNA–DNA competition values obtained among members of different RNA homology groups are as low and, occasionally, even lower than those obtained between any member and *E. coli*, thus providing further evidence of the marked heterogeneity of the genus *Pseudomonas* (Figure 1.2).

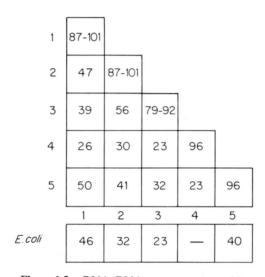

Figure 1.2. rRNA–DNA per cent competition values among the different *Pseudomonas* RNA homology groups. The figures given for inter-group comparisons are average values

A list of the species that have been thoroughly characterized by the methods that have been discussed in this chapter is presented in Table 1.2, together with some synonyms. A large number of nomenspecies has been left out from our investigations. Perhaps some of the excluded species are synonyms of some of the species described here, or could be assigned to some of our species groups. It is also likely that a number of species described in the literature, of which in many cases no representative strains have survived in collections, may represent well-defined entities which, hopefully, future

Table 1.2. List of some of the well-characterized species of the genus *Pseudomonas*

RNA homology group	Species	Synonyms	Comments
	P. aeruginosa	*P. pyocyanea,* *P. polycolor*	
	P. fluorescens	*P. marginalis,* *P. chlororaphis,* *P. aureofaciens,* *P. lemonnieri,* *P. geniculata*	Subdivided into 7 biotypes. Some of the biotypes, however, deserve independent species status.
I	*P. putida*	*P. ovalis, P. convexa*	Subdivided into 2 biotypes, one of which (A) is 'typical' *P. putida*.
	Fluorescent plant pathogens		See Table 1.4.
	P. stutzeri	*P. stanieri*	
	P. mendocina		
	P. alcaligenes		Non-fluorescent species.
	P. pseudoalcaligenes		
	P. cepacia	*P. multivorans*	
	P. marginata	*P. alliicola,* *P. gladioli* (?)	See also Table 1.4.
	P. caryophylli		
	P. pseudomallei		
II	*P. mallei*	The specific name *mallei* has been attached to several genera: *Bacillus, Acinetobacter, Loefflerella, Malleomyces,* etc.	
	P. solanacearum		See Table 1.4.
	P. pickettii		
	P. acidovorans	*P. indoloxydans,* *P. desmolytica*	
	P. testosteroni		
III	*P. facilis*	*Hydrogenomonas facilis*	Hydrogen bacterium
	P. delafieldii		
	P. saccharophila		Hydrogen bacterium
IV	*P. diminuta*		
	P. vesicularis		

Table 1.2. Continued

RNA homology group	Species	Synonyms	Comments
V	$\left\{\begin{array}{l}\end{array}\right.$ *P. maltophilia*		
	P. lemoignei		
	P. ruhlandii	*Hydrogenomonas ruhlandii*	$\left.\begin{array}{l}\end{array}\right\}$ Hydrogen bacteria
	P. flava		
	P. palleronii		

Note: The last four species have not been thoroughly analysed by nucleic acid hybridization techniques, and therefore their relation to other species of the genus is not known at present.

studies might rediscover or help circumscribe more precisely than in their original descriptions. A list of these species and a key including some of their salient phenotypic properties will appear in the forthcoming edition of *Bergey's Manual of Determinative Bacteriology* (genus *Pseudomonas*, by M. Doudoroff and N. J. Palleroni), now in press. Table 1.1 and the key of Table 1.4 in this chapter should also be consulted.

5. THE INTERNAL SUBDIVISION OF THE GENUS *PSEUDOMONAS*

5.a. The *P. fluorescens* RNA Homology Group (Group I)

The central position occupied in this group by the species *P. fluorescens* is the reason for its designation. The group is quite heterogeneous in phenotypic properties, but some of these properties can be used in particular combinations for the internal subdivision of this large cluster within the genus. The species included in this RNA homology group are the same that constitute the previously defined *P. fluorescens* DNA homology complex (Palleroni & Doudoroff, 1972). A variable degree of DNA homology links directly or indirectly the constituent species; besides, a high degree of *r*RNA–DNA homology is shared by all members of the group that have been analyzed.

The *P. fluorescens* RNA homology group includes fluorescent and non-fluorescent species. Among the fluorescent species is the type species of the genus, *P. aeruginosa*, *P. fluorescens* (sub-divided into a number of 'biotypes'), *P. putida* (including 2 biotypes), and many species of fluorescent plant pathogenic pseudomonads. The non-fluorescent species can in turn be subdivided into the stutzeri subgroup (*P. stutzeri* and *P. mendocina*), and the alcaligenes subgroup *P. alcaligenes* and *P. pseudoalcaligenes*). The G + C content of the DNA of all the organisms of the group that have been analysed ranges from

58 to 68 moles %, a span almost as wide as that of the whole genus. No important phenotypic properties are common to all members of the group, which at the same time could be used to exclude other species belonging to other RNA homology groups. Before the assignment of *P. pseudoalcaligenes* to the group (Ralston, 1972) all members had in common one important negative property, the inability to accumulate poly-β-hydroxybutyrate as carbon reserve material (Table 1.3).

Table 1.3. General properties of the saprophytic fluorescent pseudomonads[a]

A. Production of fluorescent pigment.
B. Absence of poly-β-hydroxybutyrate as carbon reserve material, and incapacity to hydrolyse extracellular poly-β-hydroxybutyrate granules.[b]
C. Oxidase reaction positive.
D. Arginine dihydrolase reaction positive.[b]
E. Substrates utilized by the majority of the strains examined: glucose, fructose, gluconate, 2-ketogluconate, pelargonate, benzoate and/or *p*-hydroxybenzoate, β-alanine, arginine, betaine and/or sarcosine, putrescine and/or spermine.
F. Substrates rarely utilized: L-rhamnose, maltose, maleate, glycollate, ethylene glycol, D-mandelate, *m*-hydroxybenzoate, phthalate, α-aminobutyrate, D-tryptophan.
G. Substrates not utilized by any strain: D-arabinose, D-fucose, starch, cellobiose, salicin, lactose, threonine, norleucine.

---

[a] The characters in this table also apply to two pathogens: *P. aeruginosa* and the strains of *P. fluorescens* biotype B included among the plant pathogens under the name *P. marginalis.*
[b] The following two correlations seem to be absolute for the entire genus: arginine dihydrolase positive strains are found *only* among those which are able to use arginine aerobically; the capacity for hydrolysis of extracellular poly-β-hydroxybutyrate is found *only* among strains capable of accumulating this compound as carbon reserve material.

a.i. The Fluorescent Species of Group I

The type species, *P. aeruginosa*, will be discussed in some detail in order to provide the reader with reference material for the species with which this book will be most concerned. The elements of the description have been taken from observations of the author and his colleagues over a number of years and also from the excellent work by Jessen (1965). *P. fluorescens* and *P. putida* will be less extensively described, in part because it is very difficult at the present moment to give a condensed account of the properties of these complex species. Other species of the genus will be given even less attention, but the interested reader can find satisfactory descriptions in the literature (Doudoroff, 1940; van Niel & Allen, 1952; Hayward, 1964; Delafield *et al.*, 1965; Redfearn *et al.*, 1966; Stanier *et al.*, 1966; Ballard *et al.*, 1968; Ballard *et al.*, 1970; Davis *et al.*, 1970; Palleroni & Doudoroff, 1971; Palleroni *et al.*, 1972; Palleroni & Doudoroff, 1972; Ralston, 1972; Ralston, Palleroni & Doudoroff, 1973).

P. aeruginosa is a very well-defined species that in most cases can be readily recognized by a limited number of simple characters. The name of the species refers to the most striking character of the cultures, a colour resembling copper rust or verdigris. The cells are rods of an average thickness of 0·5 μm and a length ranging from 1·4 to 3 μm. Most strains are motile and they typically possess one polar flagellum per cell. Among the 354 strains of Jessen's collection 20 were non-motile. One strain had an unusually high proportion of cells (15 %) with two flagella, thus deviating from the maximum limit (10 %) arrived at by Lautrop and Jessen (1964) for typically mono-trichous strains.

The production of pyocyanin and fluorescent pigments are characteristic of the species, but some strains may lack one or both of these types of pigment. Several other pigments have been reported; phenazine-1-carboxylic acid, dihydroxyphenazine-1-carboxylic acid, chlororaphin, oxychlororaphin, the red pigments aeruginosin A and B, melanin, and the chlorine-containing pyoluteorine (Takeda, 1958; Stanier *et al.*, 1966; Palleroni & Doudoroff, 1972). Some of these pigments may also be produced by biotypes of the species *P. fluorescens*.

P. aeruginosa strains have interesting characters which, in combination, are important for the diagnosis of the species. Jessen (1965) has listed the following: tendency to localized swarming from the edge of the colonies; pigmentation; characteristic smell (described by some bacteriologists as 'fruity'); presence of shining metallic patches on the surface of the colonies. This combination is present in the majority of the cultures examined, and thus it constitutes a powerful diagnostic tool. The colonies frequently spread on the surface of the agar from the point of inoculation, a phenomenon that depends on the composition of the medium (spreading is striking in media containing 'Tween' 80). Many colonies of *P. aeruginosa* are pitted.

All strains grow fast at 37 °C, a character that differentiates *P. aeruginosa* from other fluorescent pseudomonads. Some strains fail to grow at 42 °C, but 44 °C is the maximum growth temperature for many. There is no growth at 5 °C.

The organism is an absolute aerobe except in media containing nitrate; when nitrate is present in relatively high concentration in liquid cultures the reduction to nitrogen gas becomes visible. The catalase and oxidase reaction are positive. The arginine dihydrolase reaction is positive. Gelatin is hydro-lysed, while starch and poly-β-hydroxybutyrate are not. Most strains have a weak lipolytic action, as evidenced by the formation of a small amount of precipitate in media supplemented with polysorbate 'Tween' 80 and calcium salts.

The egg yolk reaction is negative. This poorly defined reaction has been attributed to the presence of lecithinases, but its complexity is shown by the fact that *P. aeruginosa* gives a negative reaction, although it has been reported as a strong lecithinase producer. Probably the reaction involves proteins of the complex medium used for the assay, and these may be destroyed by the extracellular proteases before they can participate in the reaction.

The species is nutritionally versatile, and individual strains can use about 80 or more organic compounds. Characteristic substrates of the nutritional spectrum are sebacate, geraniol, mandelate, acetamide and some hydrocarbons. All strains can grow with ammonia, valine, lysine and urea as nitrogen sources; nitrate and creatinine (but not creatine) can be used by most strains.

A concentration of 3 % sodium chloride is well tolerated in culture media, and many strains grow with 6 % salt, which is perhaps close to the maximum concentration tolerated.

Strains of *P. aeruginosa* are commonly isolated from hospital specimens, and the potential pathogenic capacity of the organism provokes preoccupation in medical practice due to its resistance to common antibiotics (see Chapter 2). The species has also been found to be phytopathogenic, and thus represents a very rare case of a microorganism pathogenic for both plant and animals. The name *P. polycolor* has been applied to strains of plant origin, although the synonymy with *P. aeruginosa* is now generally accepted. Isolation of strains from soil is easy using enrichment cultures under denitrification conditions with a variety of carbon sources. The strains are remarkably uniform both in general phenotypic properties and in nucleic acid homology (Stanier *et al.*, 1966; Palleroni & Doudoroff, 1972; Palleroni *et al.*, 1972). Although the species shares important properties with other fluorescent pseudomonads, it is clearly separate from them; the G + C content of the DNA (*ca.* 67 moles %) is higher than for any other fluorescent pseudomonad and many phages are specific for the strains of the species. Strains of *P. aeruginosa* are capable of producing specific bacteriocins (aeruginocins).

P. fluorescens is probably the most complex species of the genus *Pseudomonas* (see Table 1.2). The differentiation of *P. fluorescens* from *P. putida*, another common fluorescent saprophytic species, can be done on the basis of gelatin liquefaction, a diagnostic character used by Flügge in 1886. This property, however, is not infallible but the differentiation can also be achieved by means of other phenotypic properties (Stanier *et al.*, 1966). Among the gelatinase negative fluorescent pseudomonads are *P. putida* and many species of plant pathogenic bacteria.

P. fluorescens has been subdivided into a number of biotypes by Stanier *et al.* (1966), but the subdivision could not accommodate all the organisms of the species in the Berkeley collection. The subdivision in biotypes was primarily based on two salient properties that had been used by other authors (Fuchs, 1959), namely, capacity for denitrification and synthesis of levan from sucrose. Some other characters correlate fairly well with these two, but the delimitation of the biotypes in some cases is rather unclear.

The strains of biotype A are capable of levan synthesis but do not denitrify, and among them has been placed the type strain proposed by Rhodes (1959). A substantial level of DNA homology is shared with other biotypes, particularly with biotype B (Palleroni *et al.*, 1972). The strains assigned to biotype B are phenotypically more heterogeneous than those of biotype A, and are capable of both denitrification and levan formation. The biotype includes the species *P. marginalis*, a fluorescent plant pathogenic species. Biotype C includes denitrifiers incapable of levan formation. The biotype is clearly heterogeneous and it can be split into several subgroups that are now under study.

Two groups of organisms originally described as species, *P. chlororaphis* and *P. aureofaciens* were reduced in this classification to the category of biotypes (D and E, respectively). This decision was taken at the time because of a number of properties of these two species in common with *P. fluorescens*, but it is now clear that they can be readily differentiated from other biotypes of the species and that they perhaps deserve separate specific status. Their G + C content is higher than that of other biotypes, and they also share a high level of DNA homology (see Figure 2 of Palleroni & Doudoroff, 1972). Strains of the biotypes D and E produce very characteristic phenazine pigments.

Biotype F includes the species *P. lemonnieri*, characterized by the production of an insoluble blue pigment (Starr, Blau & Cosens, 1960). This property can be easily lost by subcultivation in laboratory media. Only two strains have been thoroughly characterized, and therefore it is difficult to assess with with some precision the relationship of this biotype to others. These two strains were until recently the only two known in culture collections, but several new strains have now been isolated and are currently under study. All strains are levan producers and denitrifiers.

Biotype G is reserved for strains incapable of denitrification and levan formation and, as often happens with groups defined on the basis of negative properties, it represents the most artificial and heterogeneous biotype of the species. Strains of other biotypes which have lost by mutation one or both of the basic properties of the primary classification, would be placed in this biotype. Phenotypic studies performed on a large number of strains of Jessen's collection seem to indicate the presence of definite clusters within this biotype (unpublished observations).

In summary, the internal subdivision of *P. fluorescens* is far from satisfactory. Several biotypes or members of biotypes may deserve species status, and much work is yet to be done before attempting to unravel the internal complexity of this species.

Of the 57 strains of fluorescent pseudomonads isolated by den Dooren de Jong (1927), 23 came from soil and only one of these liquefied gelatin. All other strains were of aquatic origin and appeared to be gelatin liquefiers.

These results carried interesting ecological implications, and the conclusion was that *P. putida* (the non-liquefier) is mainly a soil organism while *P. fluorescens* occurs predominantly in water. The collection of strains studied at Berkeley contained 41 fluorescent, gelatinase negative strains that could be assigned to *P. putida*; only one of them came from a water sample, and the origin of the remaining strains for which the history was known, was in all cases the soil. However, many *P. fluorescens* strains from our collection came from soil, and in Jessen's collection (Jessen, 1965) there is a good number of *P. putida* strains that have been isolated from water. In view of the evidence collected from the histories of a large number of *P. putida* strains, the hypothesis of the ecological specialization of the two species seems no longer tenable. It must be granted, however, that the ample opportunities for cross-contamination of natural waters with soil and vice versa would require very careful sample collecting techniques before meaningful conclusions could be drawn about the ecology of the two species.

    P. putida is a fairly well-defined species, which is somewhat surprising if one considers the fact that the differentiation from other saprophytic fluorescent *Pseudomonas* species is achieved mainly through some of its negative properties: the incapacity for gelatin liquefaction and for denitrification, and the inability to give the egg yolk reaction. None of the strains grow at 41 °C, very few grow at 4 °C, none hydrolyse 'Tween' 80 or utilize sucrose, and none of the sucrose utilizers is capable of producing levan. The species has been subdivided into two biotypes on the basis of the ability to use L-tryptophan for growth. The properties of the tryptophan utilizers (biotype B) suggest that they constitute a group in some ways more resembling *P. fluorescens* than typical *P. putida* (biotype A). DNA hybridization experiments provided additional evidence in favour of this hypothesis (Palleroni *et al.*, 1972). *P. putida* biotype A appears to be fairly homogeneous in phenotypic traits, but the DNA homology, which has been performed on 12 strains of the collection, has shown a marked internal heterogeneity. More work will have to be done on a larger collection in order to detect any significant degree of internal clustering in this species. One of the strains of *P. putida* (ATCC 12633) has been the subject of many important pieces of research on the metabolism of aromatic compounds.

    A number of fluorescent pseudomonads can be isolated from diseased plants, constituting a large group whose taxonomic structure is still highly unsatisfactory. Some gelatin liquefiers (*P. marginalis*) are now assigned to *P. fluorescens*. Both this species and *P. cichorii* are oxidase positive, while most other species are oxidase negative. The oxidase negative organisms constitute a very complex assemblage including many nomenspecies that have been named on the basis of the host of origin. Most of them will be lumped together under one name, *P. syringae*, in the forthcoming edition of *Bergey's Manual of Determinative Bacteriology*; however, it is becoming clear that this may be a gross oversimplification as work on these organisms progresses in various laboratories, and several of the clusters that can be differentiated on the basis of DNA homology, selected nutritional characters and host range, may well deserve independent specific rank. These and other plant pathogenic bacteria will be discussed in more detail below in this chapter.

a.ii. The Non-fluorescent Species of Group I

Four non-fluorescent species of the RNA homology group I are arranged into two subgroups, the stutzeri and the alcaligenes subgroups. The alcaligenes subgroup was originally defined on purely phenotypic grounds by Stanier *et al.* (1966), who noticed some resemblance between these organisms and the fluorescent pseudomonads. Recently a number of strains have been added to the collection and a better circumscription of the taxon has been achieved (Ralston, 1972). In this subgroup, *P. alcaligenes* is represented by three strains which are not very similar to one another; *P. pseudoalcaligenes*, on the other hand, has many more representative strains, and these are linked by a fair degree of DNA homology. DNA homology can also be demonstrated between this species and members of the stutzeri subgroup. Some strains of *P. pseudoalcaligenes* are the only members of the *P. fluorescens* RNA homology group that are able to accumulate poly-β-hydroxybutyrate. Little is known about the ecology of the alcaligenes subgroup; many strains have been isolated from specimens of clinical origin, but it seems likely that they are casual contaminants of such materials.

P. stutzeri has been known for a long time to soil bacteriologists and recently a new species isolated from soil in Mendoza, Argentina, could be placed relatively closely to this species (Palleroni *et al.*, 1970). While *P. stutzeri* is internally very heterogeneous, *P. mendocina* is remarkably uniform. *P. stutzeri* is very common in soils and it has been isolated frequently even from desert soils of very low organic matter content. The widespread occurrence probably accounts for the fact that a fair number of strains could also be isolated from clinical materials. Simple enrichment cultures under conditions of denitrification are procedures that commonly yield *P. stutzeri* and occasionally *P. mendocina* as non-fluorescent denitrifying species. *P. stutzeri* is frequently recognized by the characteristically wrinkled colonies on the plates (van Niel & Allen, 1952). One of the recommended procedures based on denitrification enrichment (van Iterson, 1902) uses tartrate as carbon source which, however, does not seem to be a suitable substrate for the pure cultures of the species that can be obtained from the enrichment.

One indication of the internal heterogeneity of *P. stutzeri* is given by the wide range of G + C values of the species (61 to 66 moles %); this is, in fact, the widest range for a single species in the whole genus. Not even *P. fluorescens* or the fluorescent plant pathogens, which include entities equivalent to independent species, are so variable in DNA base composition. Mandel (1966) has tried to solve the problem by the creation of a new species, *P. stanieri*, reserved for the strains with the lower G + C content, but no clear-cut phenotypic differences seem to correlate with this division. The inclusion of several strains of Argentine origin in the *P. stutzeri* collection has erased the discontinuity in the GC scale suggestive of the presence of only two species (Palleroni *et al.*, 1970).

5.b. The pseudomallei–cepacia RNA Homology Group (Group II)

This large assembly of aerobic pseudomonads is interesting because of the fact that with the exception of one species (*P. pickettii*) all members are pathogens of animals and plants. The strains of *P. pickettii* studied by Ralston *et al.* (1973) are of clinical origin, but the possible role of this species as a pathogen is doubtful and has not been explored. Two of the species of the group, *P. pseudomallei* and *P. mallei*, are the most virulent animal pathogens in the genus.

The members of this large RNA homology group have few phenotypic characters in common, the most important of which is the ability to accumulate poly-β-hydroxybutyrate as carbon reserve material. As in the case of the *P. fluorescens* RNA homology group, the members of this group are also related to one another directly or indirectly by a certain level of DNA homology.

Among the species of the group one of the most remarkable is *P. cepacia* (*P. multivorans*). Some of the strains of this species are the most nutritionally versatile in the genus *Pseudomonas*. It is at present not clear why this species, so well endowed with the capacity of using a large variety of organic compounds for growth, is not isolated more frequently from natural materials; this is an interesting ecological problem that deserves further consideration. The species is not very homogeneous internally in phenotypic properties or in DNA homology. These two experimental approaches do not correlate very well, and it is at present impossible to suggest any basis for the internal subdivision of the species.

The phytopathogen *P. marginata* includes at present another species, *P. alliicola* because a close relationship between the two species has been demonstrated (Ballard *et al.*, 1970). These two names may be dropped altogether in the future since it is proposed that the name *P. gladioli* may have priority (Hildebrand, Palleroni & Doudoroff, 1973). A third species of plant bacteria in this group, *P. caryophylli*, is rather distant from the other members of the group.

P. pseudomallei and *P. mallei* are the agents of the animal diseases melioidosis and glanders, respectively. These serious diseases can be transmitted to man (Redfearn & Palleroni, 1973). *P. mallei* has been assigned to the genus *Pseudomonas* because of its clear phylogenetic relationship to *P. pseudo-mallei* (Redfearn *et al.*, 1966; Rogul *et al.*, 1970). *P. mallei* has never been isolated except from animal hosts while *P. pseudomallei* may be normally a free-living organism. This last species is nutritionally very versatile and probably plays a significant role in the process of mineralization of organic matter. Interesting aspects of the ecology of these two organisms have been discussed elsewhere (Redfearn *et al.*, 1966; Redfearn & Palleroni, 1973).

The last two species of the group, *P. solanacearum* and *P. pickettii*, are related to one another by DNA homology. *P. solanacearum* is a plant pathogen that has been known for a long time and is considered as one of the most serious of bacterial phytopathogens of tropical regions. Its name indicates a relation to plants belonging to the family *Solanaceae*, but plants of many other botanical families are similarly susceptible. A subdivision of the species into four biotypes was proposed by Hayward (1964) but cannot be fully supported by phenotypic or DNA data. However, at least two groups appear to be evident within the species (Palleroni & Doudoroff, 1971). The recently described species *P. pickettii* (Ralston *et al.*, 1973) is a very uniform species representing a link between *P. solanacearum* and other members of the group.

5.c. The acidovorans RNA Homology Group (Group III)

This RNA homology group is at present constituted by the two species of the acidovorans DNA homology group, *P. acidovorans* and *P. testosteroni*, by the two species of the facilis–delafieldii DNA homology group, and by the species *P. saccharophila*. These five species can thus be subdivided into three DNA homology groups that share with one another no detectable DNA homology, and therefore this is the first RNA homology group composed by various independent DNA homology groups.

P. acidovorans and *P. testosteroni* resemble one another phenotypically and are among the best defined and most easily recognizable species of the genus (Stanier *et al.*, 1966). In spite of having different DNA base composition, the two species share a low but easily detectable level of DNA homology. A number of important biochemical properties differentiate these two species from most of the species we have considred so far.

P. facilis, *P. delafieldii* and *P. saccharophila* have some phenotypic resemblance, but DNA homology can only be demonstrated between the first two species (Ralston, Palleroni & Doudoroff, 1972). The homology is very high, and this is one of the most convincing arguments against the assignment of *P. facilis* to a different genus (*Hydrogenomonas*). *P. saccharophila* was discovered by Doudoroff (1940) and has been used for some outstanding research in carbohydrate metabolism.

5.d. The diminuta RNA Homology Group (Group IV)

This small group contains two species, *P. diminuta* and *P. vesicularis*, that are unique morphologically because of the short wavelength of the single polar flagellum, and also physiologically due to the requirements for organic growth factors and their inability to use nitrate as a nitrogen source (Ballard *et al.*, 1968). A low level of DNA homology can be demonstrated between the two species.

5.e. The *P. maltophilia–Xanthomonas* RNA Homology Group (Group V)

Several important phenotypic properties single out *P. maltophilia* from most *Pseudomonas* species: the oxidase reaction is negative, methionine is required for growth in mineral media, and nitrate is not used as nitrogen source. It is also the only known species of the genus capable of using lactose as carbon sources. Most strains of the Berkeley collection have a clinical origin, but the species seems to be widespread in nature because it has also been isolated from water, milk and hydrocarbon enrichments. *r*RNA–DNA competition experiments have shown that this species is related to *Xanthomonas* species. The meaning of these unexpected results is very difficult to assess at the present moment. *Xanthomonas* species are plant pathogens capable of producing special carotenoid-like pigments which, on superficial examination, seem to have little in common with *P. maltophilia*.

5.f. Miscellaneous Species

In this section will be presented a brief discussion of several species whose relationships to other well-characterized species of the genus have not been explored until now. *P. lemoignei* is a well-defined species which is represented by a single strain. The species is very easily identified because of its very restricted nutritional spectrum. Among the very few organic compounds that can be utilized for growth is the polymer poly-β-hydroxybutyrate. The original description of the species (Delafield *et al.*, 1965) included the inability to grow at 41 °C, but this has recently been found to be a mistake (R. Gherna, personal communication). The G + C content of the DNA of this organism is in the lower limit of the *Pseudomonas* range (58 moles %), and the DNA does not show homology with that of other species of the genus.

The remaining species of well-characterized *Pseudomonas*, *P. ruhlandii*, *P. flava* and *P. palleronii* are hydrogen bacteria (Davis *et al.*, 1970). The last two species produce carotenoid pigments.

5.g. The Plant Pathogenic Species of *Pseudomonas*

The distribution of plant pathogens among the different species groups of *Pseudomonas* is far from random. The number of species described is enormous but many of them cannot be recognized from their original descriptions and type cultures are not available in many cases. Therefore, many named species could not be reexamined in recent taxonomic investigations. Even so, we can perhaps state at present that two important groups, the fluorescent and pseudomallei–cepacia groups include practically all the

Table 1.4. Key for the differentiation of the phytopathogenic species of *Pseudomonas*

I. Produce fluorescent pigment. Do not accumulate poly-β-hydroxybutyrate. The GC content of the DNA ranges from 58 to 61 moles % (the only exception is *P. aeruginosa*, with *ca.* 67 moles %).	
A. Arginine dihydrolase reaction positive.	
1. Produce pyocyanin; grow at 41 °C	*P. aeruginosa (syn. P. polycolor)*
2. Do not produce pyocyanin; do not grow at 41 °C	Some strains of *P. fluorescens* biotype B (syn. *P. marginalis*)
B. Arginine dihydrolase reaction negative.	
1. Oxidase reaction positive	*P. cichorii* (syn. *P. papaveris?*)
2. Oxidase reaction negative	*P. syringae*[a]
II. Do not produce fluorescent pigment. Accumulate poly-β-hydroxybutyrate. The GC content of the DNA ranges from 65 to 68 moles %.	
A. Use arginine and betaine for growth. Grow at 41 °C.	
1. Denitrify. Arginine dihydrolase reaction positive	*P. caryophylli*
2. Do not denitrify. Arginine dihydrolase reaction negative	
a. Use mesaconate and D($-$)tartrate as carbon sources; do not use levulinate, *m*-hydroxybenzoate or tryptamine	*P. marginata* (syn. *P. alliicola, P. gladioli?*)
b. Do not use mesaconate or D($-$)tartrate; use levulinate, *m*-hydroxybenzoate and tryptamine for growth	*P. cepacia* (syn. *P. multivorans*)
B. Do not use betaine or arginine for growth. Do not grow at 41 °C	*P. solanacearum*

[a] This collective species include a large number of nomenspecies. The synonyms of *P. syringae* and of some of the species mentioned in the second list below, are the following: *P. avenae, P. barkeri, P. cerasus, P. citrarefaciens, P. citriputealis, P. hibisci, P. holci, P. matthiolae, P. nectarophila, P. papulans, P. prunicola, P. punctulans, P. rimafaciens, P. sojae, P. spongiosa, P. trifoliorum, P. vignae, P. viridifaciens, P. utiformica.* The following species have also been included as synonyms of *P. syringae*, although there are indications that some of them may deserve independent specific rank: *P. angulata, P. aptata, P. coronafaciens, P. delphinii, P. dysoxyli, P. garcae, P. glycinea, P. helianthi, P. lachrymans, P. mellea, P. mori, P. morsprunorum, P. panacis, P. phaseolicola, P. pisi, P. savastanoi, P. sesami, P. tabaci, P. tonelliana* (Dr. D. C. Hildebrand, personal communication).

phytopathogenic species of the genus. This fact is obviously important, since plant pathogenicity was not one of the characters that was taken into consideration for the primary subdivision; therefore this distribution may indicate that the subdivision presented above probably approaches a natural classification of the genus.

Table 1.4 shows a key for the differentiation of the phytopathogenic species of *Pseudomonas* which is given with the idea of presenting the names of most accepted species and some of their synonyms, together with some of the useful characters for their differentiation. It should be remembered that, like any other determinative key, the present one has been drawn from the properties of a given collection of organisms, and therefore its structure may change radically when other strains or new species are included. Some phytopathogenic species do not seem to correspond to any of the species mentioned in the key, but their precise location cannot be given at present on the basis of their published descriptions.

Taxonomic work based on the approach of Stanier *et al.* (1966) has been performed on the fluorescent plant pathogens by Misaghi and Grogan (1969) and by Sands, Schroth and Hildebrand (1970); their work should be consulted for more details. The paper by Lelliott, Billing and Hayward (1966) also contains very valuable information on a limited range of characters of a large collection of organisms of the fluorescent group.

6. CONCLUSIONS AND PREDICTIONS

The classification of the species of *Pseudomonas* presented in this chapter is still incomplete with respect to some of the species groups, for which more work on a larger collection of organisms will be necessary. The study of additional strains will perhaps result in a more precise circumscription of the taxa and the creation of new subgroups. Species such as *P. alcaligenes*, now represented by only three strains that are quite dissimilar to one another, may be brought into sharper focus; on the other hand, more strains assignable to poorly defined clusters within species such as *P. fluorescens* or *P. stutzeri* may help for a better understanding of these complex species. This rather optimistic view of *Pseudomonas* taxonomy is not shared by all workers in the field, some of whom seem to doubt the power of present taxonomic approaches for the resolution of the formidable complexity of some of the groups (Rhodes, 1971).

At present it is difficult to assign to conventional taxonomic hierarchies the groups defined by ribosomal RNA–DNA competition experiments. The first obvious conclusion is that these groups may be equivalent to independent genera and that the general definition of the genus as presented should be applied to a suprageneric category. In view of our poor understanding of bacterial taxonomic hierarchies, particularly of those above the generic level, it may be totally unwarranted now to try to define such a taxon. If *Pseudomonas* were to be divided into several genera, these genera would share many basic morphological and physiological properties and the differentiation would only be feasible by means of a few seemingly unimportant characters in addition to the results of tests that are still far from the reach of the practical taxonomist. The diminuta group may be an exception for which a separate generic category might be created on the basis of morphological and physiological traits. Aside from this case, there are no important phenotypic characters that are shared by all members of a given RNA homology group, that at the same time are not shared by one or more species of other groups and, as we have seen, the DNA base composition of each group is almost as variable as for the whole genus as presently defined.

Early in the work in which the author has participated, in an attempt at establishing practical criteria for the differentiation of species of *Pseudomonas*, the concept of 'ideal phenotype' was introduced (Stanier *et al.*, 1966). The ideal phenotype consisted of a constellation of a limited number of properties permitting the differentiation of a given species from all others that had been characterized by the same methods. While the ideal phenotype is a reasonable tool within a given collection, its general usefulness may be questioned. As any other key system, the ideal phenotype describes the species as accurately as the collection from which it was abstracted, and a given constellation of characters may lose much of its value when more strains are added to the collection or more species are included in the comparison.

From the extensive phenotypic characterization of the species discussed in this chapter, some characters were given more weight than others for the classification, even though in this visual comparison of the data the main characteristics of the total phenotype were always kept in mind. Later on, the phenotypic data were subjected to numerical analysis, and the similarity (S_J) and matching (S_{SM}) coefficients were in some cases arranged according to procedures of cluster analysis (Palleroni *et al.*, 1972). The results confirmed, with few exceptions, the early assignment of the species into subgroups and the data of the DNA–DNA hybridization experiments. Unfortunately, all the evidence gathered so far seems to be insufficient for an evaluation of the precise circumscription of the taxa, considering that we still know very little about the roles played by the exchange of genetic material or by the forces acting on different organisms which have participated of the same natural niches. These and other problems will have to be resolved through considerable work before the answer to speciation in the genus *Pseudomonas* can be attempted on a solid basis. For the time being, the best we can do is to base the taxonomic system on information obtained in various fields of research, since any one line taken independetly is likely to lead to some misevaluation of relationships. In practice, classification may be performed on the basis of phenotypic characters, but other studies at the biochemical and genetic levels may provide valuable additional evidence since often these approaches reach a finer resolving power than the simple phenotypic characterization. Comparative studies involving data from different approaches have not yet been performed on a sufficiently large scale in members of this vast genus, but is is only by means of a balanced contribution of studies on the various aspects of the biology of these organisms that the road will eventually be open to a natural classification of the species of *Pseudomonas*.

7. BIBLIOGRAPHY

Amako, K., Yasunaka, K. & Takeya, K. (1970) *Journal of General Microbiology*, **62**, 107.
Ambler, R. P. & Murray, S. (1973) *Biochemical Society Transactions*, **1**, 107.
Ambler, R. P. & Wynn, M. (1973) *Biochemical Journal*, **131**, 485.
Ballard, R. W., Doudoroff, M., Stanier, R. Y. & Mandel, M. (1968) *Journal of General Microbiology*, **53**, 349.
Ballard, R. W., Palleroni, N. J., Doudoroff, M., Stanier, R. Y. & Mandel, M. (1970) *Journal of General Microbiology*, **60**, 199.
Baumann, L., Baumann, P., Mandel, M. & Allen, R. D. (1972) *Journal of Bacteriology*, **110**, 402.
Bendich, A. J. & McCarthy, B. J. (1970) *Proceedings of the National Academy of Sciences U.S.A.*, **65**, 349.
Bradley, D. E. (1967) *Bacteriological Reviews*, **31**, 230.
Bradley, D. E. (1972) *Genetical Research*, **19**, 39.
Brenner, D. J. & Cowie, D. B. (1968) *Journal of Bacteriology*, **95**, 2558.
Brenner, D. J., Martin, M. A. & Hoyer, B. H. (1967) *Journal of Bacteriology*, **94**, 486.
Calhoun, D. H., Carson, M. & Jensen, R. A. (1972) *Journal of General Microbiology*, **72**, 581.
Cánovas, J. L., Ornston, L. N. & Stanier, R. Y. (1967) *Science*, **156**, 1695.
Carrick, L. & Berk, R. S. (1971) *Journal of Bacteriology*, **106**, 250.
Clarke, P. H. (1972) *Journal of General Microbiology*, **71**, 241.
Cohen, G. N., Stanier, R. Y. & Le Bras, G. (1969) *Journal of Bacteriology*, **99**, 791.
Crawford, I. P. & Yanofsky, C. (1971) *Journal of Bacteriology*, **108**, 248.
Datta, P. (1969) *Science*, **165**, 556.
Davis, D. H., Stanier, R. Y., Doudoroff, M. & Mandel, M. (1970) *Archiv für Mikrobiologie*, **70**, 1.
Delafield, F. P., Doudoroff, M., Palleroni, N. J., Lusty, C. J. & Contopoulou, R. (1965) *Journal of Bacteriology*, **90**, 1455.
den Dooren de Jong, L. E. (1927) *Bijdrage tot de kennis van het mineralisatie-process*, Nijgh & Van Ditmar, Rotterdam.
Dickerson, R. E. (1971) *Journal of Molecular Biology*, **57**, 1.
Doudoroff, M. (1940) *Enzymologia*, **9**, 59.
Dworkin, M. & Foster, J. W. (1956) *Journal of Bacteriology*, **72**, 646.
Enatsu, T. and Crawford, I. P. (1971) *Journal of Bacteriology*, **108**, 431.
Feist, C. F. & Hegeman, G. D. (1969) *Journal of Bacteriology*, **100**, 869.
Flügge, C. (1886) *Die Mikroorganismen*, 2 Aufl., F. C. W. Vogel, Leipzig.
Fuchs, A. (1959) *On the Synthesis and Breakdown of Levan by Bacteria*, Waltam, Delft.

Fuerst, J. A. & Hayward, A. C. (1969a, b) *Journal of General Microbiology*, **58**, 227, 239.
Garibaldi, J. A. (1967) *Journal of Bacteriology*, **94**, 1296.
Gryder, R. M. & Adams, E. (1969) *Journal of Bacteriology*, **97**, 292.
Hayward, A. C. (1964) *Journal of Applied Bacteriology*, **27**, 265.
Heinrichsen, J. (1972) *Bacteriological Reviews*, **36**, 478.
Heumann, W. & Marx, R. (1964) *Archiv für Mikrobiologie*, **47**, 325.
Hildebrand, D. C., Palleroni, N. J. & Doudoroff, M. (1973) *International Journal of Systematic Bacteriology*, **23**, 433.
Hoffman, H. P., Geftic, S. C., Heymann, H. & Adair, F. W. (1973) *Journal of Bacteriology*, **114**, 434.
Hoyer, B. H., McCarthy, B. J. & Bolton, E. T. (1964) *Science*, **144**, 959.
Jensen, R. A., Nasser, D. S. & Nester, E. W. (1967) *Journal of Bacteriology*, **94**, 1582.
Jessen, O. (1965) *Pseudomonas aeruginosa and Other Green Fluorescent Pseudomonads. A Taxonomic Study*, Munksgaard, Copenhagen.
Johnson, J. L. & Ordal, E. J. (1968) *Journal of Bacteriology*, **95**, 893.
King, E. O., Ward, W. K. & Raney, D. E. (1964) *Journal of Laboratory Clinical Medicine*, **44**, 301.
Kingsbury, D. T. (1967) *Journal of Bacteriology*, **94**, 870.
Kingsbury, D. T., Fanning, G. R., Johnson, K. E. & Brenner, D. J. (1969) *Journal of General Microbiology*, **55**, 201.
Lautrop, H. (1965) *Publication of the Faculty of Sciences, University J.E. Purkyne*, Ser. K., **35**, 322.
Lautrop, H. & Jessen, O. (1964) *Acta Pathologica et Microbiologica Scandinavica*, **60**, 588.
Lelliott, R. A., Billing, E. & Hayward, A. C. (1966) *Journal of Applied Bacteriology*, **29**, 470.
Lessie, T. G. & Neidhardt, F. C. (1967) *Journal of Bacteriology*, **93**, 1800.
McCarthy, B. J. & Bolton, E. T. (1963) *Proceedings of the National Academy of Sciences U.S.A.*, **50**, 156.
Mandel, M. (1966) *Journal of General Microbiology*, **43**, 273.
Misaghi, I. & Grogan, R. G. (1969) *Phytopathology*, **59**, 1436.
Ornston, M. K. & Ornston, L. N. (1972) *Journal of General Microbiology*, **73**, 455.
Palleroni, N. J., Ballard, R. W., Ralston, E. & Doudoroff, M. (1972) *Journal of Bacteriology*, **110**, 1.
Palleroni, N. J. & Doudoroff, M. (1971) *Journal of Bacteriology*, **107**, 690.
Palleroni, N. J. & Doudoroff, M. (1972) *Annual Review of Phytopathology*, **10**, 73.
Palleroni, N. J., Doudoroff, M., Stanier, R. Y., Solánes, R. E. & Mandel, M. (1970) *Journal of General Microbiology*, **60**, 215.
Palleroni, N. J., Kunisawa, R., Contopoulou, R. & Doudoroff, M. (1973) *International Journal of Systematic Bacteriology*, **23**, 333.
Palleroni, N. J. & Stanier, R. Y. (1964) *Journal of General Microbiology*, **35**, 319.
Quayle, J. R. (1972) *Advances in Microbial Physiology*, **7**, 119.
Queener, S. F. & Gunsalus, I. C. (1970) *Proceedings of the National Academy of Sciences U.S.A.*, **67**, 1225.
Ralston, E. (1972) *Some Contributions to the Taxonomy of the Genus* Pseudomonas, Thesis, University of California, Berkeley.
Ralston, E., Palleroni, N. J. & Doudoroff, M. (1972) *Journal of Bacteriology*, **109**, 465.
Ralston, E., Palleroni, N. J. & Doudoroff, M. (1973) *International Journal of Systematic Bacteriology*, **23**, 15.
Redfearn, M. S. & Palleroni, N. J. (1973) 'Melioidosis. Glanders', In *Diseases Transmissible from Animals to Man*. (Eds. W. T. Hubbert, W. F. McCulloch and P. I. Schurrenberger), Academic Press.

Redfearn, M. S., Palleroni, N. J. & Stanier, R. Y. (1966) *Journal of General Microbiology*, **43**, 293.

Rhodes, M. E. (1959) *Journal of General Microbiology*, **21**, 221.

Rhodes, M. E. (1971) *Journal of General Microbiology*, **69**, xi.

Robert-Gero, M., Poiret, M. & Stanier, R. Y. (1969) *Journal of General Microbiology*, **57**, 207.

Rogul, M., Brendle, J. J., Haapala, D. K. & Alexander, A. D. (1970) *Journal of Bacteriology*, **101**, 827.

Rosenfeld, H. & Feigelson, P. (1969) *Journal of Bacteriology*, **97**, 697.

Sands, D. C., Schroth, M. N. & Hildebrand, D. C. (1970) *Journal of Bacteriology*, **101**, 9.

Schlesinger, S., Scotto, P. & Magasanik, B. (1965) *Journal of Biological Chemistry*, **240**, 4331.

Sneath, P. H. A. (1960) *Iowa State Journal of Science*, **34**, 243.

Stanier, R. Y. (1968) In *Chemotaxonomy and Serotaxonomy* (Systematic Association Special Volume), **2**, 201.

Stanier, R. Y., Palleroni, N. J. & Doudoroff, M. (1966) *Journal of General Microbiology*, **43**, 159.

Stanier, R. Y., Wachter, D., Gasser, C. & Wilson, A. C. (1970) *Journal of Bacteriology*, **102**, 351.

Starr, M. P., Blau, W. & Cosens, G. (1960) *Biochemische Zeitschrift*, **333**, 328.

Stephenson, M. (1930) *Bacterial Metabolism*; 2nd ed. 1939; 3rd ed. 1949, Longmans, Green & Co. Ltd., London.

Suzuki, A. & Goto, M. (1971) *Bulletin of the Chemical Society (Japan)*, **44**, 1869.

Takeda, R. (1958) *Hakko Kogaku Zasshi*, **36**, 281.

van Iterson, G. (1902) *Proceedings of the Section of Sciences. Kroninklijke Akademie van Wetenschappen* (Amsterdam), **5**, 148.

van Niel, C. B. & Allen, M. B. (1952) *Journal of Bacteriology*, **64**, 413.

Whittenbury, R., Phillips, K. C. & Wilkinson, J. F. (1970) *Journal of General Microbiology*, **61**, 205.

Yamamoto, T. (1967) *Journal of Bacteriology*, **94**, 1746.

Young, H. O., Chao, F., Turnbill, C. & Philpott, D. E. (1972) *Journal of Bacteriology*, **109**, 862.

CHAPTER 2

Ecological Importance of
Pseudomonas aeruginosa:
Medical Aspects

EDWARD J. L. LOWBURY

1. INTRODUCTION

Microbial species are classified for medical purposes as pathogenic and non-pathogenic, the former sometimes including avirulent as well as virulent strains. This subdivision is convenient in distinguishing organisms that cause specific communicable diseases, such as tuberculosis, from the saprophytes and most of the commensals that multiply harmlessly on the surfaces of the human body. Saprophytes and commensals can, however, cause severe pathological effects when they gain access to tissues with poor antimicrobial defences, or to patients whose resistance is impaired by illness or by certain forms of medical treatment: such infections are commonly called 'opportunistic'. Adverse and potentially fatal immediate effects can also occur when large numbers of bacteria are introduced into the bloodstream with contaminated infusion fluids or blood, even though the bacteria thus introduced do not multiply in the patient's body and the condition is not, in the strict sense, an 'infection'.

Virtually any microorganism that can, under some circumstances, grow in human tissues can act as an opportunist, but there are certain species of bacteria, viruses, fungi and protozoa which are particularly associated with this type of infection. Such organisms are characterized by resistance to a wide range of antibiotics which are usually active against the more pathogenic bacteria; with this biological advantage, they emerge by selection in hospitals where antibiotics are much used. Of the bacteria showing these properties, *Pseudomonas aeruginosa* has played a particularly important role. Though virtually harmless towards healthy, uninjured subjects (except in infancy), it appears to be more pathogenic than, for example, *Staphylococcus aureus* in the anterior chamber of the eye and more prone than other bacteria to cause fatal infection in patients with severe burns. It presents special hazards in patients requiring instrumentation (e.g. catherization) which can introduce bacteria to susceptible tissues; also in lesions containing necrotic tissue (e.g. burns) or fluids (e.g. urine, CSF) in which the organisms can grow rapidly, from small inocula to potentially invasive numbers, without interference from the natural defences. *P. aeruginosa* produces a variety of toxins and enzymes that make it a potentially dangerous invader if the natural defences are breached; its ability to grow as a saprophyte in water with minimal nutrient additives also makes it a particularly important potential contaminant of infusion fluids and other sterile solutions.

As a result of improvements in treatment, many patients who would formerly have died from causes other than infection (e.g. shock following severe burns) are today kept alive, but with enhanced susceptibility and more prolonged exposure to the risk of infection. Modern treatment of severely ill patients often involves the use of equipment (e.g. respiratory ventilators) which is difficult to sterilize or disinfect and more liable to contamination with *P. aeruginosa* than with, say, *S. aureus* or *Haemophilus influenzae*, since the latter do not multiply (and many strains will not survive for long) in moist inanimate environments. The same patients may be receiving antibiotics which lead to the selection of a variety of resistant opportunist bacteria, and some will also be receiving steroid hormones, immunosuppressive drugs and other treatments which render them more susceptible to infection.

For such reasons *P. aeruginosa* and certain other Gram-negative bacilli (especially *Klebsiella* spp, *Proteus* spp and *Serratia marcescens*) have in recent years come to play a more important role in hospital infection (Barrett *et al.*, 1968; Adler *et al.*, 1971). These changes are associated with the use of a wide range of antibiotics and, in infant nurseries, with the selective action of hexachlorophane applications used in prophylaxis against *S. aureus* (Forfar *et al.*, 1968; Light *et al.*, 1968). *P. aeruginosa* has, however, been recognized as a potentially dangerous pathogen from the early days of bacteriology; the clinical features and variety of such infections were well illustrated in reviews by Fraenkel (1912, 1917), Epstein and Grossman (1933) and Stanley (1947). In the last few years there has been some evidence of a diminished incidence and severity of *P. aeruginosa* sepsis (e.g. Moyer *et al.*, 1965; Lowbury, 1971; Noone & Shafi, 1973).

2. ISOLATION AND IDENTIFICATION OF *P. AERUGINOSA* FROM CLINICAL SOURCES

2.a. Introduction

P. aeruginosa multiplies rapidly on simple culture media, usually producing characteristic colonies, pigments and odour; it is one of the bacterial species most easily isolated from and identified in clinical specimens such as wound exudate, pus, urine, etc. In these situations other species of *Pseudomonas* are rarely found. Bacteriological sampling of the inanimate environment (surfaces, solutions, etc.) is sometimes required in epidemiological studies; for example, in attempting to trace the source of an outbreak of infection. From these sites, too, it is usually easy to isolate *P. aeruginosa*, but other species of fluorescent *Pseudomonas* are also likely to be present in them and could be mistaken for *P. aeruginosa* unless careful confirmatory tests are made. Moreover, a small proportion of strains of *P. aeruginosa* are atypical— e.g. in colony form, or in failing to produce the characteristic pigments; special methods for isolation of *P. aeruginosa* from clinical sources are therefore appropriate where this organism presents a special infective hazard to the patient.

In clinical bacteriology it is usual to inoculate specimens from wounds and miscellaneous sources on blood agar and into a liquid medium, with the addition of selective or enrichment media for the detection of particular organisms. Though *P. aeruginosa*, if abundant and typical, will normally be identified without difficulty in the simpler media, the pigments pyocyanin and fluorescein are obscured in blood agar and often not apparent in cooked meat broth; scanty or poorly pigmented growth is likely to be missed. The following methods have therefore been developed for the detection of *P. aeruginosa* in a larger proportion of specimens, including those containing small numbers of the bacteria.

2.b. Methods of Isolation and Detection

The isolation of *P. aeruginosa* from clinical sources has been facilitated by the use of a selective medium containing 0·03 % cetrimide (Lowbury & Collins, 1955) and by examination of overnight cultures on blood agar or on the selective medium under an ultraviolet lamp for the characteristic yellow–green or blue–green fluorescence (Lowbury, 1951a). These methods have been improved by the use of King's medium B (King, Ward & Raney, 1954), which enhances the production of fluorescein, as the base for cetrimide agar (Brown & Lowbury, 1965), by the examination of cultures for fluorescence in an ultraviolet viewing cabinet suitable for use in an undarkened room (Lowbury, Lilly & Wilkins, 1962), and by the development of a medium containing 0·2 mg cetrimide and 15 mg nalidixic acid per ml of base (Goto & Enomoto, 1970; Lilly & Lowbury, 1972), which is more selective and less inhibitory towards *P. aeruginosa* than media containing 0·03 % cetrimide. Chloroxylenol has also been used in a culture medium as a selective agent for *P. aeruginosa* (Gould & McLeod, 1960), but this medium has been less extensively used and studied than cetrimide agar.

Though fluorescence under ultraviolet irradiation is a very useful criterion in the detection of *P. aeruginosa*, is is poorly developed or not seen on some agar media (e.g. McConkey's); on blood agar fluorescence is usually strong, especially at the periphery of areas of confluent growth. In a mixed culture individual highly fluorescent colonies must not be assumed to be the *Pseudomonas*, because fluorescein, which diffuses rapidly through the agar medium from *P. aeruginosa* colonies, is absorbed by neighbouring colonies of *Escherichia coli*, *Proteus* spp. and other species, which then fluoresce brilliantly (Hurst & Lowbury, 1953). Detection of *P. aeruginosa* in mixed cultures by production of pyocyanin, e.g. on milk agar (Brown & Scott Foster, 1970), has also been used for the detection of *P. aeruginosa*.

Fluorescence of burns exposed under an ultraviolet lamp has recently been developed as a technique for immediate detection, without bacterial culture, of severe infection with *P. aeruginosa* (Polk *et al.*, 1969).

2.c. Identification of *P. aeruginosa*

Bacterial growth from clinical sources showing blue–green or yellow–green fluorescence after 18–24 hours' culture at 37 °C on cetrimide or cetrimide–nalidixic acid agar is virtually always *P. aeruginosa*, and almost all strains of *P. aeruginosa* give fluorescent growth on these media. In routine clinical bacteriology such a presumptive identification is commonly accepted. It is often, however, necessary to confirm that a fluorescent pseudomonad is *P. aeruginosa*, especially in epidemiological studies on strains isolated from the inanimate environment, and a relatively small range of tests is appropriate for this purpose. Tests used in the taxonomic characterization of *P. aeruginosa* are discussed by Liston *et al.* (1963). Colwell (1964) and others. Perhaps the most useful of these are:

1. Production of pyocyanin on modified Sierra medium (Wahba & Darrell, 1965) or on King's medium A (King *et al.*, 1954) or on milk agar (Brown & Scott Foster, 1970), which also shows casein hydrolysis by *P. aeruginosa*. Over 90 % of strains of *P. aeruginosa* produce this blue, chloroform-soluble, non-fluorescent pigment when grown on these special media, and no other species produce the pigment.

2. Gluconate oxidation and slime production in a liquid medium containing potassium gluconate (Haynes, 1951).

3. Growth at 42 °C,

4. Reduction of nitrate to gaseous nitrogen.

5. Growth, with production of red colonies, on 1 % tetrazolium chloride agar (Selenka, 1958).

In addition to these specific tests for *P. aeruginosa*, tests for inclusion in the genus *Pseudomonas* are appropriate in a detailed study—expecially the oxidase reaction (Kovačs, 1956), oxidative metabolism of glucose (Hugh & Leifson, 1953) and the arginine dihydrolase reaction (Thornley, 1960). Where all Gram-negative bacilli are put through a range of biochemical tests, further tests for *P. aeruginosa* can be reserved for those which are positive for *Pseudomonas*. Phillips (1969) has devised a simplified scheme for the isolation and identification of *P. aeruginosa* within 24–48 hours, the great majority of strains being identified by pyocyanin production. (See Chapter 1 for a detailed discussion of *Pseudomonas* species).

2.d. Typing of Strains

P. aeruginosa can be further subdivided by serological typing, phage typing and bacteriocin ('pyocin' or 'aeruginocin') typing methods. Typing of strains is essential in tracing the sources and routes of transfer of organisms isolated in an outbreak. The methods of typing are discussed in detail elsewhere in this book (see Chapter 5). No single method gives as much information as can be obtained by combining at least two of them. Bacteriocin and serological types, the latter determined by use of antisera to the O-antigens produced by immunizing animals with boiled suspensions (Habs, 1957), are closely correlated (Cziszar & Lanyi, 1970), but unrelated to phage types, so that a combination of phage typing with either serological or bacteriocin methods is rational. Recent technical developments, especially the preparation of phage-free pyocin extracts (Rampling & Whitby, 1972), give promise of improved bacteriocin sensitivity typing of *P. aeruginosa*. Another approach in which phage and pyocin typing methods are combined, has been developed by Farmer and Herman (1969).

3. *P. AERUGINOSA* INFECTIONS

3.a. Introduction

Clinical infections with *P. aeruginosa* include *local infection*, e.g. of wounds (especially burns), the urinary tract, the respiratory tract, the intestine, the eye and the ear, and *generalized infections* (blood-borne or 'septicaemic') arising from sites of primary local infection in patients with impaired resistance, and leading to the development of metastatic foci in various organs. In some patients with local infection, general illness may occur without evidence of blood-borne distribution of the bacteria (*toxaemia*).

The characteristic clinical and pathological features of invasive infection with *P. aeruginosa* were presented in Fraenkel's (1917) review and have been described and extended in many subsequent publications (e.g. Forkner *et al.*, 1958; Williams *et al.*, 1960; Rabin *et al.*, 1961; Speirs *et al.*, 1963; Sevitt, 1964; Teplitz *et al.*, 1964a; Teplitz, 1965). The organism invades the walls of small blood vessels, producing an inflammatory arteritis and thrombosis, with arterial occlusion and development of small areas of necrosis (infarction); septicaemic spread, which sometimes occurs, leads to the development of characteristic focal embolic lesions, appearing first as macules or papules, and these increase in size, becoming indurated and developing a black, haemorrhagic, necrotic centre, from which a heavy growth of *P. aeruginosa* can be obtained on culture; when superficial these lesions commonly ulcerate through the skin. The haemorrhagic foci and the deep 'ecthymatous' ulcers which ensue are very characteristic of *Pseudomonas* septicaemia, though not found in all cases of such infection.

Neutrophil leucocytosis and pyrexia, which are typically associated with pyogenic infections, are often found in patients with *Pseudomonas* sepsis, in some cases perhaps due to coexisting staphylococcal infection. *Pseudomonas* sepsis is, however, characterized more by a *reduced* leucocyte count and a neutropenia, which may follow an initial neutrophilic leucocytosis. Other characteristic features are hypothermia (often following initial pyrexia), a sudden fall in blood pressure, anuria and paralytic ileus. While severe coccal infections are commonly associated with mental confusion, the mental state of patients with *Pseudomonas* septicaemia is usually clear (Tumbusch *et al.*, 1962; Polk & Stone, 1972). Cardiac arrhythmia, progressive jaundice and neurological abnormalities are also described (Forkner *et al.*, 1958).

The prognosis for patients who develop *Pseudomonas* septicaemia is poor. Some authors report a very high (sometimes 100 %) mortality (e.g. Liedberg *et al.*, 1954; Markley *et al.*, 1957; Forkner *et al.*, 1958); others have reported variable proportions of survivors (e.g. Jones *et al.*, 1966). Even when antibiotics highly active against the bacteria are used as soon as possible after the condition is diagnosed, treatment may fail because irreparable damage has by then already occurred (Jones *et al.*, 1966), or even because the destruction of bacteria by the antibiotic floods the circulation with endotoxin capable of causing a fatal reaction. Tumbusch and his colleagues (1962) have suggested that endotoxin released into the circulation may cause a local Schwartzman reaction in tissues sensitized by blood-borne organisms. A characteristic sign which may appear in severe invasive infections with *P. aeruginosa* is verdoglobinuria, i.e. the excretion of urine with a green fluorescent pigment derived from haemoglobin and clearly differentiated from fluorescein and pyocyanin (Stone *et al.*, 1963).

While a small proportion of patients colonized by *P. aeruginosa* develop severe sepsis, with the signs described above, a much larger proportion of patients in whom burns or other wounds become colonized with *P. aeruginosa* show no obvious pathological effects of such colonization, though local effects (e.g. rejection of skin grafts) may sometimes occur (Jackson *et al.*, 1951).

3.b. Pathogenesis

b.i. Animal Infections

Pathological changes similar to those found in invasive human infections can be obtained by inoculation of cultures of *P. aeruginosa* into animals. Intravenous or intraperitoneal injection of virulent *P. aeruginosa* into mice is followed by the rapid development of septicaemia, from which the animals die usually within 24–48 hours; or, with smaller dosage (intravenous), by a progressive infection, characterized by multiple abscesses of the kidneys (Gorrill, 1952). A gradual, progressive infection, with development of characteristic signs (loss of weight, pyrexia followed by hypothermia, focal lesions in kidneys, liver, spleen, and septicaemia) leading to death usually within 3–10 days, can be obtained by inoculation of burns in rats or mice (Teplitz *et al.*, 1964a,b; Jones *et al.*, 1966; Jones & Lowbury, 1966). Similar results were obtained on inoculation of *P. aeruginosa* into excised wounds (Walker *et al.*, 1964).

P. aeruginosa has been found to cause spontaneous outbreaks of pneumonia in the guinea pig (D'Ascani & Venturi, 1958). In rabbits, intradermal injection leads to the development of large papular or pustular lesions showing the characteristic histological changes; the rabbits do not die (Fox & Lowbury, 1953b; Teplitz, 1965).

The lesions which appear in these animals show microscopic changes similar to those found in human *Pseudomonas* sepsis. For example, Teplitz (1965) inoculated *P. aeruginosa* on the full skin thickness burns of the rat occupying 20 % of the body surface. Over 90 % of the animals died within 15 days. Autopsies performed immediately after death showed *P. aeruginosa* in blood, spleen and liver cultures. Eighty-five per cent of the rats developed metastatic septic lesions—round, circumscribed, haemorrhagic necrotic lesions in lungs, liver, kidney and spleen. Histopathological changes included dense infiltration of the media and adventitia of arteries and veins with Gram-negative bacilli (*Pseudomonas* vasculitis) and perivascular cuffing of arteries and veins by closely packed bacilli. The perivascular bacilli in larger arteries were sometimes sharply confined by the external elastic membrane; this layer and the basement membrane of the vessel appeared to present a barrier to bacterial penetration, but most of the vessels which were involved were small vessels without an external elastic membrane. Colonization of the vessel walls appeared to be a preliminary stage towards invasion of the blood stream. There was little leucocytic infiltration and often no evidence of thrombosed blood in the lumen. The lesions recall earlier descriptions by Hitschmann and Kreibig (1897) and Fraenkel (1917).

b.ii. Virulence

Strains of *P. aeruginosa* vary greatly in their virulence for mice, as shown by the LD_{50} of an intraperitoneal injection or by the mortality of mice on inoculation of small fresh burns with cultures of *P. aeruginosa*; by this latter technique it was shown that some strains (in a dose of about 7×10^8 bacteria) caused a mortality of 70–80 %, with septicaemia, while other strains did not increase the mortality above that (approximately 5 %) found in mice which were burned but not infected (Jones *et al.*, 1966; Jones & Lowbury, 1966). Strains found avirulent when applied to burns caused death on intraperitoneal injection, but a larger challenge dose was required to produce a fatal infection with these strains (Jones, 1970). Strains isolated from the bronchial mucus of patients with cystic fibrosis of the pancreas or with bronchiectasis often produce mucoid colonies which may be encapsulated (Sonnenschein, 1927; Cetin *et al.*, 1969; Diaz *et al.*, 1970); mucoid variants usually emerge in the bronchus of patients at first colonized by typical rough colonies of *P. aeruginosa* (Doggett *et al.*, 1971). Bacterial virulence is enhanced by the possession of a mucoid envelope which protects the organism against phagocytes and antibiotics; it is common, however, for virulent strains which cause infection of burns to produce non-mucoid colonies.

b.iii. Toxic factors

P. aeruginosa produces a variety of toxins, enzymes and pigments, some of which undoubtedly contribute to the pathogenic properties of the bacteria.

Liu and his colleagues (Liu *et al.*, 1961; Liu & Mercer, 1963; Liu, 1964; Atik *et al.*, 1968) have studied the role of various fractions obtained from cultures of *P. aeruginosa* in defined media containing α-alanine, aspartic acid, glutamic acid, glucose and salts; they have concluded that *P. aeruginosa* differs from other Gram-negative bacilli in owing most of its pathogenic activity to extracellular products—in particular, to toxic components of the extracellular slime—and not to endotoxins. Using cellophane-covered plates and a synthetic medium, Liu obtained protease, lecithinase and haemolysin from *P. aeruginosa*. Lethal effects could be obtained by inoculation of mice with extracellular slime and, to a smaller extent, with lecithinase and proteinase. Intradermal injection of the enzymes into rabbits caused severe local lesions. It seemed unlikely that sufficient quantities of lecithinase or of haemolysin could be produced in human tissues to have pathogenic effects, because more glucose is required for their production than is present in the tissues. Proteolytic enzymes, however, could be produced in high concentration from media containing no glucose, and these enzymes may therefore be a more important factor in pathogenesis. Their activity, moreover, is enhanced by lactic acid which tends to accumulate in injured tissues (Thomas & Stetson, 1949). Kabota and Liu (1971) found that the injection of *P. aeruginosa* culture into ligated loops of intestine caused a large accumulation of fluid; they attribute the effect to an enterotoxin probably distinct from haemolysin, proteinase, lecithinase and lethal toxin.

Atik and his colleagues (1968) showed that an exotoxin prepared from a non-proteolytic strain of *P. aeruginosa* (P-A-103) caused two types of reactions in dogs:

1. An immediate anaphylactoid reaction from which the animals recovered.

2. A late hypotensive reaction which persisted until the animals died in about 24 hours.

The circulatory and biochemical changes which occurred in these animals resembled those found in endotoxin shock, but the authors attribute the effects to an exotoxin because no toxic effects were obtained (as would be expected from Gram-negative bacilli producing endotoxin) on intraperitoneal injection of suspensions of *P. aeruginosa* washed in acid to remove the extracellular slime. Other workers (e.g. Homma & Uehara, 1971; Rubio & Lopez, 1972), however, have described endotoxins of *P. aeruginosa* and discuss their probable role in clinical infection. Chün-Hsiang *et al.* (1964) found similar pathological changes in animals which died from endotoxin shock and in patients who died with *P. aeruginosa* septicaemia. Extensive burns rendered animals more sensitive both to endotoxins and to bacterial infection: sublethal doses of endotoxins greatly enhanced the lethality of *P. aeruginosa*.

Mull and Callahan (1965) have studied the role of elastase, one of the proteinases of *P. aeruginosa*. By destroying the elastic laminae of blood vessels in which the organisms are aggregated, the enzyme may facilitate invasion of the bloodstream. Collagenase is also produced by some strains of *P. aeruginosa* (Mandl, 1961), and probably exerts a pathogenic effect by damaging the walls of small arteries (Diener *et al.*, 1973); in mice, intranasal instillation of an enzyme preparation led to confluent pulmonary haemorrhage, and severe haemorrhage was also caused by intraperitoneal and intravenous injection.

Carney and Jones (1968) studied the biological properties of culture filtrates of virulent and avirulent strains of *P. aeruginosa*, grown in a synthetic medium described by Liu (1964). The non-dialysable residue of concentrates from 20 litre batches of 5-day cultures were fractionated by gel filtration through Sephadex G200. Ultraviolet absorption at 280 nm of eluted fractions revealed three peaks with different moleculer size (F1, F3 and F4); fractions, F1, F3 and F4 were tested for toxicity, enzyme activity and immunogenic properties. Fractions from both virulent (P14) and avirulent (P2AB) strains showed similar properties. The F1 fraction was more lethal on intraperitoneal injection in mice than F3 and F4 fractions, and it was present in at least as high a concentration in equivalent amounts of extracts from both strains P14 and P2AB. The F3 fraction showed proteolytic activity, and more of this enzyme was produced by the virulent than by the avirulent strain. From this evidence it seems possible that the pathogenic properties of *P. aeruginosa* involve the action of enzymes (F3) which help the bacteria to invade the tissues, and of a separate lethal toxin (F1); though the proteolytic enzymes apparently do not themselves have important lethal properties, they may determine virulence by assisting invasion of the tissues by the bacteria.

Small amounts of *hydrogen cyanide* are produced by *P. aeruginosa* both in culture and in the animal body after death (Patty, 1921). Though these are unlikely to reach toxic levels, cyanide produced by *P. aeruginosa* colonizing extensive burns might contribute something to the overall pathogenic effects; it might also have medico-legal implications.

Pyocyanin has been shown to have toxic effects on tissue cultures of fibroblasts (Schoental, 1941) and epithelial cells (Cruickshank & Lowbury, 1953); the concentration of pyocyanin causing necrosis of epithelial cells was much smaller than that commonly found in the exudate of infected burns, so that infection with *P. aeruginosa* could be expected to interfere with healing and, in particular, with the 'take' of split skin grafts. Intradermal inoculation in animals of pyocyanin caused no local lesions like those caused by cultures of *P. aeruginosa* (Fox & Lowbury, 1953b) or by enzymes; it did not kill mice on intraperitoneal injection (Liu *et al.*, 1961). Stewart-Tull and Armstrong (1972) have reported inhibition of oxygen uptake by mouse liver mitochondria in presence of chloroform-soluble products of *P. aeruginosa*, including pyocyanin; they suggest that the release of these compounds into the bloodstream from infected burns might be a factor in the pathogenesis of *Pseudomonas* infection.

Wretland and others (1973) compared the production of lipase, esterase, protease, haemolysin, DNAase, RNAase, lecithinase, elastase and staphylolytic enzyme by 50 strains of *P. aeruginosa* from patients with bacteraemia and 148 strains from carriers and non-bacteraemic infections. The majority of strains produced these factors, and they were produced by a similar proportion in each of the two groups of strains examined.

3.c. Immunity

The role of natural and acquired immunity is particularly relevant in assessing mechanisms of pathogenesis of opportunist organisms. Numerous studies have been made in recent years on the immunological properties of *P. aeruginosa*, and most of these have been concerned with the application of such knowledge to prophylaxis and treatment by vaccines and antisera. These clinical applications are considered elsewhere (Brown, 1975). We shall consider here natural resistance and the immunological response to infection in human subjects.

c.i. Antibodies

Antibodies to *P. aeruginosa* in normal subjects are described by Lilley and Bearup (1928) and by Gibson (1930). Sandiford (1937) reported the presence of agglutinins to *P. aeruginosa* in urinary tract infections with these organisms. Fox and Lowbury (1953a) showed, in the serum of normal subjects, agglutinins to a range of serotypes of *P. aeruginosa*. They showed also that higher titres were present in patients whose burns were infected with *P. aeruginosa*. Increase in titre was associated with the duration of infection and the extent of the burns, and the highest titres were against the strain and serotype colonizing the patient's burns.

The relevance of these findings in relation to host defence is hard to assess. High titres of antibody were sometimes found in the serum of patients dying with *P. aeruginosa* infection, suggesting that these antibodies had little or no protective value. However, rabbits immunized with a range of serotypes of *P. aeruginosa* were found to have acquired some specific immunity, as shown by reduced local pathogenic effects of cultures injected intradermally (Fox & Lowbury, 1953b). Mice could be passively protected with specific antisera given concurrently with intraperitoneal challenge by *P. aeruginosa*, but if there was a delay of 48 hours in giving the antiserum it had no protective effect (Jones & Lowbury, 1966). The high titre of *Pseudomonas* agglutinin in certain patients who subsequently died with *Pseudomonas* septicaemia was found at a stage when irreparable damage was likely to have been caused by the infection. That such rapid effects can occur was shown in a patient who died in spite of successful elimination of *P. aeruginosa* from the blood-stream by chemotherapy (Jones *et al.*, 1966).

Invasion of *P. aeruginosa* into tissues and blood-stream is likely to occur when tne level of antibodies is reduced by disease, e.g. in hypogamma-globulinaemia (Speirs *et al.*, 1963) and leukaemia; also in severely burned patients who shortly after injury are shown to have reduced primary response to antigens (Alexander & Moncrief, 1966) and a lower level of immuno-globulins (Arturson *et al.*, 1969; Ritzmann *et al.*, 1969). Treatment with cytotoxic and immunosuppressive drugs also facilitates invasion by *P. aeruginosa*.

Graber and others (1961) found that mice were protected against intra-peritoneal challenge with *P. aeruginosa* when inoculated with undiluted serum from burned patients who had high haemagglutinin titres against the same infecting strain. Jones and Lowbury (1965) tested the serum of 12 healthy persons for agglutinins to two serotypes of *P. aeruginosa*. The serum of one subject, which showed no detectable agglutinin to either strain, failed to protect burned mice passively against either strain of *P. aeruginosa* inoculated on the burned area. Of the other sera, 8 showed agglutinin titres of between 1/2 and 1/80 against one strain (B4) and 9 showed titres between 1/6 and 1/110 against the other strain (P14). The highest titre against each strain was found in the same serum, and injection of this gave fairly good protection to burned mice against both strains. Serum from a burned patient with an agglutinin titre of 1/256 gave complete protection against the infecting strain.

Though fresh serum has bactericidal action, possibly due to the action of properdin (Pillemer *et al.*, 1954), against some strains of *P. aeruginosa* (e.g. Colebrook *et al.*, 1960), specific immunity to *P. aeruginosa* does not depend on this effect. For example, Fox and Lowbury (1953b) found no evidence of bactericidal action in sera from immunized rabbits against strains of *P. aeruginosa* towards which the rabbits showed increased resistance; they did, however, find opsonic action in serum from immunized rabbits, indicating that the protective action of serum was mediated through phagocytes. The leucopenia which occurs in severe *Pseudomonas* infection can therefore be expected to interfere with the protective action of antibodies. Liu and Mercer (1963) found that the antibody to the extracellular slime had a protective function and was associated with agglutinins. Early protection, obtained in mice within 48 hours of the first immunizing dose of a vaccine (Jones, 1971), is associated with an absence of agglutinin, and apparently with the early rise in IgM; the subsequent rise in agglutinins is associated with the more delayed appearance of IgG (Jones *et al*, 1971, 1972).

c.ii. Phagocytes

Cellular resistance to infection involves the ingestion of bacteria by poly-morphonuclear neutrophils and other phagocytic cells and the destruction of the bacteria inside the cells. The two processes are separate, and sometimes bacteria survive after being ingested. Phagocytosis of some bacteria, e.g. capsulated virulent strains, may be dependent on antibodies (opsonins) active against the bacteria, or on serum with complement but without specific antibodies; or it may occur without the stimulus of factors in the environment. A number of factors, including phagocytins, bactericidin, low pH and hydro-gen peroxide, determine the bactericidal action of phagocytes.

Alexander and Wixson (1970) found that the ability of neutrophils to kill ingested bacteria was deficient in patients with severe burns; cyclic variations in neutrophil function appeared to be present in normal subjects as well as in burned patients. The number of neutrophils as well as their ability to destroy *P. aeruginosa* was reduced in patients with burns; this deficiency in phagocytic defence combined with the reduced antibody response described above renders the severely burned patient particularly vulnerable during the first days after injury.

Studies with an immunofluorescent technique have shown poor leucocyte response and rapid invasion of hair follicles and subcutaneous tissue by a *virulent* strain (not by avirulent strains) of *P. aeruginosa* (Carney *et al.*, 1973). Since passive immunization prevents invasive infection, it is probable that this is achieved, in part, though neutralization of toxic factors of *P. aeruginosa* which interfere with the leucocytic response. These data and the fact that protective antisera often have no bactericidal activity suggest that phagocytic action aided by opsonin is the principal mechanism of protection against *P. aeruginosa* (see Jones & Dyster, 1973).

The role of various antigens, and in particular of the F1 fraction, is discussed elsewhere (Brown, 1974). The F1 fraction has been shown to have outstanding immunogenic powers, causing active immunity in mice injected with extremely small quantities of a lyophilized F1 preparation (Dr. R. J. Jones, personal communication). This may have some relevance to the usual insusceptibility of healthy persons to *P. aeruginosa*.

3.d. Sites and Patterns of Infection with *P. aeruginosa*

P. aeruginosa produces characteristic lesions, with the histological features described above, in any living tissue in any part of the body which it succeeds in invading from the slough of burns, umbilical stump or other adjacent areas of colonization. In some patients these local changes are associated with septicaemia or pyaemia, and the development of metastatic lesions in the skin, the meninges, the endocardium and elsewhere. Generalized illness may, however, occur without evidence of blood-borne dissemination of the bacteria, presumably through absorption of toxins (e.g. Teplitz *et al.*, 1964a).

Burns and Wounds. *P. aeruginosa* commonly appears as a predominant organism colonizing burns in hospital (Colebrook, 1960; Lowbury 1960); the incidence of colonization at any particular time varies with the extent of the burned area (Lowbury & Fox, 1954). In most patients colonization is not associated with obvious pathogenic effects, but the ability of *P. aeruginosa* to interfere with skin grafts and, in consequence, to delay healing was demonstrated in a controlled trial of topical chemoprophylaxis with polymyxin cream (Jackson *et al.*, 1951).

Such local pathogenic effects, which are usually inconspicuous and much less severe than those caused by *Streptococcus pyogenes*, are of small importance compared with the invasive infections that occur in a small proportion of patients with burns; this is a hazard mainly of extensive full skin thickness burns (Jackson *et al.*, 1951; Liedberg *et al.*, 1954; Tumbusch *et al.*, 1962), but it may, on occasion, occur in debilitated patients with smaller burns (Jones *et al.*, 1966). In the 1960's *P. aeruginosa* infection was widely recognized as the most important cause of death in severely burned patients. Though the hazard of infection is much smaller in burns which do not involve the full thickness of the skin, studies with arteriography have shown that deep partial skin thickness burns may be converted to full skin thickness lesions as the result of infection with *P. aeruginosa* (Order *et al.*, 1965a,b). Unlike partial skin thickness burns, full skin thickness lesions require skin grafts because they will not heal spontaneously. The poor cellular inflammatory response in the devitalized burn with invasive *Pseudomonas* infection may be responsible for the poor 'take' of skin grafts in some burns.

Chemotherapy for *Pseudomonas* infection of burns presents a number of problems which are discussed elsewhere (see Brown, 1975, Chapter 5). Special emphasis is therefore placed on prophylaxis by various methods, including a first line of defence against contamination of the burns (by topical chemoprophylaxis, isolation and skin grafting) and a second line of defence against invasion from the burn surface (by chemotherapy and immunological methods) (Lowbury, 1967). The success of such methods has led, in some centres, to a significant reduction in fatalities from *P. aeruginosa* infection (Lindberg *et al.*, 1965; Cason & Lowbury, 1968; Bull, 1971); at the same time other organisms, including *Klebsiella* spp. and fungi, have emerged in some centres as the main opportunist pathogens of burned patients (Pruitt & Curreri, 1971; MacMillan *et al.*, 1972).

P. aeruginosa has played a relatively smaller role as a cause of sepsis in traumatic and operation wounds than in burns. Surveys by the Public Health Laboratory Service (1960) and the National Research Council (1964) showed the predominance of *S. aureus* in septic wounds, though in the American report 13 % of the septic wounds yielded *P. aeruginosa*. More recent surveys (e.g. Bruun, 1970) show an increased relative proportion of Gram-negative bacilli in post-operative infection, but *P. aeruginosa* has not been a predominant organism in this group. Outbreaks of *Pseudomonas* infection of wounds can occur through the use of contaminated materials leading to infection of soft tissues and bone (Sussman & Stevens, 1960). *P. aeruginosa* is a prominent cause of osteomyelitis (Pulaski, 1964). Johanson (1968) has described infection with *P. aeruginosa* following puncture wounds of the foot in 11 children.

Skin. P. aeruginosa is commonly found among the bacterial flora of eczematous exudates. Its pathogenic role in such lesions is usually unimportant when compared with its potential role in extensive burns, but local inflammation and exudation may occur (Selwyn, 1965; Noble & Savin, 1966). Extensive pyoderma has been reported in a patient with non-eczematous dermatitis (Hall *et al.*, 1968). Prolonged or frequent immersion of the hands or feet in water predisposes towards infection of the macerated epithelium by *P. aeruginosa*, e.g. in the green nail syndrome (Goldman & Fox, 1944) and in tropical immersion foot syndrome (Taplin *et al.*, 1965; Taplin & Zaias, 1966) (see also Infection of Ears).

In patients with immunodeficient disease or severe burns, characteristic ecthymatous ulcers may appear on normal skin, associated with pyaemic transfer of *P. aeruginosa* from the site of primary infection (Speirs *et al.*, 1963; Sevitt, 1964).

In infants, who are especially susceptible to infection, lesions in the mouth and throat may be caused by *P. aeruginosa* (Jacobs, 1964).

Urinary Tract. Infections of the urinary tract can be divided into *primary infections*, in which there is no previous or predisposing illness, and *secondary infections*—i.e. relapses or reinfections after treatment, or infections that follow surgical treatment or instrumentation (catheterization or endoscopy). Primary infections are almost always caused by *Escherichia coli* derived from the patient's intestinal flora, and usually sensitive to all or most antimicrobial drugs. Secondary infections are caused by bacteria acquired from external sources, including multiresistant strains transmitted from other patients, and from various inanimate environmental sources, some of which originally acquired the organism from patients in the ward. Amongst these miscellaneous coliform bacilli *Proteus mirabilis*, *P. aeruginosa* and *Klebsiella* spp., are commonly represented (Gould, 1968). (For details, see the section below on Mechanisms of Transmission).

Pseudomonas infection of the urinary tract may remain localized in the bladder (cystitis) or it may ascend through the ureters to the pelvis of the kidneys, from which it may invade the kidney tissue, causing pyelonephritis; this serious and potentially fatal type of urinary tract infection is much more commonly caused by endogenous infections with *E. coli* than by *P. aeruginosa*.

In septicaemic infections with *P. aeruginosa*, local lesions may occur in the kidney similar to those found in experimental infections of animals (Teplitz *et al.*, 1964b; Teplitz, 1965); such lesions develop into cortical or corticomedullary pyelonephritis without involvement of the inner medulla (papillae).

Transient bacteraemia, which may be associated with septic shock, is a well known hazard of urethral instrumentation and of irrigation with contaminated fluids; *P. aeruginosa* is one of the Gram-negative bacilli occasionally responsible for this condition (Last, Harrison & Marsh, 1966).

Myerowitz and his coworkers (1972) have found strains of *E. coli* and a strain of *P. aeruginosa* in urine of patients with urinary tract infection possessing antigens cross-reactive with capsular polysaccharide of *Streptococcus pneumoniae* types I and III, *Neisseria meningitidis* groups C and H, and *Haemophilis influenzae* type b. These antigens, which resemble *E. coli*

K antigens, may confer characteristics on the bacteria, allowing them to cause renal parenchymal infection.

Respiratory Tract. P. aeruginosa is not a pathogen of the normal bronchial lining, and primary *Pseudomonas* bronchitis or tracheitis probably never occurs. There has, however, been an increasing reported incidence of *Pseudomonas* infection of the respiratory tract, all of it associated with pre-existing lesions, surgical instrumentation and the use of nebulizers, anti-biotics or steroids (Goslings, 1963; Pierce *et al.*, 1966; Barson, 1971). Such cases include cystic fibrosis and bronchiectasis (Burns & May, 1968) and those treated with tracheostomy in intensive care units (Phillips & Spencer, 1965; Sutter *et al.*, 1966a). In cystic fibrosis *P. aeruginosa* causes a particularly severe and intractable bronchopneumonia, the organism usually appearing at a later stage of the disease than *S. aureus* and *H. influenzae*. Burns and May (1968) have shown that *Pseudomonas* antibodies (precipitins and agglutinins) appear in the serum of a proportion of patients with cystic fibrosis and *P. aeruginosa* in the sputum; they regard the development of antibodies to *P. aeruginosa* colonizing the respiratory tract as evidence of the pathogenic involvement of the organism, and this view is supported by the fact that antibodies in the serum are usually associated with the presence of mucoid strains of *P. aeruginosa*, which predominate in the more severely ill patients (Diaz *et al.*, 1970). The clinico-pathological appearances in *Pseudomonas* bronchopneumonia are characteristic, with massive infiltration of the bacteria into the walls of arterioles and venules, but minimal inflammatory response. There is a diffuse necrotic change around the vessel walls which eventually rupture, causing haemorrhage into the adjacent alveoli (Barson, 1971). Tillotson and Lerner (1968) describe nodular bronchopneumonic lesions which developed into haemorrhagic abscesses in a series of patients with *Pseudomonas* pneumonia.

The sources of infection (see below) include mechanical ventilators, nebulizers, suction apparatus and the hands of nurses; airborne contamina-tion is unimportant (Lowbury *et al.*, 1971). Improved design and disinfection of equipment, the use of disposable suction catheters, the use of vaporizers rather than nebulizers for humidification of inhaled air and the wearing of gloves by nurses in intensive care units help to reduce the hazard of respiratory *Pseudomonas* infection.

Infections of the Eye. P. aeruginosa can cause conjunctivitis, necrotizing infection of the eyelids and dacryocystitis (see Stanley, 1947). The most important types of infection, however, are those which involve the chambers of the eye. If *P. aeruginosa* gains access to the anterior chamber it causes a severe type of hypopyon ulcer, which is likely to develop into panophthalmitis (Duke-Elder, 1970). Bignell (1951) found that almost all the severe corneal ulcers, sometimes leading to early loss of an eye, were caused by *P. aeruginosa*, and this organism is cited as a particularly dangerous pathogen to the

injured eye by Burns and Rhodes (1961), Crompton (1962) and Allen (1963);
Burns and Rhodes (1961) describe fatal eye infections in premature infants.
The anterior chamber is peculiarly susceptible; hence opportunists, including
Bacillus subtilis, can cause endophthalmic infection. Crompton (1962) refers
to experiments in the rabbit in which severe infection with loss of an eye
followed injection of as few as 50 cells of *P. aeruginosa* into the anterior
chamber; to achieve a similar effect with *S. aureus* 9000 cells had to be injected.

Bacteria may enter the anterior chamber of the eye as a result of injuries
or corneal disease, and also through breaches in asepsis during operations.
The sources of *P. aeruginosa* causing infection of eyes include solutions used
for irrigation and eye drops; e.g. solutions of fluorescein, used to locate the
presence of a corneal ulcer or abrasion. Contamination at the site of injury,
followed by corneal ulceration, may lead to the appearance of pus in the
anterior chamber and the development of acute panophthalmitis, necessitating evisceration of the eye. Operations exposing the anterior chamber (e.g.
for removal of cataract) present a particular hazard. In one hospital irrigation
with contaminated saline solution led to an outbreak of endophthalmitis, a
single phage type and serotype of *P. aeruginosa* being isolated from all the
infected eyes and from one batch of 'sterile' saline used in the operating
theatre (Ayliffe *et al.*, 1966). Allen (1959) reported 5 patients who suffered the
loss of an eye with *P. aeruginosa* endophthalmitis due to preoperative use of
a contaminated solution of 1/100 benzalkonium chloride, and other infections
in which the sources were a detergent solution used for immersion of cataract
knives, a sulphonamide solution instilled prophylactically into eyes with
foreign bodies, and saline used to irrigate the anterior chamber. Corneal
scarring, with diminished vision, has followed the use of contaminated
atropine drops after extraction of a lens (Crompton, 1962). *Pseudomonas*
conjunctivitis is also reported in a patient with intact cornea who used
contaminated polocarpine eye drops for the treatment of glaucoma (Stobie,
1961).

The hazard of eye infection with *P. aeruginosa* has in recent years been
reduced by better precautions in surgery and dispensing, including the use of
single dose containers of sterile solutions for instillation into eyes during
endophthalmic operations or when the cornea is damaged.

Infections of the Ear. P. aeruginosa is commonly found in the exudate of
external otitis. The moisture of the external auditory meatus favours their
growth in this site, especially in tropical climates (Saltzman, 1963; Mawson,
1967). Perochondritis, a painful infection of the auricle complicating otitis
externa, is particularly associated with *P. aeruginosa* (Shambaugh, 1959).
P. aeruginosa may also be found in the exudate of otitis media, where it
usually appears as a secondary invader from the external meatus after
successful treatment of a primary infection by Gram-positive cocci (Shambaugh, 1959).

Central Nervous System. Meningitis caused by *P. aeruginosa* is well recognized as a potential hazard of contamination during lumbar puncture and neurosurgical operations, by direct extension from infected ear, sinus or mastoid process, or through haematogenous transfer (Stanley, 1947; Knight *et al.*, 1952; Nunn & Wellman, 1960). Such infections are often fatal. Barrie (1941), Smith and Smith (1941) and Garrod (1946) cited contamination of lumbar puncture equipment on rinsing with supposedly sterile water as a special hazard of infection with Gram-negative bacilli, including *P. aeruginosa.*

In a neurosurgery centre, Ayliffe and his colleagues (1965) described a series of infections, including meningitis, urinary tract infection and wound infection, caused by a single epidemic type of *P. aeruginosa.* The outbreak was eventually traced to a shaving brush occasionally used for preoperative depilation of the scalp; the contaminant had apparently survived disinfection of the scalp, after depilation, with 0·5 % chlorhexidine in 70 % ethanol.

Meningitis and cerebral abscess arising from haematogenous spread of *P. aeruginosa* are occasionally found in patients with *Pseudomonas* septicaemia. They may also occur after penetrating head wounds (Botterell & Magner, 1945).

Endocarditis. Acute bacterial endocarditis is an occasional complication of *Pseudomonas* septicaemia (Rabin *et al*, 1961; Sevitt, 1964).

Other Sites. Osteomyelitis and arthritis may occur through invasion of bones and joints with *P. aeruginosa* from adjacent wounds or burns (e.g. Bishop, 1938). Clinical infection can occur in the alimentary tract during the course of generalized infection; ulcers, which may appear in the stomach or intestinal wall, are similar to the ecthymatous ulcers of the skin.

P. aeruginosa is one of the organisms that have been isolated from the lesions of pseudomembranous enterocolitis (Reiner *et al.*, 1952; Kay *et al.*, 1958). These lesions have usually occurred in patients under antibiotic therapy and occasionally after operations; the role of bacterial colonization in pseudomembranous enteritis is uncertain, as the pathology of the condition is complex. Outbreaks of severe diarrhoea due to *P. aeruginosa* in infants (especially premature infants) have been described; for example by Falcao *et al.* (1972). These workers describe two outbreaks, in one of which the strain was traced to an oxygen bubbler that had been contaminated by a nursery worker carrying the organism in her gut.

Typhoid-like infections with *P. aeruginosa* in the tropics are described by Pons (1927).

Infection in Infancy. *P. aeruginosa* is recognized to have a particularly important pathogenic role in infant nurseries of maternity hospitals; e.g. the outbreaks cited above (Falcao *et al.*, 1972) in which diarrhoea was a predominant feature. Neonates, and in particular premature infants and those suffering from illness, are more susceptible than older children or adults, because their antibody-forming mechanism is inadequate.

P. aeruginosa infection in infants includes infection of lungs, urinary tract, intestinal tract, skin and eyes; in some cases local infection leads to systemic invasion with septicaemia, meningitis, brain abscess, endocarditis and circulatory collapse (Hoffman & Finberg, 1955; Neter & Weintraub, 1955; Jacobs, 1964; Paul & Marget, 1964).

Infection is acquired from a variety of sources. Cole, Thom and Watrasiewicz (1971) found *P. aeruginosa* in the faeces of 8 % of infants during the first six months of life, from which self-infection was possible, though the presence of *P. aeruginosa* in the intestine was not associated with a higher incidence of intestinal infection. External sources, however, are more commonly implicated, in particular suction apparatus. Rogers (1960) found that catheters and suction apparatus used to clear the respiratory tract became contaminated with *P. aeruginosa* from weak antiseptic solutions.

Bassett, Thompson and Page (1965), Rubbo, Gardner and Franklin (1966) and Drewett and his colleagues (1972) all found inadequately disinfected suction apparatus to be a source from which infants acquired *P. aeruginosa.* These authors stress the need for more reliable disinfection by avoidance of disinfectants known to be ineffective against *P. aeruginosa,* by the use of disposable or boilable mucus extractors in place of aspirators which are difficult to disinfect, and by the use of subatmospheric steam for disinfection of equipment (e.g. suction apparatus) which will not stand heat sterilization (Drewett *et al.,* 1972). Other potential sources of *P. aeruginosa* in infants are humidifiers (nebulizers), incubators (Laursen, 1962), water baths, feeds and feeding bottles (Ayliffe *et al.,* 1970), taps and sinks (Wilson *et al.,* 1961), sponges, hands and solutions. In general, the sources which actually bring large numbers of the organisms into contact with susceptible tissues (e.g. aspirators, solutions) are more important than remote or sparse sources, such as sinks or floors.

Obstetrics. P. aeruginosa presents much greater hazards to infants than to mothers in obstetric practice. It may, however, be a predominant organism in septic abortion (e.g. Durcieux & Brett, 1967) and could be an opportunist pathogen when it colonizes the puerperal uterus in an immunodeficient woman. It was not among the bacteria found in high vaginal swabs of patients with puerperal pyrexia (Calman & Gibson, 1953).

Septicaemia. Speirs, Selwyn and Nicholson (1963) described *Pseudomonas* septicaemia arising spontaneously in an apparently healthy child who was found to have congenital hypogammaglobulinaemia. Immunodeficiency through disease (leukaemia, burns, diabetes) or through treatment (corticosteroids, immunosuppressive agents, radiation) are the conditions required for development of septicaemia, the most severe form of invasive infection with *P. aeruginosa.* The characteristic focal lesions and their appearance in different anatomical sites have been described above. *Pseudomonas* septicaemia is usually the result of invasion from a burn or some other local site of infection, but it may also be due to contaminants introduced at the site of intravenous infusion or catheterization (Moran *et al.*, 1965).

Some authors have found *P. aeruginosa* septicaemia invariably fatal even with polymyxin therapy (e.g. Tumbusch *et al.*, 1962). The arrival of gentamicin in 1963 and carbenicillin in 1966 improved the prospects, but the prognosis is still grave, and it may be too late to arrest the disease by the time a diagnosis is made.

4. MECHANISMS OF TRANSMISSION

4.a. Introduction

The mode of transfer of infectious disease depends on a number of factors, including the normal habitat of the pathogens, their ability to grow or to survive in moist or dry environments outside the body, their presence in infective sites of patients and in carrier sites of nurses and other members of staff, the pathogenicity of bacteria which have been exposed to drying and other adverse influences, and the effectiveness of measures used in hospitals to prevent the transfer of infection.

Infection with *P. aeruginosa* is almost restricted to patients in hospital. Hospital infection may be acquired from external sources (exogenous infection) by the airborne route or by contact with human or inanimate vectors, or it may be endogenous infection ('self-infection') by bacteria present in sites of carriage—on the patient's skin, respiratory tract or intestinal tract. Bacteria are sometimes acquired first on these sites of carriage from the hospital environment from which they are transferred, at a second stage, to sites of infection.

P. aeruginosa, like other Gram-negative bacilli, is readily killed by evaporation of its suspending fluid medium. There is considerable variation between strains, but with most strains more than 90 % of the bacteria, and often many more, are killed during the evaporation of suspensions allowed to dry on cover slips (Lowbury & Fox, 1953); those which have survived the effects of drying, however, continue to survive in the dried state as do *S. aureus* and other Gram-positive cocci, which are much less vulnerable to desiccation. *P. aeruginosa* can usually be found in fragments of burn eschar in the dust of burns wards where infected patients are nursed (Lowbury & Fox, 1954; Hurst & Sutter, 1966) and such residues can, under certain conditions, be transferred by air to other patients (see below). By contrast with its poor survival on drying, *P. aeruginosa* can survive well in aqueous suspension; while all the cells of *S. aureus* died within 48 hours when kept suspended at room temperature in deionized water, a strain of *P. aeruginosa* suspended in the same medium survived for over 5 weeks (Pettit & Lowbury, 1968). Moreover, *P. aeruginosa* can multiply in water containing minimal nutrient additives, so that large numbers of the bacteria are likely to be present in moist environments which become contaminated with them.

4.b. Sources

The relative importance of different sources and reservoirs of *P. aeruginosa* varies with the clinical situation; some examples have already been given in relation to specific sites and types of patient.

b.i. Human Sources

The proportion of normal subjects carrying *P. aeruginosa* in faeces is low. Lányi, Gregačs and Adam (1966) found 1·7 % carriers among 7650 subjects examined; higher proportions have been found by others—e.g., 3–4 % (Lowbury & Fox, 1954; McLeod, 1958) and 11–12 % (Ringen & Drake, 1952; Shooter *et al.*, 1966; Sutter *et al.*, 1967); Sutter *et al.* (1967) found *P. aeruginosa* in numbers ranging from less than 5 to 100,000 per gram of faeces. Volunteers who swallowed a culture of *P. aeruginosa* carried the organism in faeces for up to six days, but for longer periods if treated with ampicillin (Buck & Cooke, 1969). Because of the greater exposure in hospital to reservoirs of the organism and the selective influence of antibiotic therapy, higher proportions are found in hospital patients, e.g. 38 % in a surgical ward (Shooter *et al.*, 1966) and 20 % in a burns unit (Lowbury & Fox, 1954). At autopsy 36 % of patients were shown by Stoodley and Thom (1970) to have *P. aeruginosa* in the faeces.

Though the hands of nurses working in wards with infected patients often carry *P. aeruginosa* (e.g. Lowbury & Fox, 1954; Lowbury *et al.*, 1971; Falcao *et al.*, 1972), the skin, mouth and respiratory tract are not normal sites of colonization; *P. aeruginosa* inoculated on to the surface of the skin died rapidly through the effects of evaporation (Ricketts *et al.*, 1951). The moist skin of the axilla and perineum were reported by Ružička (1898) as sites from which *P. aeruginosa* could be isolated in some normal subjects; increased humidity may lead to the colonization of the skin of infants by the organism (Hoffman & Finberg, 1955). Premature infants have been found to carry *P. aeruginosa* on the skin and in faeces much more often than they are normally carried by adults (Neter & Weintraub, 1955). Sutter *et al.* (1966b) found the organism in 6/350 samples of saliva from patients attending a dental clinic.

The most important human sources are, undoubtedly, infected wounds, urine and lesions producing exudate; these moist sources contaminate dressings, bedclothes, nurses' hands and equipment, on which the organism may be transferred to other patients.

b.ii. Inanimate Reservoirs

Because *P. aeruginosa* is highly sensitive to drying but survives and multiplies in water, the important inanimate sources of the organism are moist environments. A number of these have already been mentioned in relation to individual types of patient. Contaminated fluids which have caused infection with *P. aeruginosa* include eye drops and other ophthalmic solutions (Thygeson, 1948; Ridley, 1958; Ayliffe *et al.*, 1966); lotions, detergent antiseptics (in particular quaternary ammonium compounds) and chloroxylenol when inadequately dispensed, e.g. in cork-stoppered bottles, or in bottles replenished without adequate disinfection, or due to inactivation by cotton swabs (Lowbury, 1951b; Plotkin & Austrian, 1958); dilute phenolic solutions (Bean & Farrell, 1967; Simmons & Gardner, 1969), hand creams (Ayliffe *et al.*, 1968) and steroid creams (Noble & Savin, 1966); moisture in suction apparatus, respiratory ventilators, nebulizing humidifiers (Grieble *et al.*, 1970) and incubators (e.g. Bassett *et al.*, 1965; Phillips & Spencer, 1965); in shaving brushes used for preparing the scalp before brain surgery (Ayliffe *et al.*, 1965) and nail brushes (Ayliffe *et al.*, 1968); also in miscellaneous items, including floor cloths, dish mops and sponges, and in sinks and taps (Whitby & Rampling, 1972; Teres *et al.*, 1973). Food has been found to carry *P. aeruginosa* in hospital kitchens, the same type appearing subsequently in the faeces of patients in the wards (Shooter *et al.*, 1969, 1970; Kominos *et al.*, 1972). Surgical instruments (e.g. resectoscopes and Bigelow evacuators) and mucus extractors have transmitted infection when disinfected with inadequate preparations (Rogers, 1960; Moore & Forman, 1966).

Dry reservoirs are relatively unimportant, because so few of the pseudomonads survive drying. However, *P. aeruginosa* can be found regularly in the dust of wards where patients (especially burns cases) infected with *P. aeruginosa* are treated. Cellulose wadding used for packing plaster bandages is another dry source which has been found to cause infection (Sussman & Stevens, 1960). Dry dust carrying *P. aeruginosa* can be dispersed by air, and large numbers may be found in the air during the dressing of burns (Lowbury & Fox, 1954; Barclay & Dexter, 1968).

The patterns of infection with *P. aeruginosa* are shown in many reports from hospitals. Examples of epidemic infection are the outbreaks in ophthalmic surgery and neurosurgery described above in which a single serotype and phage type of the organism was isolated from all the infected patients and from a source (eye drops; shaving brush). Endemic infection occurs in burns units, intensive care units and urological wards where there are continuing sources of infection and a continuing supply of fresh targets for infection; in such infections a number of different phage and serotypes of *P. aeruginosa* are usually found to cause infection (e.g. Davis *et al.*, 1969). Endemic infection may also be endogenous, as in leukaemic children treated with cytotoxic drugs and successfully screened against exogenous infections by protective isolation in ultraclean isolation rooms or isolators, with sterile food and other supplies (Jameson *et al.*, 1971).

P. aeruginosa infections in a burns ward have been found in most cases to be acquired from other patients in the ward (Lowbury & Fox, 1954; Sutter *et al.*, 1967; Liljedahl *et al.*, 1972). In a burns unit with two wards (Davis *et al.*, 1969) it was found that strains of *P. aeruginosa* acquired by burns were usually of the same serotypes as those previously found in the burns of other patients in the *same* (not the other) ward.

4.c. Routes of Transfer

Much evidence, including the rarity of airborne *P. aeruginosa*, points to the minimal importance of airborne infection. In a burns unit, however, large numbers of *P. aeruginosa* may be dispersed during the removal of dressings. In a controlled trial (Lowbury, 1954) comparable groups of patients had their dressings changed in a plenum ventilated dressing room and in the same room without mechanical ventilation; there was a significantly lower incidence of infection with *P. aeruginosa* in the patients dressed in the plenum ventilated room, from which it could be inferred that *P. aeruginosa* was sometimes transferred by air after a dressing of one patient to the burn of the next patient who had his dressings changed in the dressing station.

In other situations, however, airborne infection with *P. aeruginosa* appears to be very rare, and even in a burns ward, except under the circumstances described above, contact transfer is more important. This was shown by a controlled trial in which comparable groups of patients with burns were nursed for periods up to three weeks in a plastic isolator with an open top that protected them only against.*contact* contamination from the hands and clothing of nurses and others, in an 'air-curtain' isolator that protected them only against *airborne* contamination, and in the open ward. The acquisition of *P. aeruginosa* by the burns was identical (50 %) in the patients who were treated in the 'air-curtain' isolator and in those treated in the open ward, but none of the patients treated in the open-topped plastic isolator acquired *P. aeruginosa* (Lowbury *et al.*, 1971). Nearly half of the nurses in the Burns Unit were found to carry *P. aeruginosa* on their hands while on duty (Lowbury & Fox, 1954), and this would seem to be a particularly important vector (see also Kominos *et al.*, 1972). Other Gram-negative bacilli were as commonly acquired (probably by self-infection) in the plastic isolator as in the two other environments.

5. COMMENTS

The clinical importance of ecological factors is well illustrated by *P. aerugi-nosa*. Though a weak pathogen, its ability to survive and grow under conditions adverse to more strongly pathogenic species gives it the potentialities of an opportunist. The ability to grow well at body temperature and the possession of a range of toxic factors give it an advantage over many other opportunists. Its predilection for moist environments and its relative resistance to various disinfectants determine the mechanisms by which it is often transferred to hosts. These factors also indicate the aseptic and hygienic methods needed to prevent such transfer. Moreover the resistance of *P. aeruginosa* to most antibiotics and the uncertainty of success in treatment make it important to have effective methods of prophylaxis.

P. aeruginosa became prevalent because methods which were effective against *Strep. pyogenes* and *S. aureus* were ineffective against the *Pseudomonas*; also because of the longer survival, under modern treatment, of debilitated, opportunist-prone patients and because of the use of measures that rendered patients more susceptible to opportunist attack. In recent years there has been a widespread reduction in the frequency and severity of *Pseudomonas* infection, due probably to better protection of the patients who were formerly most at risk—e.g. in burns which have been successfully protected with 0·5 % silver nitrate solution and other forms of topical chemoprophylaxis, by better control of urinary tract infection, and by effective prevention of outbreaks. With smaller numbers of infected patients the problems of preventing cross-infection are less insurmountable. There are still, however, important hazards, especially in neonates, in cytotoxic therapy and in burns treated under adverse conditions—e.g. in developing countries, or where there have been delays in starting treatment. The organism is still widespread, and disuse or misuse of effective measures could lead to increased incidence of *Pseudomonas* sepsis. The virulence of strains may well decline due to fewer opportunities for colonizing human tissues, but there is no evidence that this has contributed to the recent decline in *Pseudomonas* sepsis reported from some centres.

6. BIBLIOGRAPHY

Adler, J. L., Burke, J. P. & Finland, M. (1971) *Archives of Internal Medicine*, **127**, 460.
Alexander, J. W. & Moncrief, J. A. (1966) *Archives of Surgery*, **93**, 75.
Alexander, J. W. & Wixson, D. (1970) *Surgery, Gynecology and Obstetrics*, **130**, 431.
Allen, H. F. (1959) *Transactions of the American Ophthamological Society*, **57**, 377.
Allen, H. F. (1963) in *Infectious Diseases of the Conjunctiva and Cornea*, St. Louis, p. 80.
Arturson, G., Högeman, C. F., Johanson, S. G. O. & Killander, J. (1969) *Lancet*, i, 546.
Asay, L. D. & Koch, R. (1960) *New England Journal of Medicine*, **262**, 1062.
Atik, M., Liu, P. V., Hanson, B. A., Amini, S. & Rosenberg, C. F. (1968) *Journal of the American Medical Association*, **205**, 134.
Aycliffe, G. A. J., Barry, D. R., Lowbury, E. J. L., Roper-Hall, M. J. & Martin Walker, W. (1966) *Lancet*, i, 1113.
Ayliffe, G. A. J., Brightwell, K., Collins, B. J. & Lowbury, E. J. L. (1968) *Lancet*, ii, 1117.
Ayliffe, G. A. J., Collins, B. J. & Petit, F. (1970) *Lancet*, i, 559.

Ayliffe, G. A. J., Lowbury, E. J. L., Hamilton, J. G., Small, J. M., Asheshov, E. A. & Parker, M. T. (1965) *Lancet*, **ii**, 365.

Barclay, T. L. & Dexter, F. (1968) *British Journal of Surgery*, **55**, 197.

Barrett, F. F., Casey, J. I. & Finland, M. (1968) *New England Journal of Medicine*, **278**, 5.

Barrie, H. J. (1941) *Lancet*, **i**, 242.

Barson, A. J. (1971) *Archives of Diseases in Childhood*, **46**, 55.

Bassett, D. C. J., Thompson, S. A. S. & Page, B. (1965) *Lancet*, **i**, 781.

Bean, H. S. & Farrell, R. C. (1967) *Journal of Pharmacology*, 19/Suppl. 1833.

Bignell, J. L. (1951) *British Journal of Opthalmology*, **35**, 419.

Bishop, W. A. (1938) *Journal of Bone & Joint Surgery*, **20**, 216.

Botterell, E. H. & Magner, D. (1945) *Lancet*, **i**, 112.

Brown, M. R. W. (1975) *Resistance of Pseudomonas aeruginosa*, John Wiley, London.

Brown, M. R. W. & Scott Foster, J. H. (1970) *Journal of Clinical Pathology*, **23**, 172.

Brown, V. I. & Lowbury, E. J. L. (1965) *Journal of Clinical Pathology*, **18**, 752.

Bruun, J. (1970) *Post-operative Wound Infection*, Bergen.

Buck, A. C. & Cooke, E. M. (1969) *Journal of Medical Microbiology*, **2**, 521.

Bull, J. P. (1971) *Lancet*, **ii**, 1133.

Burns, M. W. & May, J. R. (1968) *Lancet*, **i**, 270.

Burns, R. P. & Rhodes, D. H. (1961) *Archives of Ophthalmology*, Chicago, **65**, 517.

Calman, R. M. & Gibson, J. (1953) *Lancet*, **ii**, 649.

Carney, S. A., Dyster, R. E. & Jones, R. J. (1973) *British Journal of Dermatology*, **88**, 539.

Carney, S. A. & Jones, R. J. (1968) *British Journal of Experimental Pathology*, **49**, 395.

Cason, J. S. & Lowbury, E. J. L. (1968) *Lancet*, **i**, 651.

Cetin, E. T., Toreci, K. & Ang, O. (1969) *Journal of Bacteriology*, **89**, 1432.

Chün-Hsiang, L., Chien-Yin, H., Hsing-Min, C. & Chun, S. (1964) *Chinese Medical Journal*, **83**, 779.

Cole, A. P., Thom, A. R. & Watrasiewicz, K. (1971) *Lancet*, **ii**, 1155.

Colebrook, L. (1960) *A New Approach to the Treatment of Burns and Scalds*, Fine Technical Publications, London.

Colebrook, L., Lowbury, E. J. L. & Hurst, L. (1960) *Journal of Hygiene*, Cambridge, **58**, 357.

Colwell, R. R. (1964) *Journal of General Microbiology*, **37**, 181.

Crompton, D. O. (1962) *Australasian Journal of Pharmacology*, Oct. 30, p. 1030.

Cruickshank, C. N. D. & Lowbury, E. J. L. (1953) *British Journal of Experimental Pathology*, **34**, 583.

Cziszar & Lanyi, B. (1970) *Acta Microbiologica Academiae Scientarium Hungaricae*, **17**, 361.

D'Ascani, E. & Venturi, D. (1958) *Zooprofilassi*, **18**, 617. Cited by Wilson and Miles (1964).

Davis, B., Lilly, H. A. & Lowbury, E. J. L. (1969) *Journal of Clinical Pathology*, **22**, 634.

Diaz, F., Mosovitch, L. L. & Neter, E. (1970) *Journal of Infectious Diseases*, **121**, 269.

Diener, B., Carrick, L. & Berk, R. S. (1973) *Infection and Immunity*, **7**, 212.

Doggett, R. G., Harrison, G. M. & Carter, R. E. (1971) *Lancet*, **i**, 236.

Drewett, S. E., Payne, D. J. H., Tuke, W. & Verdon, P. E. (1972) *Lancet*, **i**, 946.

Duke-Elder, S. (1970) *Parsons' Diseases of the Eye*, 15th ed., Churchill, London, p. 384.

Durcieux, R. & Brett, A. J. (1967) *Revue Francaise de Gynecologie et d'Obstetriques*, **62**, 451.

Epstein, J. W. & Grossman, A. B. (1933) *American Journal of Disease in Childhood*, **46**, 132.

Falcao, D. P., Mendonca, C. P., Scrassolo, A. & de Almeida, B. B. (1972) *Lancet*, **ii**, 38.

Farmer, J. J. III & Herman, L. G. (1969) *Applied Microbiology*, **18**, 760.

Forfar, J. O., Gould, J. C. & MacCabe, A. F. (1968) *Lancet*, **ii**, 177.

Forkner, C. E., Frei, C. E., Edgcomb, J. H. & Utz, J. P. (1958) *American Journal of Medicine*, **25**, 877.

Fox, J. E. & Lowbury, E. J. L. (1953a) *Journal of Pathology and Bacteriology*, **65**, 519.

Fox, J. E. & Lowbury, E. J. L. (1953b) *Journal of Pathology and Bacteriology*, **65**, 533.

Fraenkel, E. (1912) *Zeitschrifft für Hygiene und Infektionskrankheiten*, **72**, 486.

Fraenkel, E. (1917) *Zeitschrifft für Hygiene und Infektionskrankheiten*, **84**, 369.

Garrod, L. P. (1946) *British Medical Bulletin*, **4**, 106.

Gibson, H. J. (1930) *Journal of Hygiene, Cambridge*, **30**, 337.

Goldman, L. & Fox, H. (1944) *Archives of Dermatology and Syphilology*, **49**, 136.

Gorrill, R. H. (1952) *Journal of Pathology and Bacteriology*, **64**, 857.

Goslings, W. R. O. (1963) in *Infection in Hospitals*, (Eds. R. E. O. Williams & R. A. Shooter), Blackwell, Oxford, p. 21.

Goto, S. & Enomoto, S. (1970) *Japanese Journal of Microbiology*, **14**, 65.

Gould, J. C. (1968) in *Urinary Tract Infection*, (Eds. F. O'Grady & W. Brumfitt), Oxford University Press, p. 43.

Gould, J. C. & McLeod, J. Q. (1960) *Journal of Pathology and Bacteriology*, **79**, 295.

Graber, C. D., Cummings, D., Vogel, E. H. & Tumbusch, W. T. (1961) *Texas Reports on Biological Medicine*, **19**, 268.

Grieble, H. G., Colton, F. R., Bird, T. J., Toigo, A. & Griffiths, L. G. (1970) *New England Journal of Medicine*, **282**, 531.

Habs, I. (1957) *Zeitschrifft fur Hygiene und Infektionskrankheiten*, **144**, 218.

Hall, J. H., Callaway, J. L., Tindall, J. P. & Graham Smith, J. (1968) *Archives of Dermatology*, **97**, 312.

Haynes, W. C. (1951) *Journal of General Microbiology*, **5**, 939.

Hitschmann, C. & Kreibig, K. (1897) *Wiener Klinische Wochenschrifft*, **10**, 1093; quoted by Fraenkel (1912).

Hoffman, M. A. & Finberg, L. (1955) *Journal of Pediatrics*, **46**, 627.

Homma, J. Y. & Uehara, T. (1971) *Japanese Journal of Experimental Medicine*, **41**, 593.

Hugh, R. & Leifson, E. (1953) *Journal of Bacteriology*, **66**, 24.

Hurst, L. & Lowbury, E. J. L. (1953) *Journal of Clinical Pathology*, **5**, 359.

Hurst, V. & Sutter, V. L. (1966) *Journal of Infectious Diseases*, **116**, 151.

Jackson, D. M., Lowbury, E. J. L. & Topley, E. (1951) *Lancet*, **ii**, 137.

Jacobs, J. (1964) *Postgraduate Medical Journal*, **40**, 590.

Jameson, B., Gamble, D. R., Lynch, J. & Kay, H. E. M. (1971) *Lancet*, **i**, 1034.

Johanson, P. H. (1968) *Journal of the American Medical Association*, **204**, 262.

Jones, R. J. (1970) *British Journal of Experimental Pathology*, **51**, 53.

Jones, R. J. (1971) *British Journal of Experimental Pathology*, **52**, 100.

Jones, R. J. & Dyster, R. E. (1973) *British Journal of Experimental Pathology*, **54**, 416.

Jones, R. J., Hall, M. & Ricketts, C. R. (1972) *Immunology*, **23**, 889.

Jones, R. J., Jackson, D. M. & Lowbury, E. J. L. (1966) *British Journal of Plastic Surgery*, **19**, 43.

Jones, R. J., Lilly, H. A. & Lowbury, E. J. L. (1971) *British Journal of Experimental Pathology*, **52**, 264.

Jones, R. J. & Lowbury, E. J. L. (1965) *Lancet*, **ii**, 623.

Jones, R. J. & Lowbury, E. J. L. (1966) in *Research in Burns*, (eds. J. Wallace & J. Wilkinson), Livingstone, Edinburgh, p. 474.

Kabota, Y. & Liu, P. V. (1971) *Journal of Infectious Diseases*, **123**, 97.
Kay, A. W., Richards, R. L. & Watson, A. J. (1958) *British Journal of Surgery*, **46**, 45.
King, E. O., Ward, M. K. & Raney, D. E. (1954) *Journal of Laboratory and Clinical Medicine*, **44**, 301.
Knight, V., Hardy, R. C. & Negrin, J. (1952) *Journal of the American Medical Association*, **149**, 1395.
Kominos, S. D., Copeland, C. & Grosiak, B. (1972) *Applied Microbiology*, **23**, 309.
Kovačs, N. (1956) *Nature, London*, **178**, 703.
Lányi, B., Cregačs, S. & Adam, T. (1966) *Acta Microbiologica Academiae Scientarium Hungaricae*, **13**, 319.
Last, P. M., Harrison, P. A. & Marsh, J. A. (1966) *Lancet*, **i**, 74.
Laursen, H. (1962) *Acta Obstetrica Gynecolgica Scandanavica*, **41**, 254.
Liedberg, N. C. F., Reiss, E. & Artz, C. P. (1954) *Surgery, Gynecology & Obstetrics*, **99**, 151.
Light, I. J., Sutherland, J. M., Cochrane, M. L. & Sartorius, J. (1968) *New England Journal of Medicine*, **278**, 1243.
Liljedahl, Malmborg, A. S., Mystrom, B. & Sjöberg, L. (1972) *Journal of Medical Microbiology*, **5**, 473.
Lilley, A. B. & Bearup, A. J. (1928) *Medical Journal, Australia*, **i**, 362.
Lilly, H. A. & Lowbury, E. J. L. (1972) *Journal of Medical Microbiology*, **5**, 151.
Lindberg, R. B., Moncrief, J. A., Switzer, W. E., Order, S. E. & Mills, W. (1965) *Journal of Trauma*, **5**, 601.
Liston, J., Wiebe, W. & Colwell, R. R. (1963) *Journal of Bacteriology*, **85**, 1061.
Liu, P. V. (1964) *Journal of Bacteriology*, **88**, 1421.
Liu, P. V., Abe, Y. & Bates, J. L. (1961) *Journal of Infectious Diseases*, **108**, 218.
Liu, P. V. & Mercer, C. B. (1963) *Journal of Hygiene, Cambridge*, **61**, 485.
Lowbury, E. J. L. (1951a) *Journal of Clinical Pathology*, **4**, 66.
Lowbury, E. J. L. (1951b) *British Journal of Industrial Medicine*, **8**, 22.
Lowbury, E. J. L. (1954) *Lancet*, **i**, 292.
Lowbury, E. J. L. (1960) *British Medical Journal*, **1**, 994.
Lowbury, E. J. L. (1967) *British Journal of Plastic Surgery*, **20**, 211.
Lowbury, E. J. L. (1971) *Proceedings of the Royal Society of Medicine*, **64**, 986.
Lowbury, E. J. L., Babb, J. R. & Ford, P. M. (1971) *Journal of Hygiene, Cambridge*, **69**, 529.
Lowbury, E. J. L. & Collins, A. G. (1955) *Journal of Clinical Pathology*, **8**, 47.
Lowbury, E. J. L. & Fox, J. E. (1953) *Journal of Hygiene, Cambridge*, **51**, 203.
Lowbury, E. J. L. & Fox, J. E. (1954) *Journal of Hygiene, Cambridge*, **52**, 403.
Lowbury, E. J. L., Lilly, H. A. & Wilkins, M. D. (1962) *Journal of Clinical Pathology*, **15**, 339.
Lowbury, E. J. L., Thom, B. T., Lilly, H. A., Babb, J. R. & Whittall, K. (1971) *Journal of Medical Microbiology*, **3**, 39.
McLeod, J. W. (1958) *Lancet*, **i**, 394.
Macmillan, B. G., Law, E. J. & Holder, A. (1972) *Archives of Surgery*, **104**, 509.
Mandl, I. (1961) *Advances in Enzymology*, **23**, 163.
Markley, K., Gurmendi, G., Chavez, P. M. & Bazan, A. (1957) *Annals of Surgery*, **145**, 175.
Mawson, S. R. (1967) *Diseases of the Ear*, Arnold, London, p. 229.
Moore, B. & Forman, A. (1966) *Lancet*, **ii**, 929.
Moran, J. M., Atwood, R. P. & Rowe, M. I. (1965) *New England Journal of Medicine*, **272**, 554.
Moyer, C. A., Brentano, L., Gravens, P. L., Margraf, H. W. & Monafo, W. W. (1965) *Archives of Surgery*, **90**, 812.

Mull, J. D. & Callahan, W. S. (1965) *Experimental and Molecular Pathology*, **4**, 567.
Myerowitz, R. L., Schneerson, R., Robbins, J. B. & Turck, M. (1972) *Lancet*, **ii**, 250.
National Research Council (1964) *Annals of Surgery*, **160**, Suppl. No. 2.
Neter, E. & Weintraub, D. H. (1955) *Journal of Pediatrics*, **46**, 280.
Noble, W. S. & Savin, J. A. (1966) *Lancet*, **i**, 347.
Noone, P. & Shafi, M. S. (1973) *Journal of Clinical Pathology*, **26**, 140.
Nunn, S. L. & Wellman, W. E. (1960) *Medical Clinics of North America*, **44**, 1075.
Order, S. E., Mason, A. D., Walker, H. L., Lindberg, R. F., Switzer, W. E. & Moncrief, J. A. (1965a) *Surgery, Gynecology & Obstetrics*, **120**, 983.
Order, S. E., Mason, A. D., Walker, H. L., Lindberg, R. B., Switzer, W. E. & Moncrief, J. A. (1965b) *Journal of Trauma*, **5**, 62.
Patty, P. A. (1921) *Journal of Infectious Diseases*, **29**, 73.
Paul, R. & Marget, W. (1964) *German Medical Monthly*, **9**, 119.
Pettit, F. & Lowbury, E. J. L. (1968) *Journal of Hygiene, Cambridge*, **66**, 393.
Phillips, I. (1969) *Journal of Medical Microbiology*, **2**, 9.
Phillips, I. & Spencer, G. (1965) *Lancet*, **ii**, 1325.
Pierce, A. K., Edmondson, E. B., McGee, G., Ketchersid, J., Loudon, R. G. & Sanford, J. P. (1966) *American Revue of Respiratory Diseases*, **94**, 309.
Pillemer, L., Blum, L., Lepow, I. H., Ross, O. A., Todd, E. W. & Wardlaw, A. C. (1954) *Science*, **120**, 279.
Plotkin, S. A. & Austrian, R. (1958) *American Journal of Medical Science*, **235**, 621.
Polk, H. C. & Stone, H. H. (1972) in *Contemporary Burn Management*, (Eds. H. C. Polk & H. H. Stone), Little, Brown & Co., Boston, p. 303.
Polk, H. C., Ward, C. G., Clarkson, J. G. & Taplin, D. (1969) *Archives of Surgery*, **98**, 292.
Pons, R. (1927) *Annals of the Institute Pasteur*, **41**, 1338.
Pruitt, B. A. & Curreri, P. W. (1971) *Archives of Surgery*, **103**, 461.
Pulaski, E. J. (1964) *Surgical Infections*, Charles C. Thomas, Springfield, p. 214.
Rabin, E. R., Graber, O. D., Vogel, E. H., Finkelstein, R. A. & Tumbusch, W. J. (1961) *New England Journal of Medicine*, **265**, 1225.
Rampling, A. & Whitby, J. L. (1972) *Journal of Medical Microbiology*, **5**, 305.
Reiner, L., Schlesinger, M. J. & Miller, G. M. (1952) *Archives of Pathology, Chicago*, **54**, 39.
Ricketts, C. R., Squire, J. R. & Topley, E. (1951) *Clinical Science*, **10**, 89.
Ridley, F. (1958) *British Journal of Ophthalmology*, **42**, 641.
Ringen, L. M. & Drake, C. H. (1952) *Journal of Bacteriology*, **64**, 841.
Ritzmann, S. E., McClung, C., Falls, D., Larson, D. L., Abston, S. & Goldman, A. S. (1969) *Lancet*, **i**, 1152.
Rogers, K. B. (1960) *Journal of Applied Bacteriology*, **23**, 532.
Rubbo, S. D., Gardner, J. F. & Franklin, S. C. (1966) *Journal of Hygiene, Cambridge*, **64**, 121.
Rubio, N. & Lopez, R. (1972) *Applied Microbiology*, **23**, 211.
Ružička, S. (1898) *Zentrallblatt für Bakteriologie*, **24**, 11.
Saltzman, M. (1963) *Clinical Medicine*, **70**, 559.
Sandiford, B. R. (1937) *Journal of Pathology and Bacteriology*, **44**, 567.
Schoental, R. (1941) *British Journal of Experimental Pathology*, **22**, 137.
Selepka, F. (1958) *Archives of Hygiene, Berlin*, **142**, 569.
Selwyn, S. (1965) *Journal of Hygiene, Cambridge*, **63**, 59.
Sevitt, S. (1964) *Acta Chirurgica Plastica*, **6**, 173.
Shambaugh, G. E. (1959) *Surgery of the Ear*, 2nd ed., Saunders, Philadelphia.

Shooter, R. A., Cooke, E. M., Gaya, H., Kumar, P., Patel, N., Parker, M. T., Thom, B. T. & France, D. R. (1969) *Lancet*, **i**, 1227.
Shooter, R. A., Cooke, E. M., Rousseau, S. A. & Breaden, A. L. (1970) *Lancet*, **ii**, 226.
Shooter, R. A., Walker, K. A., Williams, V. R., Horgan, G. M., Parker, M. T., Asheshov, E. H. & Bullimore, J. F. (1966) *Lancet*, **ii**, 1331.
Simmons, N. A. & Gardner, D. A. (1969) *British Medical Journal*, **ii**, 668.
Smith, W. & Smith, M. M. (1941) *Lancet*, **ii**, 783.
Sonnenschein, C. (1927) *Zentralblatt für Bakteriologie, Parasitenkunde und Infektionskrankheiten, Originale*, **104**, 365.
Speirs, C. F., Selwyn, S. & Nicholson, D. N. (1963) *Lancet*, **ii**, 710.
Stanley, M. M. (1947) *American Journal of Medicine*, **2**, 253, 347.
Stewart-Tull, D. E. S. & Armstrong, A. V. (1972) *Journal of Medical Microbiology*, **5**, 67.
Stobie, P. J. (1961) quoted by Crompton, D. A. (1962).
Stone, H. H., Martin, J. D. & Graber, C. D. (1963) *Annals of Surgery*, **159**, 991.
Stoodley, B. J. & Thom, B. T. (1970) *Journal of Medical Microbiology*, **3**, 367.
Sussman, M. & Stevens, J. (1960) *Lancet*, **ii**, 734.
Sutter, V. L., Hurst, V., Grossman, M. & Calonje, R. (1966a) *Journal of the American Medical Association*, **197**, 854.
Sutter, V. L., Hurst, V. & Landucci, A. D. J. (1966b) *Journal of Dental Research*, **45**, 1800.
Sutter, V. L., Hurst, V. & Lane, C. W. (1967) *Health Laboratory Science*, **4**, 245.
Taplin, D. & Zaias, N. (1966) *Military Medicine*, **131**, 814.
Taplin, D., Zaias, N. & Rebell, G. (1965) *Archives of Environmental Health*, **11**, 546.
Teplitz, C. (1965) *Archives of Pathology*, **80**, 297.
Teplitz, C., Davis, D., Mason, A. D. & Moncrief, J. A. (1964a) *Journal of Surgical Research*, **4**, 200.
Teplitz, C., Davis, D., Walker, H. L., Raulston, G. L., Mason, A. D. & Moncrief, J. A. (1964b) *Journal of Surgical Research*, **4**, 217.
Teplitz, C., Raulston, G. L., Walker, H. L., Mason, A. D. & Moncrief, J. A. (1964) *Journal of Infectious Diseases*, **114**, 75.
Teres, D., Schweers, P., Bushnell, L. S., Hedley-Whyte, J. & Feingold, D. S. (1973) *Lancet*, **i**, 415.
Thomas, L. & Stetson, C. A. (1949) *Journal of Experimental Medicine*, **89**, 461.
Thornley, J. (1960) *Journal of Applied Bacteriology*, **23**, 37.
Thygeson, P. (1948) *Californian Medicine*, **69**, 18.
Tillotson, J. R. & Lerner, A. M. (1968) *Annals of Internal Medicine*, **68**, 295.
Tumbusch, W. T., Vogel, E. H., Butkiewicz, C. D., Graber, C. D., Larson, D. L. & Mitchell, E. T. (1962) in *Research in Burns*, (Ed. C. Artz), Saunders, Philadelphia, p. 235.
Wahba, A. H. & Darrell, J. H. (1965) *Journal of General Microbiology*, **38**, 329.
Walker, H. L., Mason, A. D. & Raulston, G. L. (1964) *Annals of Surgery*, **160**, 297.
Whitby, J. L. & Rampling, A. (1972) *Lancet*, **i**, 15.
Williams, R., Williams, E. D. & Hyams, D. E. (1960) *Lancet*, **i**, 376.
Wilson, M. G., Nelson, R. C., Philips, L. H. & Boak, R. A. (1961) *Journal of the American Medical Association*, **175**, 1146.
Wretlind, B. L., Heden, L., Sjöberg, L. & Wadström, T. (1973) *Journal of Medical Microbiology*, **6**, 91.

CHAPTER 3

Wall and Membrane Structures in the Genus *Pseudomonas*

PAULINE M. MEADOW

The outer layers of *Pseudomonas aeruginosa* like those of other Gram-negative bacteria are made up of several discrete macromolecular components whose organization to form the complete wall is by no means fully understood. It is difficult to separate the individual components either physically or functionally and it is more than likely that the interrelationship between them plays a major part in the phenotypic properties of the bacterium. However some at least of the individual components have been separated and analysis of these together with electron microscopy has allowed us to distinguish some order in the structure of isolated walls.

1. MEMBRANE AND WALL ORGANIZATION

Some structure can be seen in thin sections of *P. aeruginosa* (Plate 3.1). This section was fixed in glutaraldehyde and osmium tetroxide and post-stained in uranyl acetate and lead citrate. The bacterium is bounded by two wavy double-track membranes about 7 nm wide separated from each other by an electron-transparent region in which there is no clearly differentiated structure. The inner of these membranes is probably the cytoplasmic

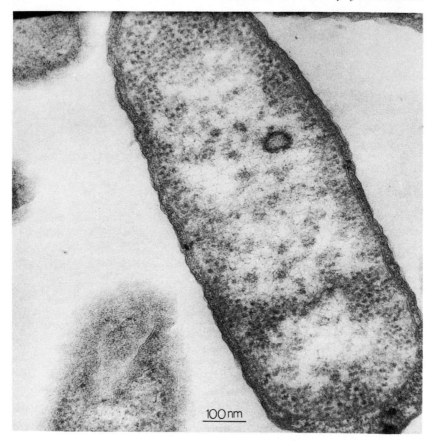

Plate 3.1. Thin section of *P. aeruginosa* (ATCC 9027) showing two double-track outer membranes. The bar represents 100 nm. (Courtesy of Dr F. W. Adair, Ciba Geigy, Summitt, New Jersey, U.S.A.)

membrane while the regions outside this may be considered as the wall. A clearer idea of the number of layers in the wall can be derived from freeze-etched studies (Lickfield *et al.*, 1972; Gilleland *et al.*, 1973). By studying bacteria with and without cryoprotective agents these two groups of workers have been able to distinguish as many as nine different layers in the walls and membranes of *P. aeruginosa* (Figure 3.1). Five of these layers (L_1, L_3, L_5, L_7 and L_9) are electron dense and some indication of their structure can be gained from the electron microphotograph of freeze-etched *P. aeruginosa* shown in Plate 3.2. These organisms were not protected with glycerol and the outer layer seen (the convex cell wall layer $\widehat{CW}$) is thought to correspond to L_3 as shown in Figure 3.1. Similar studies of organisms freeze-etched in the absence of glycerol showed a much smoother outer layer probably signifying

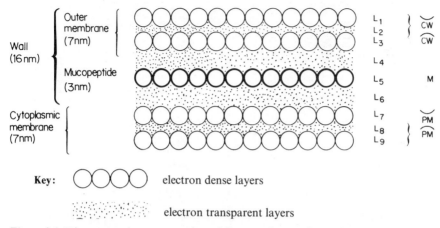

Figure 3.1. Diagrammatic representation of the outer layers of *Pseudomonas aeruginosa* as revealed by freeze-etch and thin-section electron microscopy. (Modified from Lockfield *et al.*, 1972 and Gilleland *et al.*, 1973.)

a layer containing a high lipid concentration which could be L_1. In Plate 3.2 fibrils can be seen extending from the outer wall layers to the concave cytoplasmic membrane (P̃M) but whether these are mucopeptide fibrils or artefacts of the preparations is not known. The mucopeptide layer (M) can be seen in profile between the concave cell wall and the concave cytoplasmic membrane in the cell on the left of the photograph. From the freeze-etched pictures it appears that both the outer membrane and the cytoplasmic membrane split down the middle revealing the underlying components as might be expected from a bilayer comprising lipid–protein–lipid.

In chemical terms the electron microphotographs could be interpreted as shown in Figure 3.2. The outer membrane consists of phospholipid and lipopolysaccharide arranged to form a mosaic or a lipid bilayer (L_1–L_3 in Figure 3.1). The antigenic side-chains of the lipopolysaccharide probably project outwards into the surrounding environment since they can be detected serologically in intact organisms. In addition to these two components in the outer membrane there are also proteins some of which may be lipoproteins and glycoproteins. Some of these can be seen in the L_2 layer in the freeze-etched profiles as a regular array of globular sub-units about 7 nm in diameter. They are released from isolated walls of *P. aeruginosa* by treatment with EDTA which solubilizes a protein–lipopolysaccharide complex composed of these globular sub-units which aggregate to form rodlets 20–25 nm in length. The material solubilized by EDTA consists of about 80 % protein and 30 % lipopolysaccharide (Roberts, Gray & Wilkinson, 1970), and causes lysis of the bacteria unless they are protected by 0·55 M sucrose. The depleted walls show a different appearance by electron microscopy but can be restored partially by the addition of magnesium which allows some of the material

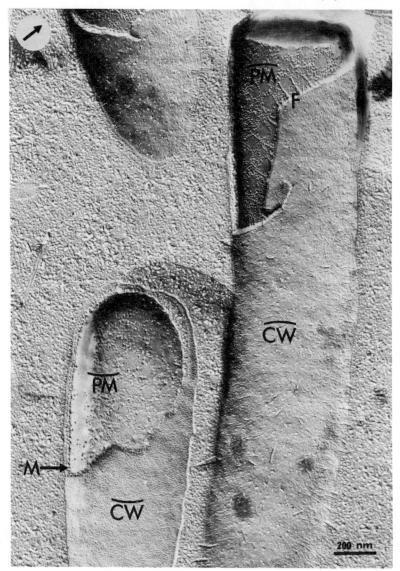

Plate 3.2. Freeze-etched *P. aeruginosa* OSU 64 after cryoprotection with glycerol. Four fracture layers can be seen; the convex outer wall layer, C̄W̄: convex cytoplasmic membrane, P̂M̂: concave outer wall layer, C̃W̃; and concave cytoplasmic membrane, P̄M̄. Fibrils run between the convex outer layer and the convex cytoplasmic membrane. A profile mucopeptide layer, M, can be seen between C̃W̃ and P̄M̄. A closely packed layer of rodlets 20–25 nm long makes up part of the C̄W̄ layer. The bar represents 200 nm. (Reproduced with permission from Gilleland *et al.* (1973) *Journal of Bacteriology*, **113,** 417–432)

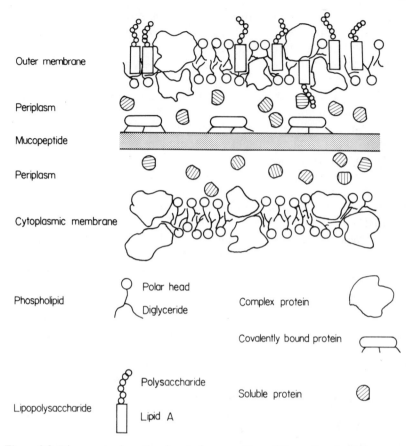

Figure 3.2. Diagram to show the chemical composition of the wall and membrane layers of *P. aeruginosa*

solubilized (85 % of the protein) to reaggregate on the membranes in the L_2 layer (Gilleland *et al.*, 1973). The restored bacteria do not however recover their viability completely, although they become osmotically stable, suggesting that some specific conformation or orientation in the L_2 layer is required for the proteins to be functional. Whether this function is enzymic or structural or both remains to be seen. Beneath the outer membrane is the so-called periplasmic space from which some proteins can be released by osmotic shock or removal of the outer membrane. Other proteins appear to be covalently linked either to the outer membrane or to the mucopeptide component. Somewhere in the periplasmic region is the rigid mucopeptide layer which may be associated either with the outer membrane or with the cytoplasmic membrane as will be discussed later. Finally there is the cytoplasmic membrane probably consisting, like the outer membrane, of a phospholipid protein mosaic or bilayer. Although the evidence for the complete organization of the outer layers of *Pseudomonas aeruginosa* is incomplete it is compatible with this type of structure.

More direct evidence of the bacterial envelope structure is however available for a related organism, a marine pseudomonad studied by MacLeod and his colleagues (Forsberg, Costerton & MacLeod, 1970a,b). This organism requires sodium ions for growth and metabolism and other cations for maintenance of cellular integrity. When it is washed with a solution containing sodium, potassium and magnesium before resuspending in 0·5 M sucrose, or if it is washed with 0·5 M sodium chloride, about 5 % of its dry weight is released into the medium as non-dialysable material. Thin sections however showed no change in appearance and the authors concluded that the material must be non-staining loosely bound material which was not required for viability. Further washing with sucrose caused the outer double-track membrane to fragment and non-sedimentable material, presumably corresponding to the lipopolysaccharide–protein layers (L_1–L_3), was released into the medium. Further washing released a third fraction which may have been derived from the periplasmic region (L_4) since it differed in both hexosamine and protein content from either of the other two layers. At this stage the organism was still rod shaped and contained all the diaminopimelic acid and muramic acid of the original bacterium. It was also osmotically stable suggesting that the mucopeptide layer was still intact and it was described as a mureinoplast. Treatment with lysozyme solubilized all the diaminopimelic acid and muramic acid and converted it into a spherical osmotically sensitive protoplast bounded only by the cytoplasmic membrane. This instability of the walls of the marine pseudomonad has proved very useful in studying the different components of the isolated wall using non-degradative techniques. In most Gram-negative bacteria the outer layers can be removed only after more drastic treatment such as hot 45 % phenol, trichloroacetic acid or sodium dodecylsulphate, which makes sequential analysis almost impossible. While the walls of this marine organism are particularly unstable there is some evidence that the walls of *P. aeruginosa* itself are less stable than those of many other species. For example Roberts, Gray and Wilkinson (1967) found that lipopolysaccharide was released during preparation of walls from *P. aeruginosa* by a Braun homogenizer. It is possible that the walls of all pseudomonads are more readily disaggregated than those of many other species and that the marine pseudomonad provides an extreme example of this phenomenon.

2. CYTOPLASMIC MEMBRANE

The dissolution of the outer layers of the marine pseudomonad by washing under controlled conditions of ionic strength and pH has enabled MacLeod and his colleagues to isolate the cytoplasmic membrane free of other contaminating layers (Martin & MacLeod, 1971). This is the first analysis of cytoplasmic membranes isolated from a Gram-negative bacterium. In Gram-positive bacteria the wall and membranes can be fairly readily separated since membranes can be isolated from protoplasts prepared by lysozyme treatment of whole organisms. In most Gram-negative bacteria the osmotically fragile forms resulting from lysozyme treatment of whole bacteria are still contaminated with lipopolysaccharide and lipoprotein. These have to be removed from the membranes by phenol extraction or by the use of proteinases and other agents which may attack the cytoplasmic membrane as well. Martin and MacLeod estimated that the membranes isolated from the marine pseudomonad occupied about 12 % of the dry weight of the organism which is comparable to the calculations in Gram-positive bacteria of about 10–20 %. They were able to isolate the membranes from stable washed protoplasts of the marine pseudomonad by disintegration in the French pressure cell followed by differential centrifugation. After treatment with ribonuclease and deoxyribonuclease, the membranes were washed and collected by centrifugation. The procedure was therefore completely non-degradative. Electron microscopy, separation on sucrose density gradients and chemical analysis showed the membranes to be essentially free from cytoplasmic and wall contaminants. The chemical composition of the membranes from this marine pseudomonad was very similar to the analyses which have been reported for Gram-positive organisms which makes it seem likely that all bacterial membranes have similar structures (Table 3.1). The relatively

Table 3.1. Membrane composition (% dry weight)

	Marine pseudomonad[a]	Gram-positive bacteria[b]
Protein	62·8	41–75
Lipid	30·5	9–36
Carbohydrate	2·0	0·2–20
Polyisoprenoid	0·01	0·01

[a] From Martin & Macleod (1971).
[b] From Salton (1967); From Bodman & Welker (1969).

high protein content of the pseudomonad membranes may reflect their metabolic versatility and may be characteristic of other pseudomonads including *P. aeruginosa*. Crude membrane preparations of *P. aeruginosa* contained about 60 % protein (Hancock & Meadow, 1969) but it is difficult to know whether they were all covalently bound in the membrane.

While some of the membrane proteins may be structural, others must include the many enzymes known to be located in the membrane fractions (for reviews see Salton, 1967; 1971). The cytoplasmic membrane is the main permeability barrier of the bacterium and is involved in active transport. It is also the site of oxidative phosphorylation and of chromosome replication and attachment. It must play a part in septum formation and in the biosynthesis of wall polymers, in protein synthesis and DNA replication; and it is involved in some way in aeruginocin tolerance. In addition a number of specific enzyme functions have been located in the membrane fractions of *P. aeruginosa* and some of these will be discussed by Clarke and Ornston (Chapter 8).

One of the difficulties in studying the proteins found in the cytoplasmic membrane, is that of separating it from the outer membrane. In *Escherichia coli* Schnaitman (1971) has shown that 'Triton' X-100 solubilizes 15–23 % of the protein from partially purified walls. On the other hand treatment of partially purified membrane fractions with this detergent solubilized 60–80 % of the protein suggesting that the cytoplasmic membrane proteins are particularly susceptible to 'Triton' attack. Whether this technique can be applied to *P. aeruginosa* remains to be seen but until cytoplasmic membrane proteins can be clearly separated from the other proteins in the bacterial envelope it is difficult to define them more accurately.

The same technical problems limit study of the phospholipid composition of the cytoplasmic membranes of *P. aeruginosa*. In the marine pseudomonad the major lipid components of the cytoplasmic membranes were phospholipids (comprising 80 % of the total lipid) and were indistinguishable from the phospholipids isolated from the intact organism which would include those in the outer membrane layers as well (Martin & MacLeod, 1971). The phospholipids extracted from crude membrane preparations of *P. aeruginosa* appear very similar to those of the marine pseudomonad. About 90 % of the lipid in these preparations occurred as phospholipid and, as in the marine pseudomonad, phosphatidyl ethanolamine was the major component with smaller amounts of phosphatidyl glycerol, diphosphatidyl glycerol and phosphatidyl glycerol phosphate (Hancock & Meadow, 1967; 1969). Although there was some variation in phospholipid composition of *P. aeruginosa* during growth on different media, phosphatidyl ethanolamine remained the major component as in *E. coli* and other Gram-negative bacteria (Randle, Albro & Dittmer, 1969).

It would appear then that the cytoplasmic membranes of *P. aeruginosa* are very like those isolated from Gram-positive bacteria both in structure, composition and function though they must differ to some extent in their complement of enzymes and other proteins. These may vary from one species to another and even with the same species as in the aeruginocin-tolerant mutants discussed by Holloway (Chapter 4). The recent discovery in *P. aeruginosa* of invaginations of the membrane forming mesosome structures provides more evidence of the similarity of cytoplasmic membranes in all species of bacteria (Hoffmann *et al.*, 1973).

3. MUCOPEPTIDE

Moving outwards from the cytoplasmic membrane the next structural component is probably the mucopeptide though there may be a periplasmic region (L_6 in Figure 3.1) between these two layers. There was initially some doubt about the presence of a mucopeptide layer in *Pseudomonas aeruginosa* since the usual techniques for demonstrating this component in Gram-negative bacteria were unsuccessful and the peculiar instability of pseudomonad walls suggested limiting amounts of mucopeptide. However Forsberg *et al.* (1972) were able to detect this layer in thin sections of plasmolysed intact cells and also in mureinoplasts (bacteria from which the outer membrane layers had been removed) of their marine pseudomonad. In this organism the mucopeptide layer was much thinner than that found in other Gram-negative bacteria; it occupied only 1 % of the dry weight of the bacterium. Furthermore it appeared to be associated more closely with the cytoplasmic membrane than with the outer double membrane. Whether this is a general characteristic of pseudomonads is difficult to know particularly as the freeze-etched studies discussed earlier suggest separation of the mucopeptide layer from both the cytoplasmic membrane and the outer layers.

Whatever its position the mucopeptide layer certainly exists in *P. aeruginosa*. Intact sacculi containing the mucopeptide components can be isolated from many Gram-negative bacteria after phenol extraction or treatment with sodium dodecyl sulphate. These techniques have now been applied to several species of Pseudomonadales including *P. aeruginosa* (Heilmann, 1972; Martin, Heilman & Preusser, 1972). In this way intact sacculi retaining the shape of the original bacterium were isolated from *P. aeruginosa* OSU 64. Electron microscopy of the sacculi showed that they differed from those obtained from the Enterobacteriaceae in having a much smoother surface structure. For example sacculi isolated from *E. coli* have a coarse-grained surface pattern made up of granules about 10 nm in diameter. The pseudomonad sacculi had about the same distribution of particles on the surface but they were very much smaller (4·7 nm ± 0·6 nm) and could only be seen with higher magnification (Plate 3.3a). Sacculi isolated from another member of the Pseudomonadales, *Spirillum serpens*, were even simpler and appeared to contain no surface granules. Treatment of the sacculi from *P. aeruginosa* with trypsin and chymotrypsin removed the granules suggesting that they were protein in nature (Plate 3.3b). On the other hand when sacculi which had not been subjected to proteolytic enzymes were digested with lysozyme, only those peptides to be expected from pure mucopeptide were detected. These results would be compatible if the amounts of protein covalently linked to the mucopeptide were too small to be detected in the lysozyme digests, or if the protein had been split off by the chloroform which was used to remove the lysozyme before analysis of the digests

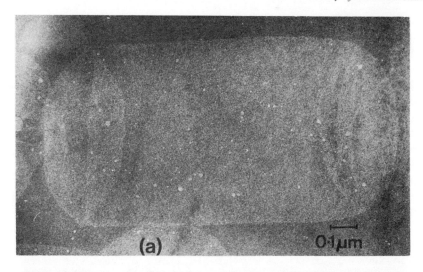

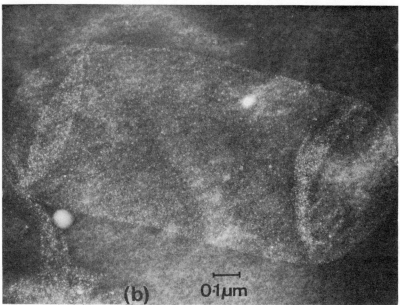

Plate 3.3. Rigid sacculi of *P. aeruginosa* OSU 64 negatively stained with potassium tungstate. (a) The sacculus is covered with (lipo)protein particles 4·7 nm in diameter. Magnification × 70,000. (b) After treatment with trypsin and chymotrypsin the surface particles have been removed exposing the smooth mucopeptide sacculus. Magnification × 70,000. (Reproduced with persmission from Martin *et al.* (1972) *Archiv für Mikrobiologie*, **83**, 332–346)

(Heilmann, 1972). In *E. coli* the mucopeptide layer is linked to the outer layers of the wall through a lipoprotein which can be removed by trypsin (Braun & Sieglin, 1970; Braun & Wolff, 1970). Although there is less covalently linked protein in *P. aeruginosa*, it may also be a lipoprotein whose function is to link the mucopeptide to the outer layers of the wall.

Chemical analysis of the isolated sacculi of pseudomonads confirmed the presence of small but detectable amounts of protein (Table 3.2). After trypsin treatment glucosamine, muramic acid, glutamic acid, alanine and diamino-pimelic acid were the major components and were present in the molar ratios $1:1:1:1\cdot73:1$. The only other amino acid present in detectable amounts was lysine (one mole for every 9 moles of diaminopimelic acid) suggesting that in *P. aeruginosa* as in *E. coli* a covalently bound protein molecule might have been attached to about 10 % of the diaminopimelic acid molecules through lysine (Heilmann, 1972). Amino acids and amino sugars made up 90 % of the weight of the mucopeptide layer after treatment with trypsin and chymotrypsin. Its simple composition is characteristic of all Gram-negative bacteria so far studied as is the presence of the *meso* isomer of diaminopimelic acid (Clarkson & Meadow, 1971). There is some cross-linking between the free amino group of the diaminopimelic acid and the terminal D-alanine as shown by the isolation of both dimers and tetramers from the lysozyme digests. The extent of cross-linking has been estimated by dinitrophenylation to be between 24 % and 30 % depending on the growth conditions (Clarkson & Meadow, 1971; Heilmann, 1972). This is very similar to the value found for other Gram-negative bacteria such as *E. coli*, for example, in which Takebe (1965) measured about 30 % cross-linking. The only other components to be detected in the purified isolated mucopeptide layers of *P. aeruginosa* were phosphorus (0·01 %) and metal ions (2·55 %). Of these sodium made up about 1·87 % of the dry weight with smaller amounts of potassium, magnesium, calcium and zinc. What role these ions play in the construction of the mucopeptide is not known since exhaustive treatment with EDTA or 8-hydroxyquinoline had no effect on the shape or composition of the mucopeptide sacculus. Its integrity was destroyed only by treatment with lysozyme (Heilmann, 1972).

One of the major functions of the mucopeptide layer appears to be to confer shape and mechanical strength on the bacterium since its loss causes protoplast formation and osmotic instability. It may however play a part in controlling the permeability of the outer layers of the bacterial wall either directly, by forming a permeability barrier, or indirectly by holding together the outer membrane layers. Indications of this role in *P. aeruginosa* Ps 18S are provided by the effect of cholate on bacteria which had been grown in the presence of concentrations of benzylpenicillin having little or no effect on the growth rate (Burman, Nordström & Bloom, 1972). Addition of cholate (15 mg/ml) to logarithmically growing bacteria caused lysis only if the bacteria

Table 3.2. Composition of isolated mucopeptide sacculi of Pseudomonadales

Recalculated from	Marine pseudomonad Forsberg et al. (1972)	P. aeruginosa Heilmann (1972)	P. aeruginosa Martin, Heilmann & Preusser (1972)	P. fluorescens Braun, Rehn & Wolff (1970)	S. serpens Martin et al. (1972)
	Molar ratios : diaminopimelic acid $= 1.0$				
Glucosamine	1.5[a]	0.92	0.86	0.88	0.86
Muramic acid	1.56[a]	0.73	0.67	0.81	0.9
Alanine	2.24	2.7	2.14	2.47	1.97
Glutamic acid	1.36	1.61	1.19	1.55	1.08
Diaminopimelic acid	1.0	1.0	1.0	1.0	1.0
Aspartic acid	tr	0.26	0.17	0.14	0.015
Threonine	0.025	0.19	0.12	0.095	0.010
Serine	0.39	0.19	0.13	0.13	0.013
Glycine	0.39	0.08	0.06	0.1	0.028
Leucine	—	0.21	—	0.105	—
Lysine	—	0.27	0.17	0.14	—
Histidine	—	—	0.03	0.1	—
Arginine	—	0.32	0.23	0.1	—

[a] Corrected for destruction during hydrolysis.

had been grown in the presence of benzylpenicillin. Cholate is known to cause lysis by distorting the cytoplasmic membrane and so allowing leakage of protein and RNA, whereas benzylpenicillin even at sublethal concentrations exerts its effect on the mucopeptide layer. It would appear therefore that the increased susceptibility of *P. aeruginosa* to cholate in the presence of benzylpenicillin is a reflexion of the increased penetration of cholate to its site of action, the cytoplasmic membrane.

From this work it is clear that the mucopeptide layer of *P. aeruginosa* retains the shape of the intact organism and that it probably comprises a single macromolecular component held together by covalent bonds linking adjacent peptide side-chains. It is substituted by a protein or lipoprotein which is smaller in size than that found in the Enterobacteriaceae but which may have a similar role in anchoring the mucopeptide either to the outer double membrane or to the cytoplasmic membrane. None of these analyses explain the apparent instability of the pseudomonad walls as compared with those of many other organisms. It is possible however that the amount of this layer per bacterium might be less than in other species. Certainly the estimate that the mucopeptide of the marine pseudomonad studied by Forsberg *et al.* (1972) occupies only 1·2 % of the dry weight of the organism is lower than that of 2 % for *E. coli* (White, Dworkin & Tipper, 1968) but this low value might be peculiar to the marine organism. There is some indication that the diaminopimelic acid content of isolated walls and whole cells of *P. aeruginosa* is lower than that of most Gram-negative bacteria which might indicate a lower mucopeptide content. It is however difficult, if not impossible, to draw valid conclusions about such figures since it is known that the composition of bacterial walls changes with growth conditions both absolutely and in the proportions of the different components (see Ellwood & Tempest, 1972). It would be necessary to isolate and analyse the walls of *P. aeruginosa* grown under a wide variety of conditions to establish the contribution of the mucopeptide layer to the overall economy of the organism.

4. PERIPLASMIC AREA

In most Gram-negative bacteria there is an area which is recognized from electron microscopy of thin sections as an unstained region. It is bounded on either side by the double membrane structures of the cytoplasmic membrane and the outer membrane layers to either of which the mucopeptide component may be attached. There is no evidence of distinct structures within this area which is known as the periplasmic region (L_4 and L_6 in Figure 3.1). Since a number of proteins can be released from Gram-negative bacteria simply by subjecting them to osmotic shock, it is assumed that this region contains a number of proteins which can be lost when the permeability of the outer layers is altered in this way. In *E. coli* a variety of enzymes have been shown to be located in this region (Neu & Chou, 1967) and some have also been characterized in *P. aeruginosa*. In an elegant series of experiments Costerton and his collaborators have studied the position and attachment of alkaline phosphatase in *P. aeruginosa* (Cheng, Ingram & Costerton, 1970; 1971). They first showed that all the bound alkaline phosphatase could be removed from *P. aeruginosa* by washing with 0·1 or 0·2 M Mg^{2+}. The release of this enzyme by magnesium appeared to have little or no effect on the viability of the bacteria but rendered the outer layers of the wall sufficiently permeable to lysozyme for it to cause spheroplast formation. The enzyme could also be released by growing the bacteria at pH 7·4 instead of 6·8. By using the Gomori reaction to study the distribution of the enzyme in thin sections, alkaline phosphatase was seen to be associated with the area between the two double membranes. Since lysozyme spheroplasts which had no mucopeptide layer retained alkaline phosphatase uniformly distributed around them, this enzyme cannot have been part of the mucopeptide. There was no evidence for the association of alkaline phosphatase with the cytoplasmic membrane and the authors concluded that alkaline phosphatase was probably associated with the lipopolysaccharide both in the periplasmic region and inside the outer double membrane, being released when this layer was disaggregated by metal ions. There is however no direct evidence to support the hypothesis. Since alkaline phosphatase was released by washing with metal ions it might be supposed that it was associated with lipopolysaccharide by ionic forces. Its release as an extracellular enzyme when the pH of the growth medium is raised might support this view since the charge of the outer layers of bacterial walls is affected by the growth conditions and particularly by the pH and ionic strength (see Ellwood & Tempest, 1972).

In many organisms the binding proteins associated with transport of amino acids and other nutrients into the bacterium are located in this periplasmic region. The outer layers of the wall are thought to function as a coarse molecular sieve allowing small molecules to pass through unhindered. The true permeability barrier is therefore the cytoplasmic membrane and specific proteins within the membrane may be required to transport material across it. Before such transport is possible however, the material may need to be concentrated in the location of the membrane and the obvious site for such proteins would be the periplasmic region. At least some such proteins in *P. aeruginosa* can be lost by osmotic shock and so are probably periplasmic components as for example the glycerol-binding protein isolated by Tsay, Brown & Gaudy (1971). For further discussion of transport into pseudomonads see Chapter 7, section 3.f, and Chapter 8, section 4.b.

5. OUTER MEMBRANE LAYER

The position of the outer double membrane layer makes it responsible for many of the properties of the intact bacterium. It must play some part in the transport of metabolites into the cell; it must have some effect on bacterio-phage and aeruginocin adsorption and it must be responsible for at least some of the antigenic and serological responses of the organism. As already discussed it appears to consist of at least two separate components lipo-proteins and a lipopolysaccharide.

5.a. Lipoprotein Components

In *E. coli* the lipoprotein attached to the mucopeptide has been isolated and characterized (Braun & Bosch, 1972; Hirashima & Inouye, 1973) but, as already discussed, the equivalent lipoprotein in *P. aeruginosa* is present in smaller amounts and has proved more difficult to isolate. Other proteins have however been isolated from *P. aeruginosa* together with lipopoly-saccharide suggesting that they may be derived from the outer membrane layers. Homma and his collaborators have studied a protein designated original endotoxic protein (OEP) which was released into the growth medium by autolysates of *P. aeruginosa* P 1-III as a complex with lipopoly-saccharide (Homma & Suzuki, 1966; Homma, 1971). The isolated protein had a molecular weight of 42,000 daltons and showed some toxic activity which was however physiologically and serologically distinct from that of the lipopolysaccharide itself. A similar protein–lipopolysaccharide complex is released from the outer membrane layers of *P. aeruginosa* by EDTA (Rogers, Gilleland & Eagon, 1969; Roberts *et al.*, 1970; Gilleland *et al.*, 1973). It contained two major protein components both of which appeared to be glycoproteins. One of these, protein A (molecular weight 43,000), may be the one studied by Homma and his colleagues. Protein B was considerably smaller with a molecular weight of about 16,000 daltons (Stinnett, Gilleland & Eagon, 1973). Whether these proteins are structural or enzymic in function remains to be seen but it might be expected that at least the enzymes concerned with the final stages of synthesis of the lipopolysaccharide and lipoproteins of the outer membrane would form part of this layer.

5.b. Lipid Components

Many of the lipids of the outer membrane are readily extractable, that is they are extracted by chloroform/methanol without prior hydrolysis and make up the bulk of the free lipids of the organism. As in the cytoplasmic membrane 90 % of these are phospholipids, the major component being phosphatidyl ethanolamine. The fatty acids in the phospholipids are similar to those found in most Gram-negative bacteria (Cho & Salton, 1966) with one major difference. In *P. aeruginosa* 18:1 is the predominant unsaturated fatty acid while 16:1 is a minor component (Hancock & Meadow, 1969). In stationary phase bacteria therefore when cyclopropane fatty acids are formed the 19 cyclopropane acid is formed in greater amounts than the 17 cyclopropane acid. The only other Gram-negative bacteria in which a preponderance of 19 cyclopropane acids have been reported are *Agrobacterium tumifaciens* (Kaneshiro & Marr, 1962) and two species of *Brucella* (Thiele, Lacave & Asselineau, 1969). Although the readily extractable fatty acid composition of *P. aeruginosa* is changed by growth conditions the overall pattern is remarkably constant, palmitic acid (16:0) representing the major component throughout the growth phase, with oleic acid (18:1) or 19 cyclopropane as the next most important component. Probably because the phospholipids of the outer layer cannot readily be distinguished from those of the inner cytoplasmic membrane we know almost nothing of their synthesis and turnover except that the cyclopropane acids appear to be synthesized from their mono-unsaturated precursors by methylation, the methyl group being donated by methionine (Hancock & Meadow, 1969).

5.c. Lipopolysaccharides

The other major component of the outer double membrane of pseudomonads, the lipopolysaccharide, has had a lot more attention partly because its chemical composition makes it readily distinguishable from other parts of the organism but also because it plays an important role in the response of the organism to its external environment. The methods used for its isolation and characterization have been developed largely from those used for the isolation of the *O*-antigenic endotoxic lipopolysaccharide in the Enterobacteriaceae the properties and biosynthesis of which have recently been well reviewed (Luderitz *et al.*, 1971). However, although the pseudomonad lipopolysaccharide has features in common with those of the Enterobacteriaceae, it has several important differences both in organization and chemical composition. As in other Gram-negative bacteria the lipopolysaccharide component can be removed from intact bacteria or isolated walls of *P. aeruginosa* by extraction with hot 45 % phenol (Clarke, Gray & Reaveley, 1967). Some of this material is apparently loosely bound since Roberts *et al.* (1967) found that it was removed even during wall preparation, while Nelson and MacLeod (1973) have shown both chemically and serologically that the lipopolysaccharide of their marine pseudomonad occurs in the periplasmic space as well as in the outer membrane.

The detailed chemical study of the *P. aeruginosa* lipopolysaccharide has been limited in general to a few laboratories, those of Homma and Egami in Tokyo, of Gray and Wilkinson in Hull and my own laboratory in London. The three groups have been working with strains of different serotype and some differences between them have emerged. Nevertheless the overall pattern of the lipopolysaccharide of *P. aeruginosa* is now becoming clear and our recent survey of a number of different serotypes suggests that in

P. aeruginosa, as in the Enterobacteriaceae, there may be a common basic lipopolysaccharide molecule which is modified in different strains. There appear to be both species and strain differences in the lipopolysaccharides of the Pseudomonadales but in *P. aeruginosa* and perhaps also in the putida groups a common basic pattern seems likely.

The endotoxic lipopolysaccharides of the Enterobacteriaceae appear to have rather high molecular weights (1.24×10^6 daltons) and there is evidence of significant heterogenicity (see Freer & Salton, 1971). The limited evidence available on the size of intact lipopolysaccharides from *P. aeruginosa* suggest they are similar. The lipopolysaccharide extracted from *P. aeruginosa* P14 by phenol and purified by differential centrifugation had a molecular weight of about 10^7 daltons, but it could be dissociated by heating with deoxycholate into sub-units of molecular weight 12,000–16,000 daltons (Ikeda & Egami, 1973). The deoxycholate treatment however also splits an amino sugar-rich fraction from the lipopolysaccharide and it is difficult to establish whether or not it caused some degradation.

As in the Enterobacteriaceae the lipopolysaccharide of *P. aeruginosa* can be split into its lipid A and polysaccharide components by mild acid hydrolysis which also partially degrades the polysaccharide part of the molecule (Fensom & Meadow, 1970). The lipid A part lacks one of the characteristic components of the lipid A of the Enterobacteriaceae, B-OH myristic acid. This is the major if not the only hydroxy acid in the lipopolysaccharides of most Gram-negative bacteria so far studied but is missing from *P. aeruginosa* 1999 (Fensom & Gray, 1969), PAC1 (Hancock, Humphreys & Meadow, 1970) and ATCC 10145, UGAVM-1 and OSU 64 (Michaels & Eagon, 1969). Furthermore we have not found this acid in any of the Habs serotypes of *P. aeruginosa* (Chester, Meadow & Pitt, 1973). Although it is a minor component in the lipopolysaccharides of some species of pseudomonads it is never the major hydroxy acid (Wilkinson, Galbraith & Lightfoot, 1973). In all the pseudomonads so far studied except *P. rubescens* 3-OH 12:0 is the major hydroxy acid and in *P. aeruginosa* lipopolysaccharide this is accompanied by 3-OH 10:0 and 2-OH 12:0 (Chester *et al.,* 1973; Fensom & Gray, 1969; Hancock, Humphreys & Meadow, 1971; Humphreys, 1971; Key, Gray & Wilkinson, 1970; Wilkinson, Galbraith & Lightfoot, 1973) (see Table 3.3). In the Enterobacteriaceae the lipid A backbone contains β 1,6- or β1,4-linked disaccharide units of glucosamine to which long-chain fatty acids are joined by ester or amide linkages to the free hydroxyl and amino groups (Adams & Singh, 1969; 1970; Gmeiner, Simon & Luderitz, 1971; Reitschel *et al.,* 1972). While there may be minor species variations it seems likely that this is the general structure of most lipids A except that of *P. diminuta* (Wilkinson *et al.,* 1973). Certainly the lipid A from *P. aeruginosa* 1999 is derived from a β 1,6-linked disaccharide of glucosamine. The hydroxyl groups of the glucosamine residues are esterified with long-chain fatty acids,

Table 3.3. Hydroxy acids of isolated *Pseudomonas* lipopolysaccharide

Species	3-OH 10:0	3-OH 12:0	2-OH 12:0	3-OH 14:0	3-OH 13:0	3-OH 11:0
P. aeruginosa	+	+ +	+	0	0	0
P. putida	+	+ +	+	0	0	0
P. acidovorans	+	+ +	+	0	0	0
P. aminovorans	+	+ +	+	0	0	0
P. syncyanea	+	+ +	+	0	0	0
P. alkaligenes	+	+ +	0	0	0	0
P. stutzeri	+	+ +	0	+	0	0
P. diminuta	0	+ +	0	+	+	0
P. pavonacea	0	+ +	0	+	+	+

+ + Major component.
 + Minor component.
 0 Not detected.

while 3-OH 12:0 is joined by amide linkage to the amino group of the glucosamine. The hydroxyl group of this hydroxy acid appears to be substituted further with 16:0 and 2-OH 12:0 (Drewry *et al.*, 1973).

Very little is known of the biosynthesis of lipid A even in the Enterobacteriaceae. Since hydroxy acids occur only in the lipid A part of *P. aeruginosa* they can be used as specific markers to study the biosynthesis of the lipid A region. The fact that the hydroxy acid composition does not vary much during growth on different media or at different times of the growth phase suggests that they are essential components in the synthesis of the complete lipopolysaccharide (Humphreys, Hancock & Meadow, 1972). All three of the lipid A hydroxy acids of *P. aeruginosa*, 3-OH, 10:0, 2-OH, 12:0 and 3-OH 12:0, can be synthesized from the coenzyme A derivatives of the corresponding saturated fatty acid, hydroxylation apparently taking place before incorporation into lipid A (Humphreys *et al.*, 1972). The addition of these acids to the glucosamine backbone might therefore be one of the final stages in the assembly of the outer layers of the wall. The enzyme preparations used for hydroxy acid synthesis are wall/membrane preparations similar to those required for the biosynthesis of the polysaccharide part of the molecule in other species. It is interesting that hydroxy acid synthesis does not apparently require the acyl carrier protein derivatives normally used for long-chain fatty acid synthesis and this may reflect the difficulty of synthesizing complex materials outside the metabolic limits of the bacterium.

While the lipid components of the lipopolysaccharide provide us with a means of distinguishing some of the pseudomonads and are responsible at least in part for their endotoxic activity, it is the polysaccharide part which seems to have more effect on the phenotype. This component appears to play some part in the heat stable serological reactions, aeruginocin and bacteriophage sensitivities as well as antibiotic sensitivities, most of which are strain rather than species characteristics. The earlier analyses of the polysaccharide part of *P. aeruginosa* lipopolysaccharide are difficult to interpret partly because of the difficulty of isolating lipopolysaccharide free from contaminating nucleic acid and cytoplasm and partly because the work was limited to a few strains of different serotypes. It now seems likely that the lipopolysaccharides contain a region which is common to all strains but there may be strain-specific side-chains. If this is so there is some similarity with the Enterobacteriaceae in which there are core components common to all strains and side-chain polysaccharides which are strain specific (Luderitz *et al.*, 1971). At present the evidence in support of this type of structure in *P. aeruginosa* is limited but consistent.

All the *P. aeruginosa* lipopolysaccharides which have been analysed so far contained glucose, rhamnose and heptose as the neutral sugars together with galactosamine, 2-keto-3-deoxyoctonic acid and usually alanine. Other neutral sugars and amino compounds, some of which are amino sugars, have been found in some strains but not all and these could be related to the serological type of the strains. Analysis of the neutral sugar and amino compounds of the lipopolysaccharides isolated from fifteen strains of *P. aeruginosa* used for serotyping by the Habs (1957) system allowed us to divide them into nine groups which could be distinguished from each other by their chemical components (Table 3.4). Six of these contained only one serotype, two groups contained one serotype and the remaining group contained five serotypes, but it remains to be seen whether this division into groups is really a serotype characteristic or whether it represents strain differences. Additional neutral sugars or amino compounds have been reported in lipopolysaccharides of other strains of *P. aeruginosa*. For example Hanessian *et al.* (1971) found xylose together with an unknown sugar in antigenic material extracted by trichloracetic acid from one of the strains of *P. aeruginosa* they studied. Such extracts would be expected to contain lipopolysaccharide together with other components. Only two unusual amino sugars, quinovosamine and fucosamine, have been identified in isolated lipopolysaccharides of *P. aeruginosa* (Suzuki, 1969; Fensom & Gray, 1969; Suzuki, Suzuki & Fukasawa, 1970; Ikeda & Egami, 1973) although unidentified nitrogenous material was found in some strains (Fensom & Gray, 1969; Fensom & Meadow, 1970). Lipopolysaccharide from a strain of *P. aeruginosa* belonging to serogroup 6 contains valine in addition to the alanine found in all lipopolysaccharides (Dr. S. G. Wilkinson, personal communication), and this could correspond to the unknown amino compound (U1) found in the strain of serotype 6 shown in Table 3.4.

When the lipid A is split from the lipopolysaccharide by mild acid hydrolysis the polysaccharide portion is partially degraded into fractions which can be separated by chromatography on 'Sephadex' G50 or G25

Table 3.4. Arrangement of lipopolysaccharides of serological type strains of *Pseudomonas aeruginosa* into chemotypes on the basis of their major sugar and amino compound components (from Chester et al., 1973) (Components present in trace amounts have been excluded)

Habs serological type	Common components[a]	Mannose	Xylose	U1	U2	U3	U4	U567	U8	Chemotype
				|		Amino compounds				
3	+	+								I
5C	+								+	II
6	+			+	+	+	+			III
5D	+						+	+		IV
10	+				+	+	+			V
11	+				+		+			VI
1, 13	+				+	+	+	+		VII
7			+							VIII
8	+						+			
2A, 2B										
4	+					+	+	+		IX
9, 12										

[a] These were heptose, glucose, rhamnose, 2-keto-3-deoxyoctonic acid, galactosamine, alanine and glucosamine.
+ Indicates component present.

(Fensom & Meadow, 1970; Drewry, Gray & Wilkinson, 1971) (see Figure 3.3). The low molecular weight fractions (L) contained ketodeoxyoctonate and phosphate accompanied in some strains by ethanolamine or ethanolamine phosphates (Drewry, Gray & Wilkinson, 1972). The high molecular weight fractions (H) varied from one strain to another both in amount and composition, but the middle fractions (M) were remarkably constant in strains of all serotypes (Chester *et al.*, 1973). They contained glucose,

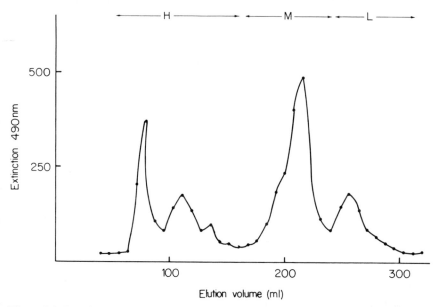

Figure 3.3 Fractionation of polysaccharide obtained from *P. aeruginosa* PAC1 lipopolysaccharide after partial degradation with a 1 % acetic acid. The material was eluted from a column of 'Sephadex' G25 and analysed for neutral sugars by the phenol–sulphuric acid method. The three regions marked represent low molecular weight material (L) the core region (M) and the high molecular weight region (H) (modified from Chester *et al.* 1973)

rhamnose, heptose, galactosamine, ketodeoxyoctonate alanine and phosphate. They may perhaps be equivalent to the core fractions in the Salmonellae. Material of similar composition has been isolated from *P. aeruginosa* P14 after heating the lipopolysaccharide with deoxycholate (Ikeda & Egami, 1973). This split off a high molecular weight fraction containing most of the amino sugars of the intact molecule leaving a lower molecular weight fraction from which the common components could be released by mild acid hydrolysis. Although there is some variation both between strains and in the same strain grown in different media there is a surprising degree of constancy in the composition of these core fractions suggesting that they may indeed be common to all strains of *P. aeruginosa.* The molar proportions show fair agreement except for glucose in PAC1 and phosphate in P14 (Table 3.5). The former is affected by growth conditions and may be part of additional substituents attached to the basic core in this strain. The low phosphate content of the core material from strain P14 may reflect its different mode of preparation. This fraction was produced by acid hydrolysis of lipopolysaccharide which had been heated with deoxycholate and some at least of the phosphate in another strain of *P. aeruginosa* has been shown to be unusually labile (Drewry *et al.,* 1971; 1972). Surprisingly, Ikeda. and Egami did not report the presence of alanine in their analyses of P14 lipopolysaccharide, but their preparations contained about 10 % protein which would have masked the alanine. We have found alanine to be present in lipopolysaccharides from all Habs serotypes of *P. aeruginosa* (Chester *et al.,* 1973) and it was found attached to the polysaccharide part in other strains studied (Naoi *et al.,* 1958; Fensom & Gray, 1969).

Table 3.5. Core polysaccharide from lipopolysaccharide of *P. aeruginosa* grown on complex media

Strain	PAC1	1999	P14
Serotype	3	2B	1
Data recalculated from	Chester & Meadow (1973)	Drewry *et al.* (1973)	Ikeda & Egami (1973)
	Molar proportions (galactosamine = 1)		
Glucose	13·0	5·2	2·3
Rhamnose	2·7	1·3	1·5
Heptose	3·0	2·3	1·4
KDO[a]	+	0·3	+
Galactosamine	1·0	1·0	1·0
Alanine	1·3	1·3	—
Phosphorus	7·0	7·3	2.6

+ Present but not estimated; − not detected.
[a] KDO = ketodeoxyoctonate.

Some idea of the relationship between the core material and the rest of the lipopolysaccharide can be gained from a mutant of *P. aeruginosa* PAC1 whose lipopolysaccharide lacks the high molecular weight peaks (H) derived from that of the parent strain. Analysis of peak M from this mutant (PAC556) showed that it also differed from its parent strain in lacking rhamnose (Fensom & Meadow, 1970) suggesting that some of the outer sugars of the core region might also be missing in this mutant. If so its lipopolysaccharide might have a site available for the addition of rhamnose. Particulate preparations of this mutant incorporated ^{14}C-rhamnose from TDP ^{14}C-rhamnose into phenol-extractable material while comparable preparations from the parent strain showed no such activity (Koval, Fensom & Meadow, unpublished). There was some indication that a butanol-soluble intermediate was formed suggesting that phospholipids or isoprene compounds may be required for polysaccharide synthesis as in the Enterobacteriaceae (Rothfield & Romeo, 1971).

Whereas the core material appears to be common to lipopolysaccharides of all strains of *P. aeruginosa* the high molecular weight material varied considerably from one strain to another. In PAC1 (serotype 3) the high molecular weight fractions contained almost half the neutral sugars of the lipopolysaccharide (Fensom & Meadow, 1970) while in strains 1999 (serotype 2B) (Drewry *et al.*, 1972) and strain P14 (serotype 1) (Ikeda & Egami, 1973) these fractions contained mainly amino sugars. It now appears that the composition of the high molecular weight fractions and indeed whether such fractions are produced at all after mild acid hydrolysis may be characteristic of the serotype. The time required to split lipid A from lipopolysaccharide by acetic acid hydrolysis varied considerably between strains of different serotypes as did the resulting polysaccharide elution profiles on 'Sephadex' chromatography (Chester *et al.*, 1973). The most complex pattern was that from serotype 3 strains as shown in Figure 3.3 and the simplest came from type 2A which had no high molecular weight material at all. The strains could also be distinguished by the proportions of components present in the different high molecular weight fractions but until many more strains have been analysed it is impossible to decide whether these are strain or serotype characteristics.

5.d. Serological Reactions

The isolated pseudomonad lipopolysaccharide, like that of Enterobacteriaceae, appears to be responsible for the heat-stable O-antigenic specificity of *Pseudomonas aeruginosa* (Adam, Kontrohr & Horvath, 1971; Chester *et al.*, 1973). The part of the lipopolysaccharide responsible for this reaction is located in the high molecular weight polysaccharide fractions obtained after acetic acid hydrolysis (Chester *et al.*, 1973). Whether this material can be equated with the side-chain region in the Enterobacteriaceae remains to be seen but the isolation of mutants with defective high molecular weight fractions or which lack this material completely and which have lost their serological specificity suggests that the relationship may be similar (Fensom & Meadow, 1970; Koval & Meadow, unpublished). None of the low molecular weight polysaccharide fractions from any of the serotypes of *P. aeruginosa* reacted with any of the antisera, but the high molecular weight materials reacted specifically with homologous antisera except for types 7, 8 and 9. Lipopolysaccharide of types 7 and 8 lacked side-chain material (Peaks H Figure 3.3) after acetic acid hydrolysis so that if side-chains existed *in vivo* they must have been degraded by this treatment. They could have been eluted with core material (M) in these strains but since this fraction showed no antigenic reactivity, there may have been too much degradation for a precipitin reaction. Type 9 lipopolysaccharide differed from all others in that high molecular weight material produced by acetic acid hydrolysis showed no serological reactivity. Reaction with type 9 antisera has however been observed not only with other fractions derived from type 9 lipopolysaccharide but from materials of intermediate molecular weight derived from strains of *P. aeruginosa* belonging to other though related serogroups (Chester & Meadow, unpublished). There appears to be heterogeneity discernable both chemically and serologically in at least some strains of *P. aeruginosa* and it will be necessary to study many more strains and fractions derived from them to establish the complete relationship between the polysaccharides and their serotype.

Changes in serological specificity induced by bacteriophages have been described by Liu (1969). He studied 13 prototype strains representing the serotype schemes of Verder & Evans (1961) and Habs (1957). These strains were tested against a number of bacteriophages and mutants resistant to these phages were isolated and typed. Nearly 20 % of the smooth colonies picked from the plaques were found to possess serotypes different both from the original types and from the strains on which the bacteriophages had been propagated. Although such changes in serotype were more common in some types than in others most of the results suggested that the changes were due to changes in the lysogenic state rather than to the selection of naturally occurring mutants by the phages. It is possible here that the introduction of different lysogenic phages allows the expression of different biochemical functions which results in the formation of altered lipopolysaccharides.

5.e. Aeruginocin and Phage Receptor Substance

By analogy with the enterobacterial lipopolysaccharides one might expect that the outer layers of *P. aeruginosa* walls would act as receptors for bacteriophages and aeruginocins. The coliphages T_3 and T_7 use the lipopolysaccharide component as their receptors while phages T_2 and T_6 apparently recognize the lipoprotein in the outer layers of the coli wall (see Rapin & Kalckar, 1971). The R-type aeruginocins have many features in common with some bacteriophages and Egami, Ikeda and their colleagues have studied the adsorption of aeruginocin R by *P. aeruginosa*. The particle has a molecular weight of about 10^7 with a structure rather like that of the tail of a T-even coliphage (Ishii, Nishi & Egami, 1965). It consists of a double hollow cylinder 120 nm and 15 nm external diameter which was shown by electron microscopy to adsorb at the tip of the core to the outer surface of aeruginocin R-sensitive bacteria (Ikeda & Nishi, 1973). Much of the aeruginocin receptor activity was solubilized when the bacteria were converted into spheroplasts by lysozyme and EDTA suggesting that the receptor activity was located in the outer layers of the bacterial envelope, outside the cytoplasmic membrane. Chemical identification of the receptor substance as a part of the lipopoly-saccharide has confirmed this deduction. The earlier work was restricted to strain P11 and an aeruginocin-resistant mutant derived from it. Aeruginocin-receptor substance was extracted from the sensitive strain with trichloracetic acid which would remove lipopolysaccharide and other materials. Equivalent preparations from the resistant strain were physically different and showed no aeruginocin-receptor activity (Ikeda & Egami, 1969). This particular strain of *P. aeruginosa* appeared to contain less than the usual amount of lipopolysaccharide and none could be extracted with phenol. The work was therefore extended to another strain (P14) from which lipopolysaccharide could be extracted readily by the phenol method. The isolated lipopoly-saccharide of molecular weight about 10^7 could be shown to inactivate aeruginocin R *in vitro*. Receptor activity was lost when the lipopolysaccharide was dissociated by heating with deoxycholate into two fractions one of which was rich in amino sugars but had little or no receptor activity. The lower molecular weight fraction (molecular weight 12,000–16,000) had no receptor activity in the presence of deoxycholate but when the latter was removed the sub-units reassociated to make polymers of molecular weight 10^5–10^6 and the aruginocin-receptor activity was restored. The amino sugar rich fraction was thus not essential for receptor activity but, since the activity of the reassociated sub-units was unstable, the authors suggest that *in vivo* the amino sugar fraction may stabilize the aggregated sub-units (Ikeda & Egami, 1973). The reassociated sub-units contained both the sugar (glucose, rhamnose and ketodeoxyoctonate) and amino sugar components (galacto-samine, glucosamine, quinovosamine and fucosamine) of the complete lipopolysaccharide despite having lost a proportion of their amino sugars. The sub-units might perhaps be equated with the core polysaccharide with fewer side-chains than occur naturally but until we know more of the structure of the lipopolysaccharide no firm conclusion can be drawn. The experiments do however show clearly that, at least *in vitro*, receptor activity can be assayed only in material of fairly high molecular weight (about 10^5 daltons).

The kinetics of aeruginocin adsorption, like those of phage adsorption, are dependent on salt concentration and temperature (Ikeda & Nishi, 1973) and there are further similarities between aeruginocin and bacteriophage sensitivities. Several *Pseudomonas* phages show immunological cross-reaction with R-type pyocins (Homma & Shionoya, 1967; Homma, Watabe & Tanabe, 1968). Furthermore Ito and Kageyama (1970) isolated a number of mutants of *P. aeruginosa* P15–16 which were resistant to various aeruginocins of the R-type and were also resistant to the immunologically-related bacteriophage PS3. From the resistance patterns of the mutants they postulated that the aeruginocin- and phage-receptor activity might be considered as specific chemical groupings projecting from the bacterial surface which could be ordered tentatively from the degree of cross-resistance to various phages and aeruginocins. Ikeda and Nishi (1973) have now shown that the isolated lipopolysaccharide which has receptor activity for aeruginocin R is also capable of inactivating phage PS5 though considerably less efficiently. The evidence to date is thus consistent with the suggestion that receptors for both aeruginocin R and some bacteriophages are located in the lipopolysaccharide part of the outer membrane layers of *P. aeruginosa*.

5.f. The Role of the Outer Layers in Antibiotic Resistance

Many antibiotics known to be highly toxic for Gram-positive bacteria are lethal to Gram-negative bacteria only at high concentrations. Cell-free systems from Gram-negative bacteria are however often as sensitive as those from Gram-positive organisms. The most likely explanations are either that intact organisms appear resistant to antibacterial compounds by destroying them with enzymes or that the outer layers of Gram-negative bacteria prevent access of antibiotics to their sites of action. Enzymes which inactivate anti-biotics are known to be produced by *P. aeruginosa* as for example those inactivating kanamycin, neomycin, streptomycin and chloramphenicol (Okamoto & Suzuki, 1965; Doi *et al.*, 1968), cephalosporin and penicillin (Sabath, Jago & Abraham, 1965; Smith, Hamilton-Miller & Knox, 1969; Roe, Jones & Lowbury, 1971). These are however not confined to *P. aeruginosa* nor do they depend on wall structure and they will not be discussed further here. There is however a growing body of evidence which suggests that the intrinsic resistance of many Gram-negative bacteria including *P. aeruginosa* is related to the composition of their walls (see Brown, 1971; Tamaki, Sato & Matsuhashi, 1971). The role of murein in preventing deoxycholate from reaching its membrane site of action has already been discussed (section 3) but it now seems likely that quantitatively the most important penetration barrier is provided by the outer membrane layers and in particular by its lipopolysaccharide and lipoprotein components. For example Hamilton (1970) has shown that conversion of *P. aeruginosa* to spheroplasts renders the organism sensitive to a number of agents such as gramicidin and cetyltri-methylammonium bromide which attack the cytoplasmic membrane. Interestingly the amount of these antibacterial agents bound by intact bacteria was unchanged when they were converted into spheroplasts probably because they can bind equally with phospholipid components on either the cytoplasmic membrane or on the outer membrane layer. Lysis resulted only when the organisms had been converted into spheroplasts in which presumably there was no outer membrane layer to prevent access of the surface-active agent to the cytoplasmic membrane.

Table 3.6. Properties of carbenicillin supersensitive mutants of *P. aeruginosa* PAC1

	Parent strain PAC1	Mutants		
		PAC7	PAC557	PAC556
Antibiotic sensitivity M.I.C. (μg/ml)				
Carbenicillin	70	30	5	5
Novobiocin	>1000	>1000	1000	5
Polymixin	20	10	10	2.5
Tetracycline	60	60	50	4
Chloramphenicol	100	100	10	10
Rifamycin	50	30	10	8
Phage type	21/68/109/119X	21/68/109	68/109	21/68
Serotype	3	3 Polyagglutinating	3	Autoagglutinating
Lipopolysaccharide (high molecular weight polysaccharide fractions separated by 'Sephadex' chromatography)				

The role of the lipopolysaccharide in determining sensitivity and resistance to certain antibiotics has been defined in chemical terms for *S. minnesota* and *E. coli*. Schlecht & Westphal (1968; 1970) studied the antibiotic sensitivities of a series of rough mutants of *S. minnesota* with known defects in lipopolysaccharide structure. Their sensitivity to erythromycin, rifamycin, actinomycin D and bacitracin increased progressively with loss of the sugar moieties from the outer regions of the core. Sensitivity to tetracyclines and penicillin derivatives could also be correlated with differences in lipopolysaccharide structure. In *E. coli* mutants isolated as supersensitive to novobiocin proved to have incomplete lipopolysaccharide structures. In particular they lacked the phosphodiester bridges which cross-link the backbone of the lipopolysaccharide in this species (Tamaki *et al.*, 1971). These mutants were also supersensitive to spiramycin and actinomycin D but their sensitivity to penicillins was unchanged. It has been suggested that the different effects of variations in polysaccharide structure on antibiotic supersensitivity of the two species may reflect differences in the detailed organization of their outer layers.

Similar studies of antibiotic supersensitive mutants of *P. aeruginosa* PAC1 suggest that their response to some antibiotics may be correlated with specific changes in lipopolysaccharide structure (Koval & Meadow, unpublished). Some of the carbenicillin supersensitive mutants isolated from *P. aeruginosa* PAC1 proved to have increased sensitivity to a number of other antibiotics whose mode of action was unrelated to the biosynthesis of the mucopeptide component. This suggested that the supersensitivity of these mutants was a reflexion of the penetrability of the carbenicillin to its size of action at the level of the cytoplasmic membrane rather than to alterations in mucopeptide structure or synthesis. The properties of some of these mutants are shown in Table 3.6. They show a variety of antibiotic sensitivities varying from those of the parent strain to those of mutant PAC556 which was particularly sensitive to all the antibiotics shown. None of the mutants had increased sensitivities to other antibiotics such as vancomycin, cephalosporin or cycloserine known to interfere with mucopeptide synthesis. On the other hand some of them were resistant to certain bacteriophages to which the parent strain was sensitive and some of them had also lost their specific serotype both of which may be determined by lipopolysaccharide structure. When the lipopolysaccharide components were isolated from the walls of the mutants, some of them could be clearly differentiated from the parent strain. The three mutants shown in the Table 3.6 all had a higher percentage of glucose in their lipopolysaccharides than did the parent strain. Since most of the glucose is thought to occur in the core fractions of partially degraded lipopolysaccharide, mutants with defective side-chain might be expected to have higher percentage glucose compostions. After hydrolysis with dilute acetic acid and separation of the polysaccharide fractions by 'Sephadex' chromatography it was clear that they were all defective in the high molecular weight (side-chain) fractions (Table 3.6). In these three mutants there appeared to be some correlation between supersensitivity to certain antibiotics and the composition of the lipopolysaccharide part of the outer membrane layers. However some other mutants showed increased sensitivity to the same antibiotics without comparable alterations to lipopolysaccharide structure. The analytical techniques we used to distinguish differences in the structure of *P. aeruginosa* lipopolysaccharides are however very crude at present and it remains to be seen whether these other supersensitive mutants will also prove to have defective lipopolysaccharides or whether they are altered in some other component of the outer layers of the wall. Certainly much more detailed work will be needed before our knowledge of the outer layers of the walls of *P. aeruginosa* reaches the level of that now established for the Enterobacteriaceae.

6. SUMMARY AND CONCLUSIONS

Although the study of the wall and membrane structure of *P. aeruginosa* has made great progress during the last few years there are still many unanswered questions particularly those involving the relationship between the composition of bacterial envelope and the phenotypic properties of the organism. The overall organization of the outer layers is pretty well established but there is some doubt about the exact size and location of the periplasmic region and it remains to be established whether the mucopeptide layer is linked to a lipoprotein in the outer membrane layer or in the cytoplasmic membrane. The gross composition of the cytoplasmic membrane of psuedomonads appears very similar to that of other organisms so far studied. However detailed work on the individual proteins of the cytoplasmic membrane of *P. aeruginosa* is still in its infancy and it may be some time before it will be possible to relate many of the enzymic metabolic capabilities of the intact organism directly to the composition and turnover of the proteins in the cytoplasmic membrane. The mucopeptide component of the wall has a very similar structure to that of other Gram-negative bacteria so far analysed. The fact that it carries less covalently linked protein or lipoprotein than many other species may however explain to some extent the fragility of the isolated pseudomonad walls. Despite the detailed studies of the effects of EDTA on *P. aeruginosa* and the identification of the lipoproteins it releases from the walls, the peculiar sensitivity of *P. aeruginosa* and a few related pseudomonads to EDTA (Wilkinson, 1967) is still not completely understood. One of the problems involved in studying this phenomenon is our current lack of knowledge of the variation of wall structure with growth conditions. Changes in sensitivity to antibacterial compounds including EDTA and polymyxin have been linked in some instances with growth under conditions of phosphate or magnesium limitation which produce measurable differences in wall composition (Brown & Melling, 1969a,b). While some progress will undoubtedly come from the analysis of bacteria grown under carefully defined conditions, as for example in the chemostat, the complex nature of the walls of *P. aeruginosa* and the fact that there are very few compounds which can be clearly shown to be limited to single macromolecular species, make detailed analysis of large batches of bacterial walls a necessary adjunct to such studies. It is to be hoped that a combination of both types of approach together with studies of mutants defective in particular components of the outer layers will give us further insight into the relationship between the structure of the cell envelope of *P. aeruginosa* and its behaviour in the external environment.

7. BIBLIOGRAPHY

Adam, M. M., Kontrohr, T. & Horvath, E. (1971) *Acta Microbiologica Academiae Scientiarum Hungaricae*, **18**, 307.

Adams, G. A. & Singh, P. P. (1969) *Biochimica et Biophysica Acta*, **187**, 457.

Adams, G. A. & Singh, P. P. (1970) *Biochimica et Biophysica Acta*, **202**, 553.

Bodman, H. & Welker, N. E. (1969) *Journal of Bacteriology*, **97**, 924.

Braun, V. & Bosch, V. (1972) *Proceedings of the National Academy of Sciences*, **69**, 970.

Braun, V., Rehn, K. & Wolff, H. (1970) *Biochemistry*, **9**, 5041.

Braun, V. & Sieglin, U. (1970) *European Journal of Biochemistry*, **13**, 336.

Braun, V. & Wolff, H. (1970) *European Journal of Biochemistry*, **14**, 387.

Brown, M. R. W. (1971) in *Inhibition and Destruction of the Microbial Cell*, (Ed. W. B. Hugo), Academic Press, London, p. 307.

Brown, M. R. W. & Melling, J. (1969a) *Journal of General Microbiology*, **54**, 439.

Brown, M. R. W. & Melling, J. (1969b) *Journal of General Microbiology*, **59**, 263.

Burman, L. G., Nordström, K. & Bloom, G. D. (1972) *Journal of Bacteriology*, **112**, 1364.

Cheng, K.-J., Ingram, J. M. & Costerton, J. W. (1970) *Journal of Bacteriology*, **104**, 748.

Cheng, K-J., Ingram, J. M. & Costerton, J. W. (1971) *Journal of Bacteriology*, **107**, 325.

Chester, I. R., Meadow, P. M. & Pitt, T. (1973) *Journal of General Microbiology*, **78**, 305.

Cho, K. Y. & Salton, M. R. J. (1966) *Biochimica et Biophysica Acta*, **116**, 73.

Clarke, K., Gray, G. W. & Reaveley, D. A. (1967) *Biochemical Journal*, **105**, 755.

Clarkson, C. E. & Meadow, P. M. (1971) *Journal of General Microbiology*, **66**, 161.

Doi, O., Ogura, M., Tanaka, N. & Umezawa, H. (1968) *Applied Microbiology*, **16**, 1276.

Drewry, D. T., Gray, G. W. & Wilkinson, S. G. (1971) *European Journal of Biochemistry*, **21**, 400.

Drewry, D. T., Gray, G. W. & Wilkinson, S. G. (1972) *Journal of General Microbiology*, **73**, viii.

Drewry, D. T., Lomax, J. A., Gray, G. W. & Wilkinson, S. G. (1973) *Biochemical Journal*, **133**, 563.

Ellwood, D. C. & Tempest, D. W. (1972) in *Advances in Microbial Physiology*, Vol. 7, (Eds. A. H. Rose & D. W. Tempest), Academic Press, London, p. 83.

Fensom, A. H. & Gray, G. W. (1969) *Biochemical Journal*, **114**, 185.

Fensom, A. H. & Meadow (1970) *FEBS Letters*, **9**, 81.

Forsberg, C. W., Costerton, J. W. & MacLeod, R. A. (1970a) *Journal of Bacteriology*, **104**, 1338.

Forsberg, C. W., Costerton, J. W. & MacLeod, R. A. (1970b) *Journal of Bacteriology*, **104**, 1354.

Forsberg, C. W., Rayman, M. K., Costerton, J. W. & MacLeod, R. A. (1972) *Journal of Bacteriology*, **109**, 895.

Freer, J. H. & Salton, M. R. J. (1971) in *Microbial Toxins*, Vol. IV, (Eds. G. Weinbaum, S. Kadis & S. J. Ajl), Academic Press, New York & London, p. 67.

Gilleland, H. E., Stinnett, J. D., Roth, I. L. & Eagon, R. G. (1973) *Journal of Bacteriology*, **113**, 417.

Gmeiner, J., Simon, M. & Luderitz, O. (1971) *European Journal of Biochemistry*, **21**, 355.

Habs, I. (1957) *Zeitschrift für Hygiene und Infectionskrankheiten*, **144**, 218.

Hamilton, W. A. (1970) *FEBS Symposium*, **20**, 71.

Hancock, I. C., Humphreys, G. O. & Meadow, P. M. (1970) *Biochimica et Biophysica Acta*, **202**, 389.

Hancock, I. C. & Meadow, P. M. (1967) *Journal of General Microbiology*, **46**, x.

Hancock, I. C. & Meadow, P. M. (1969) *Biochimica et Biophysica Acta*, **187**, 366.

Hanessian, S., Regan, W., Watson, D. & Haskell, T. H. (1971) *Nature, New Biology*, **229**, 209.

Heilmann, H. D. (1972) *European Journal of Biochemistry*, **31**, 456.

Hirashima, A. & Inouye, M. (1973) *Nature, London*, **242**, 405.

Hoffmann, H-P., Geftic, S. G. Heymann, H. & Adair, F. W. (1973) *Journal of Bacteriology*, **114**, 434.

Homma, J. Y. (1971) *Japanese Journal of Experimental Medicine*, **41**, 387.

Homma, J. Y. & Shionaya, H. (1967) *Japanese Journal of Experimental Medicine*, **37**, 395.

Homma, J. Y. & Suzuki, N. (1966) *Annals of the New York Academy of Sciences*, **133**, 508.

Homma, J. Y., Watabe, H. & Tanabe, Y. (1968) *Japanese Journal of Experimental Medicine*, **38**, 213.

Humphreys, G. O. (1971) *Ph.D Thesis: University of London*.

Humphreys, G. O., Hancock, I. C. & Meadow, P. M. (1972) *Journal of General Microbiology*, **71**, 221.

Ikeda, K. & Egami, F. (1969) *Journal of Biochemistry*, **65**, 603.

Ikeda, K. & Egami, F. (1973) *Journal of General and Applied Microbiology*, **19**, 209.

Ikeda, K. & Nishi, Y. (1973) *Journal of General and Applied Microbiology*, in press.

Ishii, S., Nishi, Y. & Egami, F. (1965) *Journal of Molecular Biology*, **13**, 428.

Ito & Kageyama, M. (1970) *Journal of General and Applied Microbiology*, **16**, 231.

Kaneshiro, T. & Marr, A. G. (1962) *Journal of Lipid Research*, **3**, 184.

Key, B. A., Gray, G. W. & Wilkinson, S. G. (1970) *Biochemical Journal*, **120**, 559.

Lickfield, K. G., Achterrath, M., Hentrich, F., Kolehmainen-Sevens, L. & Persson, A. (1972) *Journal of Ultrastructure Research*, **38**, 27.

Liu, P. V. (1969) *Journal of Infectious Diseases*, **119**, 237.

Luderitz, O., Westphal, O., Staub, A. M. & Nikaido, H. (1971) in *Microbial Toxins*, Vol. IV, (Eds. G. Weinbaum, S. Kadis & S. J. Ajl), Academic Press, London & New York, p. 145.

Martin, E. L. & MacLeod, R. A. (1971) *Journal of Bacteriology,* **105,** 1160.
Martin, H. H., Heilmann, H. D. & Preusser, H. J. (1972) *Archiv für Mikrobiologie,* **83,** 332.
Michaels, G. B. & Eagon, R. G. (1969) *Proceeding of the Society for Experimental Biology and Medicine,* **131,** 1346.
Naoi, M., Egami, F., Hamamura, N. & Homma, J. Y. (1958) *Biochemische Zeitschrift,* **330,** 421.
Nelson, J. D. & MacLeod, R. A. (1973) *Bacteriological Proceedings,* in press.
Neu, H. C. & Chou, J. (1967) *Journal of Bacteriology,* **94,** 1934.
Okamoto, S. & Suzuki, Y. (1965) *Nature, London,* **208,** 1301.
Randle, C. L., Albro, P. W. & Dittmer, J. C. (1969) *Biochimica et Biophysica Acta,* **187,** 214.
Rapin, A. M. C. & Kalckar, H. M. (1971) in *Microbial Toxins,* Vol. IV, (Eds. G. Weinbaum, S. Kadis & S. J. Ajl), Academic Press, New York & London, p. 267.
Reitschel, E. T., Gottert, H., Luderitz, O. & Westphal, O. (1972) *European Journal of Biochemistry,* **28,** 166.
Roberts, N. A., Gray, G. W. & Wilkinson, S. G. (1967) *Biochima et Biophysica Acta,* **135,** 1068.
Roberts, N. A., Gray, G. W. & Wilkinson, S. G. (1970) *Microbios,* **7–8,** 189.
Roe, E., Jones, R. J. & Lowbury, E. J. L. (1971) *Lancet,* **i,** 149.
Rogers, S. W., Gilleland, H. E. Jr. & Eagon, R. G. (1969) *Canadian Journal of Microbiology,* **15,** 743.
Rothfield, L. & Romeo, D. (1971) *Bacteriological Reviews,* **35,** 14.
Sabath, L. D., Jago, M. & Abraham, E. P. (1965) *Biochemical Journal,* **96,** 739.
Salton, M. R. J. (1967) *Annual Reviews of Microbiology,* **21,** 417.
Salton, M. R. J. (1971) *Critical Reviews in Microbiology,* **1,** 161.
Schlecht, S. & Westphal, O. (1968) *Naturwissenschaften,* **10,** 494.
Schlecht, S. & Westphal, O. (1970) *Zentralblatt für Bakteriologie, Parasitenkunde, Infecktionskrankheiten und Hygiene* (Abteilung I), **213,** 356.
Schnaitman, C. A. (1971) *Journal of Bacteriology,* **108,** 545.
Smith, J. T., Hamilton-Miller, J. M. T. & Knox, R. (1969) *Journal of Pharmacy and Pharmacology,* **21,** 337.
Stinnett, J. D., Gilleland, H. E. Jr. & Eagon, R. G. (1973) *Journal of Bacteriology,* **114,** 399.
Suzuki, N. (1969) *Biochimica et Biophysica Acta,* **177,** 371.
Suzuki, N., Suzuki, A. & Fukasawa, K. (1970) *Journal of the Japanese Biochemical Society,* **42,** 130.
Takebe, I. (1965) *Biochimica et Biophysica Acta,* **101,** 124.
Tamaki, S., Sato, T. & Matsuhashi, M. (1971) *Journal of Bacteriology,* **105,** 968.
Thiele, O. W., Lacave, C. & Asselineau, J. (1969) *European Journal of Biochemistry,* **7,** 393.
Tsay, S., Brown, K. K. & Gaudy, E. T. (1971) *Journal of Bacteriology,* **108,** 82.
Verder, E. & Evans, J. (1961) *Journal of Infectious Diseases,* **109,** 183.
White, D., Dworkin, M. & Tipper, D. J. (1968) *Journal of Bacteriology,* **95,** 2186.
Wilkinson, S. G. (1967) *Journal of General Microbiology,* **47,** 67.
Wilkinson, S. G., Galbraith, L. & Lightfoot, G. A. (1973) *European Journal of Biochemistry,* **33,** 158.

CHAPTER 4

Bacteriophages and Bacteriocins

B. W. HOLLOWAY and V. KRISHNAPILLAI

The importance of the role of plasmids in the biology of *Pseudomonas* will be stressed in other chapters of this book. Two plasmid-related structures—bacteriophages and bacteriocins—merit special consideration. Both are concerned with important aspects of the phenotype of this bacterium. The genetic determinants of both constitute an important component of the genome and both play an important role in the research tools used in investigations of *Pseudomonas*.

1. BACTERIOPHAGES

A large variety of temperate and virulent phages of *Pseudomonas* are available for study. Many if not all strains of *P. aeruginosa* are lysogenic and provided an appropriate choice of sensitive strain is made, may be used as a source of temperate phages. However, other species of *Pseudomonas* are not commonly lysogenic and for these, as well as *P. aeruginosa* virulent (or intemperate)

Table 4.1. Characteristics of *Pseudomonas* phages

Phage	Host	Morphology	Nucleic acids	G+C (%)	Virulent or Temperate	Latent period (min)	Burst Size	Mol. wt. nucleic acid ($\times 10^6$ daltons)	Other characteristics including unusual characteristics	References
ϕ-MC	*P. aeruginosa*	Icosahedral (55 nm diam. head), short tail	Double-stranded DNA	46	Temperate	45	42		Phage adsorbs to terminal ends of bacteria	Yamamoto & Chow, 1968; Chow & Yamamoto, 1969
SD1	*P. aeruginosa*	Hexagonal head (50 nm diam.) and tail (188 nm × 6·2 nm)	Double-stranded DNA	53·2	Virulent					Shargool & Townshend, 1966
B3	*P. aeruginosa*	Octahedral (52 nm) diam. head) with sheathless tail (163 nm × 8 nm), 3 short tail fibres and tail plate	Double-stranded DNA		Temperate			20–25	Ultraviolet-non-inducible: transducing: shows HCM; DNA is linear and has no cohesive ends	Holloway *et al.*, 1960; Davison, Freifelder & Holloway, 1964; Slayter *et al.*, 1964; Holloway, & van de Putte, 1968; Ritchie, 1970 (pers. comm.)
F116	*P. aeruginosa*	Octahedral (58 nm diam. head) with loosely structured and labile tail (79 nm & 8 nm)	Double-stranded DNA		Temperate			38	Ultraviolet-inducible: transducing: does not show HCM; DNA is linear and has no cohesive ends	As for phage B3

Table 4.1. (cont.)

Phage	Host	Morphology	Nucleic acids	G + C(%)	Virulent or Temperate	Latent period (min)	Burst Size	Mol. wt. nucleic acid ($\times 10^6$ daltons)	Other characteristics including unusual characteristics	References
G101	P. aeruginosa	Octahedral (66 nm diam. head), contractile tail (150 nm × 17 nm) with six tail fibres and tail plate	Double-stranded DNA		Temperate			38–41	Ultraviolet-non-inducible; transducing; shows HCM; DNA is linear and has no cohesive ends	As for phage B3
D3	P. aeruginosa				Temperate				Phage recombination has been demonstrated and also mediates phage conversion of bacterial surface antigens; shows HCM	Holloway et al., 1960; Egan & Holloway 1961; Holloway & Cooper, 1962
E79	P. aeruginosa	Octahedral (66 nm diam. head), contractile tail (150 nm × 17 nm) with six tail fibres and tail plate	Double-stranded DNA		Virulent			120 (estimate)	Non-transducing; does not show HCM	As for phage B3
Twelve phages	P. aeruginosa	Octahedral (66 nm diam. head), contractile tail (150 nm × 17 nm) with six tail fibres and tail plate	Double-stranded DNA	46–63	Virulent	35–65	10–200			O'Callaghan, O'Mara & Grogan, 1969

Table 4.1. (cont.)

Phage	Host	Morphology	Nucleic acids	G + C(%)	Virulent or Temperate	Latent period (min)	Burst Size	Mol. wt. nucleic acid ($\times 10^6$ daltons)	Other characteristics including unusual characteristics	References
ϕ-S1	*P. fluorescens*	Hexagonal (60 nm diam. head) with short tail (30 nm)			Virulent	18	65		Very broad host range (*P. aeruginosa, P. putida, P. stutzeri* and 6 other species)	Kelln & Warren, 1971
gh-1	*P. putida*	Hexagonal (50 nm diam. head) with short tail and fibres	Double-stranded DNA	57	Virulent	21	103	22·6		Lee & Boezi, 1966; 1967
23, 25F and 27	*P. putrefaciens*	Icosahedral (55–77 nm diam. head) with contractile tail or short tail (135–176 nm × 14 nm)	DNA	36·4–48·8	Virulent				Phage 27 shows HCM	Delisle & Levin, 1972a,b
Ten phages (mesophilic and psychrophilic)	*P. aeruginosa P. fluorescens P. geniculata P. putida*	Polygonal (50–80 nm diam. head) with short or long contractile tails (17–140 nm)	DNA	39·6–68·2	Virulent	Psych. phages 6–12h/ 3·5°C, $\frac{1}{2}$–1 h/ 25°C; Mesoph. phages 85–190/ 25°C 35–85/ 37°C	6–25 / 23–120 / 20–155 / 55–162		Narrow host range of mesophilic phages but broad host range of psychrophilic phages	Olsen, Metcalf & Todd, 1968

Table 4.1. (cont.)

Phage	Host	Morphology	Nucleic acids	G + C(%)	Virulent or Temperate	Latent period (min)	Burst Size	Mol. wt. nucleic acid ($\times 10^6$ daltons)	Other characteristics including unusual characteristics	References
PM2	Marine *Pseudomonas*	Icosahedral (60 nm diam.) with spikes	Double-stranded DNA	42–43	Virulent			6	Contains lipid and glycoprotein. DNA in mature phage is covalently closed and supercoiled	Franklin, 1972
Pf1, Pf2	*P. aeruginosa*	Filamentous	Single-stranded DNA		Virulent				Not sex specific (FP2±) in *P. aeruginosa* as with similar phages in *E. coli* (Hoffmann-Berling, Kaemer Knippers, 1967); narrow host range	Takeya & Amako, 1966; Minamishima *et al.*, 1968
φW-14	*P. acidovorans*	Icosahedral (85 nm diam. head) with contractile tail (140 nm)			Virulent	60	300		Appears to display lysis inhibition but super-infection decreased burst size	Kropinski & Warren, 1970
7S	*P. aeruginosa*	Spherical (25 nm diam.)	Single-stranded RNA		Virulent				Isolation of RNA phage from lysogenic *P. aeruginosa* strain is unusual; and sensitivity of free phage to RNAase is unusual	Feary *et al.*, 1963a; Feary, Fisher & Fisher, 1964

Table 4.1. (cont.)

Phage	Host	Morphology	Nucleic acids	G + C (%)	Virulent or Temperate	Latent period (min)	Burst Size	Mol. wt. nucleic acid (× 10⁶ daltons)	Other characteristics including unusual characteristics	References
PP7	*P. aeruginosa*	Spherical (25 nm diam.)	Single-stranded RNA		Virulent		36,000 (estimate)		Phage multiplication in bacterial host leads to spheroplasting which leads to lysis before phage release. Not sex specific (FP2±) in *P. aeruginosa*	Bradley, 1966; Weppelman & Brinton, 1971; Bradley, 1972a
PRR1	*P. aeruginosa*		RNA		Virulent				Plates on wide range of genera containing R factor R1822	Olsen & Shipley, 1973
φ6	*P. phaseolicola*	Polyhedral (60 nm diam. head) surrounded by membranous, compressible envelope	Double-stranded RNA	58	Virulent	120–160	125–150	9.5–16.8	Contains lipids; adsorbs to pili; RNA occurs in triple-segmented form	Semancik, Vidaver & van Etten, 1973; Vidaver, Koski & van Etten, 1973

phages may be readily isolated from soil or sewage. The techniques used for these isolations are essentially those described by Adams (1959). The properties of a range of *Pseudomonas* phages are shown in Table 4.1.

Morphologically, the range of structures is just as diverse as in other bacterial genera. Some have the typical icosahedral head, tail sheath, tail plate and tail fibres found for example in coliphage T4. As well as these, phages similar to lambda, filamentous phages and spherical phages are found (Slayter, Holloway & Hall, 1964; Takeya & Amako, 1966; Bradley, 1967; Bradley & Robertson, 1968; Lapchine *et al.*, 1969). Two types of phages utilize bacterial pili as adsorption receptors: the small RNA tail-less icosahedral (spherical) filamentous phages (see below) and a recently isolated (Bradley, 1973) phage with a hexagonal head (58 nm in diameter) and a long, non-contractile tail (186 nm in length) which appears to be unique in having this type of structure and in attaching to pili. By comparison with other phages, phage PM2, which plates on a marine pseudomonad, and phage $\phi6$ (whose host is *P. phaseolicola*) have an unusual morphology. Phage PM2 has an internal nucleocapsid surrounded by a phospholipid bilayer bridged by spikes and a final outer icosahedral shell (Franklin, *et al.*, 1972). Phage $\phi6$ has a polyhedral head surrounded by a membranous compressible envelope which appeared to elongate on attachment to pili (Vidaver, Koski & van Etten, 1973). These two phages are also the only phages reported to contain phospholipids and glycoproteins, components characteristically found in many plant and animal viruses. A range of pseudomonad phages is illustrated in Plates 4.1 to 4.4.

The nucleic acid of *Pseudomonas* phages is usually of the double-stranded DNA form ranging in size from 6-120 $\times$ 10^6 daltons. The DNA of PM2 which

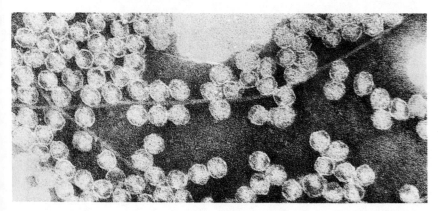

Plate 4.1. The RNA phage PP7 showing adsorption of the phage particle to a pilus of *Pseudomonas aeruginosa* ($\times$ 233,100). Photograph supplied by Dr. D. Bradley. (Reproduced from the *Journal of General Microbiology* by permission of the author and The Society for General Microbiology)

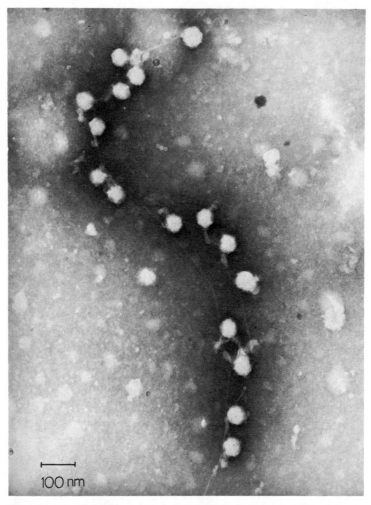

Plate 4.2. The lipid containing phage of $\phi6$ *P. phaseolicola* (Vidaver *et al.*, 1973). Note 'tails' attached to pili (the line marker indicates 100 nm). Photograph provided by Dr. M. K. Brakke. (Reproduced with permission from Vidaver, Koski and van Etten (1973) *Journal of Virology*, **11**, 799

plates on a marine pseudomonad is of particular interest. The double-stranded molecule is covalently closed and consequently is supercoiled, having 51 ± 3 superhelical turns per phage molecule. This property has been used to study the relative contributions of single-strand breaks, double-strand breaks and nucleotide damage in DNA produced by gamma radiation (van der Schans, Bleichrodt & Blok, 1972).

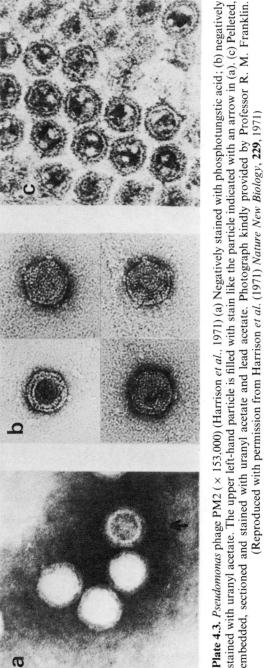

Plate 4.3. *Pseudomonas* phage PM2 ($\times$ 153,000) (Harrison *et al.*, 1971) (a) Negatively stained with uranyl acetate. The upper left-hand particle is filled with stain like the particle indicated with an arrow in (a). (c) Pelleted, embedded, sectioned and stained with uranyl acetate and lead acetate. Photograph kindly provided by Professor R. M. Franklin. (Reproduced with permission from Harrison *et al.* (1971) *Nature New Biology*, **229**, 1971)

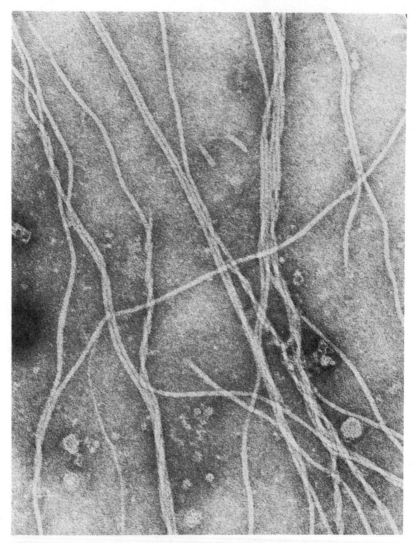

Plate 4.4. *P. aeruginosa* single-stranded DNA phage pf1 negatively stained with
2 % neutral potassium tungstate (× 133,000) Takeya & Amako (1966). Photograph kindly supplied by Professor K. Takeya and Dr. K. Amako. (Reproduced
with permission from K. Takeya & K. Amako (1966) *Virology*, **28**, 163)

Several RNA phages have been found (Feary, Fisher & Fisher, 1963a;
Bradley, 1966; Olsen & Shipley, 1973). Unlike the RNA phages of *E. coli*, only
one RNA phage shows any degree of host specificity this being PRR1 (Olsen
& Shipley, 1973) which only plates on bacterial strains possessing a particular
RP plasmid, the range including *Pseudomonas*, Enterobacteriaceae, soil

saprophytes, *Neisseria perflava* and photosynthetic bacteria. The nucleic acid of phage $\phi6$ is unusual in that it is the only known phage with double-stranded RNA (Semancik, Vidaver & Van Etten, 1973; Vidaver *et al.*, 1973) a characteristic only shared by some animal and plant viruses. Moreover its RNA exists in a triple-segmented form.

The host range of most phages is usually specific, plating only on one species. Some exceptions to this rule occur with some phages plating on such varied species as *P. aeruginosa, P. putida, P. fluorescens, P. geniculata* and *P. stutzeri* (Olsen, Metcalf & Todd, 1968; Kelln & Warren, 1971).

1.a. Lysogeny

An important aspect in the classification of bacteriophages is the property of genetic integration into the bacterial genome to establish lysogenic bacteria. Phages not having the property are referred to as intemperate, virulent or non-lysogenizing. The frequency of lysogenic strains in *P. aeruginosa* strains is probably close to 100 % (Alföldi, 1957; Hamon, Veron & Peron, 1961; Feary, Fisher & Fisher, 1963b; Patterson, 1965) in contrast to say *E. coli* where only about 10 % of strains are lysogenic (Jacob & Wollman, 1961). Polylysogeny (more than one type of prophage in a cell) also appears to be common in *P. aeruginosa* (Holloway, 1960; Holloway, Egan & Monk, 1960; Zierdt & Schmidt, 1964; Feary, Fisher & Fisher, 1964) including one example of a strain harbouring 8 and possibly 10 different prophages (Shionoya *et al.*, 1967). Lysogeny amongst other species of *Pseudomonas* appears to be very rare or undetectable (Hamon *et al.*, 1961; Patterson, 1965). Although Patterson (1965) found 3 of 8 *P. fluorescens* strains to be lysogenic, Hamon *et al.* (1961) did not find any amongst 28 *P. fluorescens* strains tested. In fact Hamon and his colleagues (1961) only found one lysogenic strain amongst a collection of 19 strains of *P. stutzeri, P. putida* and phytopathogenic *Pseudomonas* although lysogeny amongst other phytopathogenic *Pseudomonas* strains has been reported (Garrett & Crosse, 1963).

Factors which may affect the detection of lysogeny may include Host Controlled Modifications (HCM), see below, the choice of indicator strains, and other plasmids (including phage) in the genome of the sensitive strain which may prevent multiplication of the phage liberated from the lysogenic strain.

1.b. Genetics of Lysogeny

The genome of both the phage and the sensitive bacterium can influence the establishment of lysogeny. Clear plaque mutants (*c* mutants) can be readily isolated from most temperate phages and these show a reduced ability to lysogenize. A genetic analysis of such mutants in the *P. aeruginosa* phage D3 (Egan & Holloway, 1961) showed that three loci were involved, analogous to the pattern of genetic control of lysogeny in coliphage lambda (Kaiser, 1957).

The genome of *P. aeruginosa* can influence the establishment of lysogeny. Mutants with a reduced ability to be lysogenized by a range of temperate bacteriophages can be isolated by cross-streaking survivors of mutagenesis against bacteriophage (*les* mutants). The mutants behave as if the phage were virulent instead of temperate and the appearance of the streaks is markedly different from that of wild-type forms (Holloway, 1966; van de Putte & Holloway, 1968). Such mutants are also found amongst *tol* mutants of *P. aeruginosa* (q.v.) but at present the relationship between lysogenization and membrane function is not clear. Some *les* mutants also show recombination deficiency for both transduction and conjugation. The relationship of the ability to be lysogenized to other properties of the bacterial cell needs to be further investigated.

1.c. Lysogenic Conversion

The presence of the genetic material of the phage in a bacterium can affect the bacterial phenotype in a variety of ways. Apart from the ability to produce phage and be immune to lysis by superinfection with homologous phage, some lysogenic strains of bacteria have acquired other phenotypic properties (for review see Barksdale, 1959). In *P. aeruginosa* for example lysogeny with phage D3 confers a surface antigenic change and D3 can no longer adsorb to the lysogenized cells (Holloway & Cooper, 1962). Liu (1969) has shown that phage conversion occurs with a large number of serotypically different strains of *P. aeruginosa* affecting changes in surface components, the synthesis of aeruginocins, melanogenic pigments, fluorescins and proteases. It was not possible to correlate the acquisition of new phenotypic properties with lysogeny in all cases, conversion in its broadest sense, not necessarily involving the stable integration of the phage genome into the bacterial genome.

1.d. *Pseudomonas* Slime Polysaccharide Depolymerases

Pseudomonas phages specifying the synthesis of slime polysaccharide depolymerases have been described in *P. putida* (Chakrabarty, Niblack & Gunsalus, 1967) and in *P. aeruginosa* (Bartell, 1967; Bartell, Orr & Lom, 1966; Bartell & Orr, 1969b). In the latter cases 6 apparently distinct depolymerases specified by 6 different phages have been characterized and shown to belong to at least 2 distinct substrate specificity groups on the basis of enzymic activity on different slime polysaccharides (Bartell & Orr, 1969b). In the case of one of the *P. aeruginosa* phages there is good evidence that the phage genome codes specifically upon infection for the *de novo* synthesis of a unique depolymerase and not by derepression or activation of a host-coded depolymerase gene (Bartell & Orr, 1969a) and it is possible that this is also so in the case of the other depolymerases. It would be of interest to know the relationship of these polymerases to the endolysin (lysozyme) enzymes which have been identified, in say, the *E. coli* phage T4, especially since they appear to be two types—one appearing as a freely diffusible enzyme whilst the other is firmly bound to phage particles (Bartell *et al.*, 1966; Bartell & Orr, 1969b) and since their functional role in the biology of the phages is unclear.

The mucoid variant *P. aeruginosa* has an important medical significance for respiratory infections and in relation to cystic fibrosis (see Chapter 2). Martin (1972) has observed that mucoid variants occur much more frequently in association with the lysis of non-mucoid strains by certain bacteriophages. Transduction and lysogenic conversion do not appear to be satisfactory explanations of the phenomenon, but in view of the instability of both the natural and phage-induced mucoid forms (both eventually revert to the non-mucoid form) some form of plasmid as the basis of the mucoid phenotype seems likely. As the phage inducing the change are apparently non-lysogenizing, a possible explanation is that mucoid strains acquire a defective phage which achieves the status of the plasmid, and codes for the mucoid phenotype.

It is possible that the origin of such a form of plasmid may be by recombination between the phage and a resident plasmid in the bacterium. This has particular interest for the origin of the mucoid strain in human infections and as suggested by Martin, the diseases in which mucoid forms are important may be those in which the patients are liable to successive infections with different strains of *P. aeruginosa*.

The *Pseudomonas* phages continue to be very useful as a research tool in genetic, morphological and functional analyses of this genus. Transducing phages are common, and highly useful for linkage studies (see Chapter 5). The RNA phage PP7 has been used by Bradley (1972a,b,c) for the study of the functional role of *Pseudomonas* pili, and also for studies on conditions of interaction between free RNA and bacterial sphaeroplasts which lead to infectious phage assembly (Weppelman & Brinton, 1970; 1971).

1.e. Host Controlled Modification

The phenomenon of Host Controlled Modification (HCM), first recognized through effects on the plating efficiency of phages on different indicator strains has been of considerable importance in establishing the nature of procedures of genetic transfer in bacteria. HCM is detected when a phage grown in one host plates at low efficiency on a second host. Plaques isolated from the latter host show normal efficiency of plating when plated back on the original host.

The possibility of phage host range mutants giving similar plating efficiency patterns can be eliminated by appropriate controls. There are two mechanisms involved in HCM. Each bacterial strain imposes its own specificity on DNA replicating within that strain by means of methylases which produce specific patterns of methylated adenine or cytosine. This is known as modification. In addition each strain has a recognition mechanism by which it can differentiate between different patterns of DNA methylation and thus between self and non-self DNA. An endonuclease acts on the latter to inactivate its biological function—this process being known as restriction.

HCM was first detected in *P. aeruginosa* by means of phage B3 (Table 4.1) (Holloway & Rolfe, 1964) and subsequently a variety of other phages were shown to be subject to this phenomenon. However, the transducing phage F116 is not subject to restriction or modification nor are the bacterial DNA fragments which it can carry (Dunn & Holloway, 1971b). This protection of bacterial DNA against restriction is also found to occur with plasmid DNA (Chakrabarty, 1972).

Phage F126 which is related to F116 in its serological and immunity properties also shows no susceptibility to restriction enzymes. However, in transduction with F126 it is found that the bacterial DNA is not protected and is restricted when transferred into hosts with the appropriate restriction genotype. The differential nature of the restriction response to different types of DNA indicates that more than one type of restriction enzyme may occur in *P. aeruginosa* strain PAO (Krishnapillai, 1971).

HCM in *P. aeruginosa* has a varied genetic basis. Mutants with a restriction deficient Res$^-$ phenotype may be isolated directly (Rolfe & Holloway, 1968) and some, but not all may have altered modification phenotype, Mod$^-$. A large proportion of mutants resistant to *p*-fluorophenylalanine (FPA) are found to be altered in restriction and modification function (Rolfe & Holloway, 1968; Dunn & Holloway, 1971a,b). A useful but, as yet, unexplained method of obtaining a Res$^-$ phenotype in *P. aeruginosa* strain PAO is to grow it at 43 °C. The Res$^-$ phenotype persists indefinitely with growth at 43 °C, but after growth at 43 °C with subsequent multiplication at 37 °C, the restriction deficient phenotype persists for only 50–60 generations of growth at 37 °C when the normal Res$^+$ phenotype appears (Holloway, 1965).

Characteristics of other restriction phenotypes in various species of *Pseudomonas* suggest that more than one mechanism may exist to destroy the biological function of incoming DNA. In *P. putrefaciens* Delisle and Lewin (1972a,b) have described an HCM system involving the phage 27 and two different bacterial strains. There is a marked influence on both restriction and modification depending on the temperature at which the phage is propagated prior to its exposure to the HCM system. In *P. aeruginosa* multiplication of the phage CB3 is inhibited at temperature below 32 °C in some strains of bacteria but not others. By appropriate genetic crosses, components of the bacterial genome were shown to be involved as well as the temperature effect and it is evident that some mechanism other than HCM is involved (Olsen, Metcalfe & Brandt, 1968).

1.f. Location of Prophages

The genetic determinant of bacteriophage, the prophage, may either be integrated linearly into the chromosome, as in the case of lambda in *E. coli* (Campbell, 1969), or extrachromosomally as the case of P1 (Ikeda & Tomizawa, 1968). Attempts to map the genetic location of the better characterized *P. aeruginosa* phages such as B3, D3, F116 and G101 have proved fruitless to date (Holloway & Krishnapillai, unpublished data). However Krishnapillai and Carey (1972) have been able to map the prophage of phage 90 on the *P. aeruginosa* chromosome by demonstrating its linkage in conjugation to other markers (see Chapter 5). Subsequently other phages have been located on the chromosome and accurate mapping of their prophage sites is proceeding (Drs. K. Carey & V. Krishnapillai, unpublished observations, (see Figure 5.1, Chapter 5). The location in the bacterial cell of extrachromosomally located prophages is at present a matter for speculation but it is likely that, as with other plasmids, a membrane site is involved. At present, no satisfactory technique exists for demonstrating conclusively such a location for a plasmid in the bacterial cell.

2. BACTERIOCINS

2.a. Types of aeruginocins

Bacteriocins are proteins having a lethal effect on bacteria which are produced by other bacteria usually closely related to the sensitive bacteria. Substances of this type produced by *P. aeruginosa* were first identified by Jacob (1954) and given the name pyocin (after the then valid species name *P. pyocyanea*). In view of the acceptance of the species name *aeruginosa*, the term aeruginocin is more correct, but the literature shows that opinion on this nomenclature is divided, and both terms must now be accepted as synonyms. The authors prefer aeruginocin as being less liable to be confused with pyocyanin.

The production of an aeruginocin by a given strain of bacteria can usually only be determined by using a strain sensitive to the particular aeruginocin. As yet there are no rules by which such strains can be identified. If a variety of randomly selected sensitive test strains is used, it has been repeatedly shown that aeruginocinogeny is very common in strains of *P. aeruginosa*. (Hamon, 1956; Hamon *et al.*, 1961; Patterson, 1965). Indeed, the high frequency of this property has enabled aeruginocins to be effectively used in the typing of *P. aeruginosa* for epidemiological purposes (see section 3.b). In one case (Goodwin, Levin & Doggett. 1973) a strain of *P. aeruginosa* was found to be susceptible to the aeruginocin that it releases, a result which stresses the differences between self-immunity to bacteriophage lysis which is a replicative function, and immunity to bacteriocin killing which is a surface or membrane function.

Aeruginocinogenic strains can be easily identified. A loopful of an overnight broth culture of the aeruginocinogenic strain is placed on the surface of a nutrient agar plate on which has been spread 0·1 ml of an overnight broth culture of the selected sensitive strain. After overnight incubation, a zone of growth inhibition is seen around the growth of the strain producing the aeruginocin. This usually has quite a different appearance to the zone of lysis produced by a lysogenic culture spotted in the same way on a sensitive lawn.

Aeruginocins may be assayed by spotting loopfuls of serial dilutions on to the surface of a nutrient agar plate spread with a sensitive strain. The endpoint is read as the highest dilution giving either complete clearing, or showing any inhibition of growth of the sensitive lawn. Production of aeruginocins may be affected by the temperature of incubation of the aeruginocinogenic strain. For example, some strains produce significantly more aeruginocin at 32 °C (Gillies & Govan, 1966). Production at 43 °C is commonly significantly inhibited (B. W. Holloway, unpublished observations).

In general, aeruginocins show that same range of properties as the more extensively studied colicins (Reeves, 1965; 1972). Bradley (1967) proposed a classification of bacteriocins into two types. The S type, characterized by sensitivity to proteolytic enzymes and a lack of structure in electron microscopy, and the R-type bacteriocins which are not sensitive to proteolytic enzymes and show a structured appearance characteristic of components of bacteriophages (Plate 4.5).

The spontaneous production of aeruginocin can be increased by ultraviolet irradiation (Higerd, Baechler & Berk, 1967) or mitomycin C treatment of the aeruginocinogenic culture (Kageyama, 1964). Purification of aeruginocins has been achieved with a variety of techniques, typically by chemical, column and differential centrifugation procedures (Kageyama & Egami, 1962; Higerd *et al.*, 1967) and these procedures have enabled detailed characterization of a limited range of aeruginocins. Kageyama and his associates have studied aeruginocins of the R variety and shown this type of aeruginocin to be a bacteriophage tail-like particle, with a single protein type having a molecular weight about 1×10^7 daltons. All the other R-type aeruginocins were found to be very similar to each other in morphology, mode of action and immunological properties (Kageyama & Egami, 1962; Kageyama, 1964; Ito & Kageyama, 1970; Ito, Kageyama & Egami, 1970). Ishii, Nishi and Egami (1965) showed that the morphology of the aeruginocin R is that of a double hollow cylinder, 1200 Å long and 150 Å in diameter with a core and a contractile sheath around the core. Base plates and tail fibres

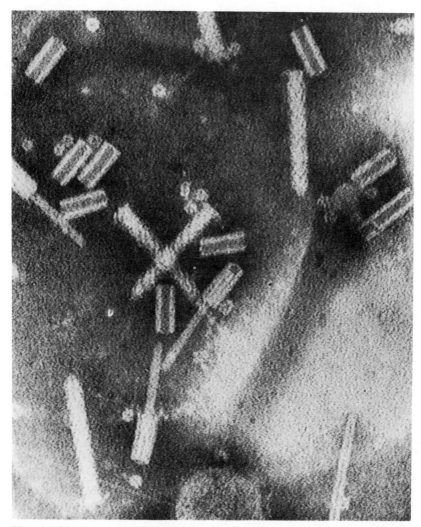

Plate 4.5. The R-type aeruginocin obtained by mitomycin induction of *P. aeruginosa* strain Götze (× 247,500) (Bradley, 1967). Extended and contracted forms are seen. (Photograph kindly provided by Dr. D. Bradley and reproduced with permission from D. E. Bradley (1967) *Bacteriological Reviews*, **31**, 230–314)

could also be observed. Kageyama (1970a,b) mapped the genetic determinant of the R2 aeruginocin at 35 minutes on the chromosome map of *P. aeruginosa*, and determined its linkage relationships to neighbouring genes by both conjugation and transduction. Subsequently two other R-type aeruginocins (R and R3) have been shown to be located at the same site (Kageyama, Shinomiya & Ohsumi, 1973).

Other R-type aeruginocins have been studied (Higerd *et al.*, 1967; 1969). Again, the aeruginocin was shown to have a marked morphological similarity to bacteriophage tails, could occur in a contracted or relaxed state with the lethal effects being produced by the relaxed form which contracts after adsorption to the bacterial cell surface. Aeruginocin 28 (Takeya *et al.*, 1969) is rod shaped, in contrast to the R-type aeruginocins, although the properties of this bacteriocin seem to be similar to the R group (Homma *et al.*, 1967) (Plate 4.6).

The structure of pyocin R has recently been studied in detail (Yui, 1971; Yui-Furihata, 1971). Methods of purification have been developed, the analysis of the sub-units carried out and it has been shown that the sub-units of the sheath are capable of polymerization (Yui-Furihata, 1972). These studies have enabled the biosynthesis and morphogenesis of R aeruginocins to be examined (Shinomiya, 1972a,b). This work is important not only for the characterization of aeruginocins, but in view of the relationship of R aeruginocins to bacteriophages (see below), they provide an ancillary study to the morphogenesis of bacteriophage.

Structures called rhapidosomes have been reported as being liberated during the growth of various bacteria, including *P. fluorescens*. A study of the pseudomonad rhapidosomes (Amako, Yasunaka & Takeya, 1970) showed that the rod-shaped particles were polymerized sheaths of bacteriocin, the evidence supporting this view being both morphological and immunological. Examination of *P. aeruginosa* rhapidosomes (Baechler & Berk, 1972) does not suggest such a close relationship with aeruginocins. These workers were unable to associate any form of biological activity with the rhapidosome. They suggest that rhapidosomes may be a product of mesosome/membrane rearrangement.

The mode of action of R aeruginocins has been investigated (Kaziro & Tanaka, 1965a,b). There was a rapid inhibition of RNA, DNA and protein synthesis similar to that found to occur with colicin K in *E. coli* (Nomura, 1967) and megacin C in *Bacillus megaterium* (Holland, 1967). In addition there were effects on ribosomes of aeruginocin-infected cells such that they were unable to carry out the polyU-dependent incorporation of ^{14}C-phenylalanine.

Aeruginocin 28 which is a rod-shaped aeruginocin of the R group, produced morphological changes in infected bacteria within 15 minutes after it is adsorbed by a sensitive cell, and Ohnishi, Takade and Takeya (1971) suggest that the site of action may be the bacterial membrane.

The other class of aeruginocins, the so called S type, are less well known. Both types seem to be equally common and both can be derived from one strain of bacteria. Fortuitously the well-characterized strain PAO of *P. aeruginosa* (Holloway, 1969) can produce both R and S aeruginocins (Ito, Kageyama & Egami, 1970). Ohkawa, Kageyama and Egami (1973) have

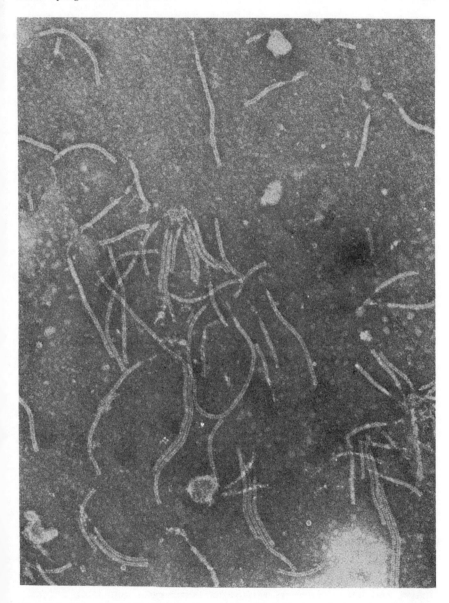

Plate 4.6. Aeruginocin 28. Negatively stained with 2 % neutral potassium tungstate (× 120,000) (Takeya *et al.*, 1967). (Photograph kindly supplied by Professor K. Takeya and Dr. K. Amako and reproduced with permission from K. Takeya *et al.* (1967) *Virology*, **31**, 167)

recently characterized the S-type aeruginocin S2 from strain PAO. Like the R-type aeruginocin, S2 can be efficiently induced by mitomycin C or u.v. The aeruginocin S2 has been shown to be a protein with a molecular weight of about 75,000. The killing action is of the single-hit variety, but the mode of action is not yet known. Holloway, Rossiter, Burgess and Dodge (unpublished observations) have studied the mode of action of a trypsin-sensitive aeruginocin named AP108 used to isolate aeruginocin-tolerant mutants. AP108 affects protein synthesis, RNA synthesis and DNA synthesis and thus shows similar characteristics to colicins of the E1 or K types which affect energy flux (Reeves, 1972).

2.b. Relationship of aeruginocins to bacteriophage

The characterization of both bacteriocins and bacteriophages has proceeded in parallel and many similarities have been pointed out by a variety of workers with different bacteriocins and bacteriophages. These similarities are primarily in terms of adsorption, structure, genetic basis and inducibility (Reeves, 1972). While these similarities should not be allowed to override the differences, it is apparent that at least in aeruginocins, the R-type aeruginocins show many properties characteristic of bacteriophages.

There are striking similarities between the morphology of R aeruginocins and phage structural components including contracted or extended sheaths, cores, base plates and tail fibres and it is difficult to escape the conclusion that what we are calling aeruginocins could well be the products of a defective prophage, particularly when one compares electron micrographs of the various R aeruginocins with those of products of conditional lethal mutants of T4 or P22 (Ishii *et al.*, 1965; Bradley, 1967; Higerd *et al.*, 1967; Homma, Goto & Shionoya, 1967; Homma & Shionoya, 1967; Takeya *et al.*, 1967; Levine, 1969; Takeya *et al.*, 1969; Ito *et al.*, 1970). Furthermore, a serological relationship between various R-type aeruginocins and certain specific temperate bacteriophages has been shown (Homma, Watabe & Kageyama, 1970).

The genetic evidence for the determinant of R2 supports this relationship. Kageyama (1970a,b) has precisely mapped the genetic determinant of R2 on the *P. aeruginosa* chromosome (see Figure 1, Chapter 5) and has pointed out the significance of this finding in terms of the relationship between bacteriophages and bacteriocins. As Kageyama points out, there is an anomaly between the lethal properties of the products of such a defective prophage and the usually accepted view that temperate phages in contrast to virulent phages are devoid of lethal properties in their tail structure. It is possible that bacteriophages may occupy any position on the spectrum between the extreme virulent phages such as T2 and the classical temperate phages such as lambda, and that R2 aeruginocins are derived by mutation to defectiveness of a bacteriophage somewhere in the middle of this range. Independently isolated R aeruginocins show remarkably similar properties and they presumably have arisen from a common bacteriophage ancestor (Kageyama *et al.*, 1973).

While it is reasonable to suppose that aeruginocins of this type have evolved from normal prophages, it is difficult to see what selective advantage they have over temperate phages, as spread of the defective prophage information within a bacterial population is entirely dependent on the host recombination mechanisms. Perhaps there has been too much attention paid to the final product of the aeruginocinogenic determinant, its structure, lethal activity and mode of action, rather than finding out what other effects such a genetic determinant can confer on the host bacterium which possess it.

As yet nothing is known of the genetic determinants of the S-type aeruginocins. Attempts to transmit these in conjugation in a manner similar to FP or RP factors have not succeeded, nor have attempts to cure aeruginocinogenicity by mitomycin C, acridines or ethidium bromide been successful (B. W. Holloway, unpublished results).

2.c. Aeruginocin-tolerant Mutants

It has been shown that mutants of *E. coli* can be isolated which are not killed by colicins but which still adsorb the colicins in much the same way as the sensitive parent strain. Such mutants are referred to as tolerant (genetic symbol *tol*) and they are providing a very useful research tool for the further understanding of bacteriocin action and membrane biology. (For references see Holloway, 1971; Reeves, 1972). Nomura (1967) pointed out that there was much circumstantial evidence to implicate the cell membrane as one step in the series of reactions initiated by bacteriocins which result in the death of a bacterium. He suggested that some mutants of bacteria which were resistant to colicins could result from alterations in the structure and function of the membrane. Conversely, *tol* mutants, may provide a means of obtaining a range of mutants affecting membrane structure and function. Experience has shown this prediction to be true. A range of such mutants has been analysed in *E. coli* and more recently, aeruginocin-tolerant mutants of *P. aeruginosa* have been isolated and characterized (Holloway *et al.*, in preparation).

Like their counterparts in *E. coli*, the *P. aeruginosa tol* mutants arise from changes in a number of genes, and show a variety of pleiotropic characteristics including alterations in membrane structure as displayed by polyacrylamide gel electrophoresis of solubilized membranes, alterations in levels of activity of membrane-bound enzymes, changes in susceptibility to toxic agents, alterations in cellular morphology and altered pattern of susceptibility to bacteriophages. Mutants of this type are important for studies on the biogenesis of membranes, the role of membranes in the cell economy and in particular the relationship of components of the genome to the membrane. In view of the role of plasmids in the *Pseudomonas* genome, this last aspect has a particular significance for the complete understanding of the genetic organization in *Pseudomonas*.

2.d. Bacteriocins of Other Species of *Pseudomonas*

Various phytopathogenic species of *Pseudomonas* have been found to produce a range of bacteriocins and can be distinguished by their host range and response to toxic agents (Garrett, Panagopoulos & Crosse, 1966; Vidaver *et al.*, 1972). It does not appear that bacteriocin typing is effective for distinguishing isolates of plant pathogens in terms of their host range.

Hamon (1956) and Hamon *et al.* (1961) have described bacteriocins (fluorocins—not to be confused with fluorescein) of *P. fluorescens* although Patterson (1965) was unable to demonstrate bacteriocinogeny for *P. fluorescens* or *P. ovalis*. *P. putida* is occasionally bacteriocinogenic (Holloway, unpublished observations) but the bacteriocins have not been further studied.

3. USE OF BACTERIOPHAGE AND BACTERIOCINS FOR EPIDEMIOLOGY

P. aeruginosa is a pathogen of considerable significance being involved in burns, wounds, urinary tract infections and more frequently in recent years, bronchopneumonia (Lancet, 1966; British Medical Journal, 1967; British Medical Journal, 1971; Lancet, 1971); such infections caused by *P. aeruginosa* showing a recent increase in both adults and children (Pierce *et al.*, 1966; Barson, 1971) (see Chapter 2). In fact, in recent years infections due to *P. aeruginosa* and some other Gram-negative bacteria have come to replace *Staphylococcus aureus* as the most important agents of nosocomial infections (McNamara *et al.*, 1967).

P. aeruginosa is particularly menacing due to its well-known resistance to chemotherapeutic agents which, in conjunction with an increasing incidence of situations where massive broad-spectrum antibiotic or immunosuppressive therapy is prescribed, has encouraged the emergence of *P. aeruginosa* as an important pathogen. Accurate epidemiological surveillance of *P. aeruginosa* in infections is imperative in view of the wide potential source of infection including faecal carriers, standard hospital equipment, disinfectant solutions and children's foods. Such surveillance is necessary as an adjunct to adequate control measures. (See Chapter 2.)

3.a. Available Techniques

Three techniques exist for the 'fingerprinting' of *P. aeruginosa*—serotyping, pyocin* typing and phage typing. The typing of *P. aeruginosa* strains for epidemiological purposes by pyocinogeny or lysogenicity patterns, where the pyocins or phages produced by tester strains are identified by a range of indicators, was first suggested by Holloway (1960) and since then a large number of investigators have developed extensive, reliable and useful typing schemes. A recent critical review of pyocin and phage typing methods is that of Vieu (1969). The typing of *P. aeruginosa* has not proved to be a simple task and the incidence of a significant number of strains untypable by any one scheme has led to the introduction of combined schemes. Since phage and pyocin typing are in more extensive use, we will primarily consider just these two but serotyping will be alluded to in relation to the two other methods.

3.b. Phage Typing

Phage typing methods for epidemiological purposes were initiated about 1960 (Gould & McLeod, 1960; Postic & Finland, 1961; Graber *et al.*, 1962). Pavlatou and Hassikou-Kaklamani (1961), using 12 phages (isolated from lysogenic bacteria or from sewage), were able to type 78 % of 174 Greek strains into 12 lysotypes. Lysotype stability was found to be somewhat variable in a small proportion of strains and this was found to be associated with the storage of cultures or after inoculation into animals (Pavlatou & Kaklamani, 1962). Typing of many separate clones of certain strains uncovered the existence of different lysotypes in strains considered pure at isolation (Pavlatou & Kaklamani, 1962). The usefulness of lysotyping for epidemiological purposes was realized by the recognition that strains isolated from the same patients or from different patients from the same environment belonged to the same lysotype although in some cases strains isolated from the same patients belonged to different lysotypes (Pavlatou & Kaklamani, 1962). Sutter and her colleagues (Sutter, Hurst & Fennell, 1965) in the United States used 18 phages (mainly from those used by Postic & Finland, 1961 and Pavlatou & Hassikou-Kaklamani, 1961) to type 317 strains on the basis of their lytic spectra. In particular they examined factors such as cultural conditions, number of phage particles applied in tests, length of storage of strains before tesing, and the reproducibility of lytic reactions. They concluded that under the appropriate standard conditions lytic spectra were essentially reproducible, and 86 % of the strains tested were typable. Meitert (1965), using 13 temperate phages was able to type 93 % of 652 Rumanian strains into 98 different lytic types. Sjöberg and Lindberg (1968)

---

* As pointed out earlier, pyocin and aeruginocin are synonymous but most epidemiologists seem to prefer pyocin, so it will be used here.

using 18 lytic phages (different to those used by Sutter *et al.*, 1965) were able to type 91 % of 667 Swedish strains from a variety of clinical sources into different lysis patterns and 243 strains from one source produced 160 different lytic patterns. Recently Bergan (1972c) in Norway has selected a new set of 24 typing phages (consisting of a primary set of 19 phages and an auxiliary set of 5) from 113 internationally available phages which had previously been used by other workers in ten different typing sets. His reasoning for so doing was to aid the standardization of typing on a world-wide basis, to develop a set which lysed as many strains as possible, and to reduce the proportion of non-typable strains. To select the new phage array, he used numerical allocation procedures to group phages according to similarities in phage lytic spectra (Bergan, 1972a,b). A comparison of the efficiency of the new set over the previous sets indicated some increase in the proportion of typable strains (95·5 %) and up to 240 lytic spectra were found with the primary set (Bergan, 1972d). There is some lack of reproducibility in phage lytic patterns when strains are tested on a number of occasions either within a few days of each other, or after storage for a few months or when typed by different observers (Sutter *et al.*, 1965; Sjöberg & Lindberg, 1968; Bergan & Lystad, 1972; Beumer *et al.*, 1972). Variation in phage typing was more extensive when lyophilized cultures were tested (Beumer *et al.*, 1972), and because of these factors, typing of strains for epidemiological purposes would appear to be more reliable by combined typing schemes (see below). However, in some instances, phage typing has been of considerable epidemiological value (Rouques *et al.*, 1969; Liljedahl *et al.*, 1972). Phage typing on the basis of lysogenicity has also been used (Holloway, 1960; Yamada, 1960; Meitert & Meitert, 1961; Feary *et al.*, 1963b; Farmer & Herman, 1969).

Lysotyping of *Pseudomonas* species other than *P. aeruginosa* has not been investigated to any extent yet, but Okabe and Goto (1963) have been able to type the phytopathogen *P. solanacerarum*, the causative agent of bacterial wilt, into about 40 lysotypes with the use of 12 virulent and temperate phages.

3.c. Pyocin Typing

Just as *P. aeruginosa* strains may be typed on the basis of lysogenicity or sensitivity to a set of phages, so may pyocins be used as a typing reagent, either by pyocinogenicity of the strains to be typed or their sensitivity to pyocin. One major advantage of pyocin typing is its logistic simplicity as opposed to typing by phages or serology, both of which require the maintenance of an adequate range of phages or antisera, respectively.

In principle the method of pyocin testing by pyocinogenicity (Gillies & Govan, 1966) involves streaking the strains to be tested across a nutrient agar plate. After 14–18 hours incubation at 32 °C, the bacterial growth is removed and the agar surface exposed to chloroform to kill residual bacteria. Indicator strains (to detect pyocin production by the strains being tested) are then streaked on the surface of the plate at right angles to the original inocula and incubation continued for a further 8 to 18 hours at 37 °C. If pyocins are produced by the strains being tested, they diffuse into the medium during the first period of incubation and exert their inhibitory activity on the indicator strains during the subsequent incubation period. The pyocin types of the tested strains are recognized from the inhibition patterns produced against the slanted indicators.

There are numerous modifications of this basic method. Taking advantage of the fact that the antibiotic mitomycin C induces the production of pyocin (Kageyama, 1964), Farmer and Herman (1969) and Tripathy and Chadwick (1971) have pyocin typed strains and the latter workers found that the incubation period for pyocin induction could be considerably shortened (32 °C for 6 hours) instead of the Gillies and Govan (1966) procedure of 32 °C for 14 hours. This has considerable logistic advantage in routine typing procedures for epidemiological purposes. It was also found that mitomycin C induction of pyocins enabled the typing of 23/28 strains which were untypable by the Gillies and Govan procedure. Gillies and Govan (1966) found that the temperature and length of incubation of the producer strains is critical. Using 32 °C for 14 hours they found that nearly twice as many strains were typable as by Wahba's (1965a) method using 37 °C for 24 hours.

Although Darrell and Wahba (1964) used pyocinogenicity against 12 indicators to type 474 strains into 13 pyocin types, Gillies and Govan (1966) were the first to apply systematically pyocinogenicity as a means of typing. They used 8 indicators to type 3227 strains and found 88 % of the strains to be typable into 36 pyocin types which contrasted with Wahba's (1965a) 10 pyocin types using 12 indicators. Bergan (1968) using the same 12 indicators as Darrell and Wahba, found 27 groups amongst a selection of Scandinavian strains. Zabransky and Day (1969) using 8 of Darrell and Wahba's (1964) indicators together with 3 others of their own were able to type 93 % of a collection of 954 isolates of U.S. origin into 15 pyocin types.

Good reproducibility in pyocin typing patterns by the Gillies and Govan (1966) procedure was shown by strains giving the same pattern of pyocinogenicity after two years in storage. Another criterion of reliability was provided by the demonstration that reiterated isolation of strains with different pyocin types from the same patient was not due to variability in the pyocin typing method but to independent infections with different strains (Gillies & Govan, 1966; Govan & Gillies, 1969). The reproducibility claimed by Gillies and Govan (1966) was questioned by Bergan (1968), and Chadwick (1972) expressed some reservations and suggested that most variations in pyocin pattern could still be intrinsically associated with the technique *per se*. Merrikin and Terry (1972) have stressed the need for rigid adherence to technical procedures when using this typing procedure.

The usefulness of this last contribution for epidemiology has also been amply shown in the U.S. where Rose and his colleagues (1971) have found it possible to subtype pyocin type 1 strains. From a purely epidemiological point of view the Govan and Gillies (1966) typing scheme has been found useful in tracing the sources of infection in a hospital environment (Teres *et al.*, 1973).

Typing by means of sensitivity patterns to a standard set of pyocins has been used only to a limited extent. Osman (1965) used 4 different pyocins to type 101 strains into 10 different inhibitory patterns. Rampling and Whitby (1972) have paid particular attention to the problem of contaminating bacteriophages in pyocin lysates which could lead to anomalies in scoring. They have been able to reduce the content of phage in mitomycin C induced lysates by ultraviolet irradiation to inactivate the phage particles, the titre of the pyocin being unaffected at the doses used. Pyocin preparations so obtained appear to be useful for routine typing purposes (Dr. A. Rampling, personal communication).

A technique of utilizing both pyocinogenicity and pyocin sensitivity simultaneously for typing has also been investigated (Farmer & Herman, 1969; Kasomson *et al.*, 1970). Farmer and Herman (1969) by the combined use of 27 indicator strains for pyocinogenicity typing (with induction of pyocins by mitomycin C to enhance pyocin titre) and 23 pyocin lysates for typing by pyocin sensitivity, found all 57 strains of *P. aeruginosa* tested by this combined method to be typable and this approach promises to be a most useful epidemiological tool.

Recently pyocin typing has been extended to mucoid strains of *P. aeruginosa* (Williams & Govan, 1973). These strains have a particular significance in that they are commonly associated with cystic fibrosis of the pancreas in children. Although initially untypable by the standard method (Gillies & Govan, 1966) the mucoid strains could be typed after mitomycin C induction of liquid cultures with the lysates being tested against standard indicators. The significance of this ability to type mucoid strains lies in the fact that both mucoid and non-mucoid strains were found to occur simultaneously in the same patient. In many cases, both strains had the same pyocin type suggesting that they are variants of the same strain and that the mucoid derivative had particular pathogenic properties related to the clinical features of cystic fibrosis.

3.d. High Incidence of Certain *P. aeruginosa* Strains Detected in Epidemiological Surveys

A fact of considerable genetic and medical significance is the observation of a very high incidence of only a few strains among the large number typed, regardless of the method of typing. For example, Table 4.2 illustrates the point from data obtained by phage or pyocin typing (see also Bergan, 1973a).

Table 4.2. Geographic distribution of lysotypes and pyocin types

A. Phage typing

Country of origin of strains	Total number of lysotypes	Number of common lysotypes	Percentage of strains contributing to common lysotypes	References
U.S.A.	87	7	44	Sutter *et al.* (1965)
Rumania	98	5	66·8	Meitert (1965)
Greece	12	4	73	Pavlatou & Hassikou-Kaklamani (1961)

B. Pyocin typing

Country of origin of strains	Percentage of typable strains belonging to Gillies & Govan (1966) pyocin types				% of strains	References
	1	3	5	10		
Scotland	34·2	25·2	5·7	2.9	68·0	Govan & Gillies (1969)
U.S.A.	52·1	7·4	3·0	11·0	73·5	Heckman *et al.* (1972)
Canada	44·7	5·3	0·3	26·8	77·1	Tripathy & Chadwick (1971)
Germany	17·6	16·2	8·6	6·7	49.1	Neussel (1971)
Australia	33·5	18·5	3·0	17·0	72·0	Tagg & Mushin (1971)
Singapore	47·0	10·0	0·0	17·0	74·0	Tagg & Mushin (1971)
Israel	65·7	4·5		11·2	81·4	Ziv *et al.* (1971)
Hungary	28·7	20·1	5·5	14·8	69·1	Csiszar & Lanyi (1970)

The high incidence of certain pyocin types has also been reported when indicators other than that used by Gillies and Govan (1966) have been used (Darrell & Wahba, 1964; Zabransky & Day, 1969; Bergan, 1968). Amongst the Gillies and Govan (1966) pyocin types, type 1 is by far the commonest type across the world and even when subtyped into its eight subtypes—the 1b, 1c and 1d forms greatly predominated (Govan & Gillies, 1969; Csiszar & Lanyi, 1970; Rose *et al.*, 1971). Predominance of certain strains has also been observed by serotyping of strains during epidemiological surveys (Lanyi, Gregačs & Adam, 1966; 1967; Muraschi *et al.*, 1966; Diaz & Neter, 1970; Lanyi, 1970; Mikkelsen, 1970; Moody *et al.*, 1972). As observed with pyocin typing, certain serotypes have a world-wide and prevalent distribution

(Muraschi *et al.*, 1966). Some patterns of distribution can also be observed with respect to the type of clinical material examined.

Taken together these facts have important implications. First that certain genotypes of *P. aeruginosa* have evolved to the point of successful persistence in the hospital and other environments and secondly that some of these have a world-wide distribution. From a clinical and epidemiological point of view, these considerations demonstrate the need of combined typing schemes for the unique identification of strains for tracing sources of infection in a hospital environment.

3.e. Combined Typing Methods

Although numerous investigators have been satisfied with the effectiveness of one or other of the three major typing methods (pyocin, phage or serology) in typing *P. aeruginosa* strains for epidemiological purposes (Gillies & Govan, 1966; Govan & Gillies, 1969; Liljedahl *et al.*, 1972; Moody *et al.*, 1972) there are others who question the validity of using a single method (Edmonds *et al.*, 1972a; Parker, 1972). Many of the uncertainties concern factors such as reproducibility of phage or pyocin typing patterns, the predominance of very few types in many epidemiological surveys and therefore difficulties in tracing isolated sources of infection. This has led to a number of investigations using combined typing methods with a view to obtaining finer subdivisions of recognition.

Wahba (1965a,b) combined the Habs (1957) serological typing method based on heat stable 'O' antigens with pyocin typing and found an increase in the typability of strains collected from the United Kingdom and 12 other countries. Twentytwo seropyocin types were recognized by the use of 14 serotypes and 10 pyocin types. However, again, as was discussed previously for pyocin types, strains tended to 'cluster' to particular serotypes and pyocin types. For example, of the 72 % of the strains which belonged to 4 of the 10 pyocin types, 84 % belonged to 4 of the 14 serotypes (Wahba, 1965a). Such a correlation was also observed in Japanese (Matsumoto, Tazaki & Kato, 1968), Hungarian (Csiszar & Lanyi, 1970) and Norwegian strains (Bergan, 1973a), the latter work also including a substantial collection of strains from world-wide sources. Csiszar and Lanyi (1970) have used an extensive serogrouping of *P. aeruginosa* based on heat-stable 'O' and heat-labile 'H' antigens into 53 serotypes and an additional 16 partially defined serotypes and were able to distinguish 165 seropyocin types when simultaneously compared with the pyocin typing scheme of Gillies and Govan (1966) and Govan and Gillies (1969). The combination of serotyping with phage typing has also been valuable in epidemiology (Meitert & Meitert, 1966; Lantos *et al.*, 1969; Bergan, 1972e). Initial serotyping followed by phage typing has increased the identification of Rumanian strains especially since the most commonly occurring serotype (comprising 64 % of the strains) was subdivisible into 29 lysotypes by phage typing although 67 % belonged to one lysotype only (Meitert & Meitert, 1966). However Bergan (1972e; 1973b) has found little correlation between serotype and lysotype nor between serotype or pyocin type with lysotype. Finally the combination of typing by pyocin, phage and serology has been utilized (Deighton, Tagg & Mushin, 1971; Edmonds *et al.*, 1972a,b; Bergan, 1973c). Although the combined procedure was useful in epidemiology, phage typing was more discriminating in determining strain relatedness (Bergan, 1973c).

The clinical importance of native *P. aeruginosa* infections, clearly establishes the need for epidemiological typing and the methods described are clearly capable of providing the necessary identification of strains. Clearly the next step is the adoption of a standard procedure which can be used internationally to provide extensive comparable data from a wide variety of situations. Any pyocins, phages or serological procedures adopted for these procedures can then can be characterized biologically, as has been done with other well-established typing systems. The hierarchical approach to these combined methods, i.e. serology, then phage typing and finally, pyocin typing, or some such sequence is highly reliable for strain identification (Deighton *et al.*, 1971; Edmonds *et al.*, 1972b; Parker, 1972).

4. SUMMARY

Clearly bacteriophages and bacteriocins are important for a variety of practical and laboratory purposes involving species of *Pseudomonas*. The contribution of their genetic determinants to the bacterial phenotype is significant and at present underestimated. The frequency of lysogeny and aeruginocinogeny in *P. aeruginosa* strains contrast with the lack of these properties in *P. putida* and *P. fluorescens*. It is possible that the differences may be related to the plasmid component of the genome of these various species and a task which now must be undertaken is the relationship between known plasmids, bacteriophages and bacteriocins. Undoubtedly what would be an essentially academic investigation will have important implications for the necessary practical uses of bacteriophages and bacteriocins.

5. BIBLIOGRAPHY

Adams, M. H. (1959) *Bacteriophages*, Interscience Publishers, New York.
Alföldi, L. (1957) *Acta Microbiologica Academiae Scientiarium Hungaricae*, **4**, 119.
Amako, K., Yasunaka, K. & Takeya, K. (1970) *Journal of General Microbiology*, **62**, 107.
Baechler, C. A. & Berk, R. S. (1972) *Microstructures*, **3**, 24.
Barksdale, L. (1959) *Bacteriological Reviews*, **23**, 202.

Barson, A. J. (1971) *Archives of Diseases in Childhood*, **46**, 55.
Bartell, P. F. (1967) *The Journal of Biological Chemistry*, **243**, 2077.
Bartell, P. F. & Orr, T. E. (1969a) *Journal of Virology*, **3**, 290.
Bartell, P. F. & Orr, T. E. (1969b) *Journal of Virology*, **4**, 580.
Bartell, P. F., Orr, T. E. & Lom, G. K. H. (1966) *Journal of Bacteriology*, **92**, 56.
Bergan, T. (1968) *Acta Pathologica et Microbiologica Scandinavica*, **72**, 401.
Bergen, T. (1972a) *Acta Pathologica et Microbiologica Scandinavica, Section B*, **80**, 55.
Bergen, T. (1972b) *Acta Pathologica et Microbiologica Scandinavica, Section B*, **80**, 89.
Bergan, T. (1972c) *Acta Pathologica et Microbiologica Scandinavica, Section B*, **80**, 177.
Bergan, T. (1972d) *Acta Pathologica et Microbiologica Scandinavica, Section B*, **80**, 189.
Bergan, T. (1972e) *Acta Pathologica et Microbiologica Scandinavica, Section B*, **80**, 351.
Bergan, T. (1973a) *Acta Pathologica et Microbiologica Scandinavica, Section B*, **81**, 70.
Bergan, T. (1973b) *Acta Pathologica et Microbiologica Scandinavica, Section B*, **81**, 81.
Bergan, T. (1973c) *Acta Pathologica et Microbiologica Scandinavica, Section B*, **81**, 91.
Bergan, T. & Lystad, A. (1972) *Acta Pathologica et Microbiologica Scandinavica, Section B*, **80**, 345.
Beumer, J., Cotton, E., Delmotte, A., Millet, M., von Grünigen & Yourassowsky, E. (1972) *Annales de l'Institut Pasteur*, **122**, 415.
Bradley, D. E. (1966) *Journal of General Microbiology*, **45**, 83.
Bradley, D. E. (1967) *Bacteriological Reviews*, **31**, 230.
Bradley, D. E. (1972a) *Genetical Research, Cambridge*, **19**, 39.
Bradley, D. E. (1972b) *Biochemical and Biophysical Research Communications*, **47**, 142.
Bradley, D. E. (1972c) *Journal of General Microbiology*, **72**, 303.
Bradley, D. E. (1973) *Virology*, **51**, 589.
Bradley, D. E. & Robertson, D. (1968) *Journal of General Virology*, **3**, 247.
British Medical Journal (1967) **4**, 309.
British Medical Journal (1971) **3**, 203.
Campbell, A. M. (1969) *Episomes*, Harper & Row, New York.
Chadwick, P. (1972) *Canadian Journal of Microbiology*, **18**, 1153.
Chakrabarty, A. M. (1972) *Journal of Bacteriology*, **112**, 815.
Chakrabarty, A. M., Niblack, J. F. & Gunsalus, I. C. (1967) *Virology*, **32**, 532.
Chow, C. T. & Yamamoto, T. (1969) *Canadian Journal of Microbiology*, **15**, 1179.
Csiszar, K. & Lanyi, B. (1970) *Acta Microbiologica Academiae Scientiarum Hungaricae*, **17**, 361.
Darrell, J. H. & Wahba, A. H. (1964) *Journal of Clinical Pathology*, **17**, 236.
Davison, P. F., Freifelder, D. & Holloway, B. W. (1964) *Journal of Molecular Biology*, **8**, 1.
Deighton, M. A., Tagg, J. R. & Mushin, R. (1971) *The Medical Journal of Australia*, **1**, 892.
Delisle, A. L. & Levin, R. E. (1972a) *Antonie van Leeuwenhoek*, **38**, 1.
Delisle, A. L. & Levin, R. E. (1972b) *Antonie van Leeuwenhoek*, **38**, 9.
Diaz, F. & Neter, E. (1970) *The American Journal of the Medical Sciences*, **259**, 340.
Dunn, N. W. & Holloway, B. W. (1971a) in *Informative Molecules in Biological Systems*, (Ed. L. Ledoux), North Holland Publishing Co., Amsterdam, p. 223.
Dunn, N. W. & Holloway, B. W. (1971b) *Genetical Research (Cambridge)*, **18**, 185.
Edmonds, P., Suskind, R. R., Macmillan, B. G. & Holder, I. A. (1972a) *Applied Microbiology*, **24**, 213.
Edmonds, P., Suskind, R. R., Macmillan, B. G. & Holder, I. A. (1972b) *Applied Microbiology*, **24**, 219.
Egan, J. B. & Holloway, B. W. (1961) *Australian Journal of Experimental Biology & Medical Science*, **39**, 9.

Farmer, J. J. & Herman, L. G. (1969) *Applied Microbiology*, **18**, 760.

Feary, T. W., Fisher, E. & Fisher, T. N. (1963a) *Biochemical Biophysical Research Communications*, **10**, 359.

Feary, T. W., Fisher, E. & Fisher, T. N. (1963b) *Proceedings of the Society for Experimental Biology and Medicine*, **113**, 426.

Feary, T. W., Fisher, E. & Fisher, T. N. (1964) *Journal of Bacteriology*, **87**, 196.

Franklin, R. M., Dalta, A. & Dahlberg (1972) *Biochimica Biophysica Acta*, **233**, 521.

Garrett, C. M. E. & Crosse, J. E. (1963) *Journal of Applied Bacteriology*, **26**, 27.

Garrett, C. M. E., Panogopoulos, C. G. & Crosse, J. E. (1966) *Journal of Applied Bacteriology*, **29**, 342.

Gillies, R. R. & Govan, J. R. W. (1966) *The Journal of Pathology and Bacteriology*, **91**, 339.

Goodwin, K., Levin, R. W. & Doggett, R. G. (1973) *Infection and Immunity*, **6**, 889.

Gould, J. C. & McLeod (1960) *The Journal of Pathology and Bacteriology*, **79**, 295.

Govan, J. R. W. & Gillies, R. R. (1969) *Journal of Medical Microbiology*, **2**, 17.

Graber, C. D., Latta, R., Vogel, E. H. & Brame, R. (1962) *The American Journal of Clinical Pathology*, **37**, 54.

Habs, I. (1957) *Zeitschrift für Hygiene*, **144**, 218.

Hamon, Y. (1956) *Annales de l'Institut Pasteur*, **91**, 489.

Hamon, Y., Veron, M. & Peron, Y. (1961) *Annales de l'Institut Pasteur*, **101**, 738.

Harrison, S. C., Caspar, D. L. D., Camerini-Otero, R. D. & Franklin, R. M. (1971) *Nature New Biology*, **229**, 197.

Heckman, M. G., Babcock, J. B. & Rose, H. D. (1972) *American Journal of Clinical Pathology*, **57**, 35.

Higerd, T. B., Baechler, C. A. & Berk, R. S. (1967) *Journal of Bacteriology*, **93**, 1976.

Higerd, T. B., Baechler, C. A. & Berk, R. S. (1969) *Journal of Bacteriology*, **98**, 1378.

Holland, I. B. (1967) *Journal of Molecular Biology*, **12**, 429.

Holloway, B. W. (1960) *The Journal of Pathology and Bacteriology*, **80**, 448.

Holloway, B. W. (1965) *Virology*, **25**, 634.

Holloway, B. W. (1966) *Mutation Research*, **3**, 452.

Holloway, B. W. (1969) *Bacteriological Reviews*, **33**, 419.

Holloway, B. W. (1971) *Australian Journal of Experimental Biology and Medical Science*, **49**, 429.

Holloway, B. W. & Cooper, G. N. (1962) *Journal of Bacteriology*, **84**, 1321.

Holloway, B. W., Egan, J. B. & Monk, M. (1960) *Australian Journal of Experimental Biology and Medical Science*, **38**, 321.

Holloway, B. W. & Rolfe, B. (1964) *Virology*, **23**, 595.

Holloway, B. W. & van de Putte, P. (1968) in *Replication and Recombination of Genetic Material*, (Eds. W. J. Peacock & R. D. Brock), Australian Academy of Science, Canberra, p. 175.

Homma, J. Y., Goto, S. & Shionoya, H. (1967) *Japanese Journal of Experimental Medicine*, **37**, 373.

Homma, J. Y. & Shionoya, H. (1967) *Japanese Journal of Experimental Medicine*, **37**, 395.

Homma, J. Y., Watabe, H. & Tanabe, Y. (1968) *Japanese Journal of Experimental Medicine*, **38**, 213.

Ikeda, H. & Tomizawa, J. (1968) *Cold Spring Harbor Symposia on Quantitative Biology*, **33**, 791.

Ishii, S., Nishi, Y. & Egami, F. (1965) *Journal of Molecular Biology*, **13**, 428.

Ito, S. & Kageyama, M. (1970) *Journal of General and Applied Microbiology*, **16**, 231.

Ito, S., Kageyama, M. & Egami, F. (1970) *Journal of General and Applied Microbiology,* **16,** 205.

Jacob, F. (1954) *Annales de l'Institut Pasteur,* **86,** 149.

Jacob, F. & Wollman, E. L. (1961) *Sexuality and the Genetics of Bacteria,* Academic Press, New York.

Kageyama, M. (1964) *Journal of Biochemistry,* **55,** 49.

Kageyama, M. (1970a) *Journal of General and Applied Microbiology,* **16,** 523.

Kageyama, M. (1970b) *Journal of General and Applied Microbiology,* **16,** 531.

Kageyama, M. & Egami, F. (1962) *Life Science,* **9,** 471.

Kageyama, M., Shinomiya, T. & Ohsumi, M. (1973) *Federation Proceedings,* (in press).

Kaiser, A. D. (1957) *Virology,* **3,** 42.

Kasomson, T., Roberts, C. E. & Panas-Ampol, K. (1970) *South East Asian Journal of Tropical Medicine and Public Health,* **1,** 391.

Kaziro, Y. & Tanaka, M. (1965a) *Journal of Biochemistry,* **58,** 357.

Kaziro, Y. & Tanaka, M. (1965b) *Journal of Biochemistry,* **57,** 689.

Kelln, R. A. & Warren, R. A. J. (1971) *Canadian Journal of Microbiology,* **17,** 677.

Krishnapillai, V. (1971) *Molecular and General Genetics,* **114,** 134.

Krishnapillai, V. & Carey, K. (1972) *Genetic Research, Cambridge,* **20,** 137.

Kropinski, A. M. B. & Warren, R. A. J. (1970) *Journal of General Virology,* **6,** 85.

Lancet, (1966) **i,** 1139.

Lancet, (1971) **i,** 1110.

Lantos, J., Kiss, M., Lanyi, B. & Völgyesi, J. (1969) *Acta Microbiologica Academiae Scientiarum Hungaricae,* **16,** 333.

Lanyi, B. (1970) *Acta Microbiologica Academiae Scientiarum Hungaricae,* **17,** 35.

Lanyi, B., Gregačs, M. & Adam, M. M. (1966/1967) *Acta Microbiologica Academiae Scientiarum Hungaricae,* **13,** 319.

Lapchine, L., Goze, A., Moillo, A. & Enjalbert, L. (1969) *Journal de Microscopie,* **8,** 503.

Lee, L. F. & Boezi, J. A. (1966) *Journal of Bacteriology,* **92,** 1821.

Lee, L. F. & Boezi, J. A. (1967) *Journal of Virology,* **1,** 1274.

Levine, M. (1969) *Annual Review of Genetics,* **3,** 323.

Liljedahl, S-O., Malborg, A-S., Nyström, B. & Sjoberg, L. (1972) *Journal of Medical Microbiology,* **5,** 473.

Liu, P. V. (1969) *Journal of Infectious Diseases,* **119,** 237.

McNamara, M. J., Hill, M. C., Balows, A. K. & Tucker, E. B. (1967) *Annals of Internal Medicine,* **66,** 480.

Martin, D. R. (1972) *Journal of Medical Microbiology,* **6,** 111.

Matsumoto, H., Tazaki, T. & Kato, T. (1968) *Japanese Journal of Microbiology,* **12,** 111.

Meitert, E. (1965) *Archives Roumaines de Pathologie Experimentale et de Microbiologie,* **24,** 439.

Meitert, T. & Meitert, E. (1961) *Archives Roumaines de Pathologie Experimentale et de Microbiologie,* **20,** 277.

Meitert, T. & Meitert, E. (1966) *Archives Roumaines de Pathologie Experimentale et de Microbiologie,* **25,** 427.

Merrikin, D. J. & Terry, C. S. (1972) *Journal of Applied Bacteriology,* **35,** 667.

Mikkelsen, O. S. (1970) *Acta Pathologica et Microbiologica Scandinavica, Section B,* **78,** 163.

Minamishima, Y., Takeya, K., Ohnishi, Y. & Amako, K. (1968) *Journal of Virology,* **2,** 208.

Moody, M. R., Young, V. M., Kenton, D. M. & Vermeulen, G. D. (1972) *The Journal of Infectious Diseases*, **125**, 95.
Muraschi, T. F., Bolles, D. M., Moczulski, C. & Lindsay, M. (1966) *Journal of Infectious Diseases*, **116**, 84.
Neussel, V. H. (1971) *Arzneimittel Forschüng*, **21**, 333.
Nomura, M. (1967) *Annual Review of Microbiology*, **21**, 257.
O'Callaghan, R. J., O'Mara, W. & Grogan, J. B. (1969) *Virology*, **37**, 642.
Ohkawa, I., Kageyama, M. & Egami, F. (1973) *Journal of Biochemistry*, **73**, 281.
Ohnishi, Y., Takade, A. & Takeya, K. (1971) *Japanese Journal of Microbiology*, **15**, 201.
Okabe, N. & Goto, M. (1963) *Annual Review of Phytopathology*, **1**, 397.
Olsen, R. H., Metcalf, E. S. & Brandt, C. (1968) *Journal of Virology*, **2**, 1393.
Olsen, R. H., Metcalf, E. S. & Todd, J. K. (1968) *Journal of Virology*, **2**, 357.
Olsen, R. H. & Shipley, P. (1973) *Journal of Bacteriology*, **113**, 772.
Osman, M. A. M. (1965) *Journal of Clinical Pathology*, **18**, 200.
Parker, M. T. (1972) *The Journal of General Microbiology*, **73**, vi.
Patterson, A. C. (1965) *Journal of General Microbiology*, **39**, 295.
Pavlatou, M. P. & Hassikou-Kaklamani, E. (1961) *Annales de l'Institut Pasteur*, **101**, 914.
Pavlatou, M. P. & Kaklamani, E. (1962) *Annales de l'Institut Pasteur*, **102**, 303.
Pierce, A. K., Edmonson, E. B., McGee, G., Ketcheroid, J., London, R. G. & Sanford, J. P. (1966) *American Review of Respiratory Diseases*, **94**, 309.
Postic, B. & Finland, M. (1961) *Journal of Clinical Investigation*, **40**, 2064.
Rampling, A. & Whitby, J. L. (1972) *Journal of Medical Microbiology*, **5**, 305.
Reeves, P. (1965) *Bacteriological Reviews*, **29**, 25.
Reeves, P. (1972) *The Bacteriocins*, Springer-Verlag, Berlin.
Rolfe, B. & Holloway, B. W. (1968) *Genetical Research* (Cambridge), **12**, 94.
Rose, H. D., Babcock, J. B. & Heckman, M. G. (1971) *Applied Microbiology*, **22**, 475.
Rouques, R., Vieu, J.-F., Mignon, F. & Leroux-Robert, C. (1969) *La Presse Medicale*, **77**, 509.
Semancik, J. S., Vidaver, A. K. & van Etten, J. L. (1973) *Journal of Molecular Biology*, **78**, 617.
Shargool, P. D. & Townshend, E. E. (1966) *Canadian Journal of Microbiology*, **12**, 885.
Shinomiya, T. (1972a) *Journal of Biochemistry*, **72**, 39.
Shinomiya, T. (1972b) *Journal of Biochemistry*, **72**, 499.
Shionoya, H., Goto, S., Tsukamoto, M. & Homma, Y. (1967) *Japanese Journal of Experimental Medicine*, **37**, 359.
Sjöberg, L. & Lindberg, A. A. (1968) *Acta Pathologica Microbiologica Scandinavica*, **74**, 61.
Slayter, H. S., Holloway, B. W. & Hall, C. E. (1964) *Journal of Ultrastructural Research*, **11**, 274.
Sutter, V. L., Hurst, V. & Fennell, J. (1965) *Health Laboratory Science*, **2**, 7.
Tagg, J. R. & Mushin, R. (1971) *The Medical Journal of Australia*, **1**, 847.
Takeya, K. & Amako, K. (1966) *Virology*, **28**, 163.
Takeya, K., Minamishima, Y., Amono, K. & Ohnishi, Y. (1967) *Virology*, **31**, 167.
Takeya, K., Minamishima, Y., Ohnishi, Y. & Amako, K. (1969) *Journal of General Virology*, **4**, 145.
Teres, D., Schweers, P., Bushnell, L. S., Hedley-Whyte, J. & Feingold, D. S. (1973) *Lancet*, **i**, 415.
Tripathy, G. S. & Chadwick, P. (1971) *Canadian Journal of Microbiology*, **17**, 829.
van de Putte, P. & Holloway, B. W. (1968) *Mutation Research* **6**, 195.

van der Schans, G. P., Bleichrodt, J. F. K. & Blok, J. (1972) _International Journal of Radiation Biology_, **23**, 133.
Vidaver, A. K., Koski, R. K. & van Etten, J. L. (1973) _Journal of Virology_, **11**, 799.
Vidaver, A. K., Mathys, M. L., Thomas, M. E. & Schuster, M. L. (1972) _Canadian Journal of Microbiology_, **18**, 705.
Vieu, J-F. (1969) _Bulletin de l'Institut Pasteur_, **67**, 1231.
Wahba, A. H. (1965a) _British Medical Journal_, **1**, 86.
Wahba, A. H. (1965b) _Zentralblatt für Bakteriologie, Parasitenkunde, Infektionskrankheitea und Hygiene_, **196**, 389.
Weppelman, R. M. & Brinton, Jr., C. C. (1970) _Virology_, **41**, 116.
Weppelman, R. M. & Brinton, Jr., C. C. (1971) _Virology_, **44**, 1.
Williams, R. J. & Govan, J. R. W. (1973) _Journal of Medical Microbiology_, **6**, 409.
Yamada, K. (1960) _Mie Medical Journal_, **10**, 359.
Yamamoto, T. & Chow, C. T. (1968) _Canadian Journal of Microbiology_, **14**, 667.
Yui, C. (1971) _Journal of Biochemistry_, **69**, 101.
Yui-Furihata, C. (1971) _Journal of Biochemistry_, **70**, 1047.
Yui-Furihata, C. (1972) _Journal of Biochemistry_, **72**, 1.
Zabransky, R. J. & Day, F. E. (1969) _Applied Microbiology_, **17**, 293.
Zierdt, C. H. & Schmidt, P. J. (1964) _Journal of Bacteriology_, **87**, 1003.
Ziv, G., Mushin, R. & Tagg, J. R. (1971) _Journal of Hygiene, Cambridge_, **69**, 171.

CHAPTER 5

Genetic Organization of *Pseudomonas*

BRUCE W. HOLLOWAY

1. INTRODUCTION

Genetic analysis of microorganisms has contributed much to microbiology and genetics. The history of bacterial genetics shows that most workers have devoted their efforts to one strain of one species, *Escherichia coli* K12, and the success of this strategy is obvious in the wealth of knowledge and the elegant experiments which continue to be associated with work on this bacterium. However, other species of bacteria have attracted the attention of geneticists, and this work has revealed to microbiologists the value of genetics to many problems involving microorganisms. *Pseudomonas* is one such example. The growth of interest over the last few years in the genetics of the two species, *P. aeruginosa* and *P. putida* has been both an intellectual stimulant and a delight to the few veteran workers in this field.

133

For both these species, systems of recombination analysis have been developed and a variety of techniques established for isolating mutants of research value to geneticists, microbiologists and biochemists. It has been shown that plasmids have an essential role in the drug resistance and biochemical versatility of these bacteria, indeed these being the very properties which have made *Pseudomonas* attractive to so many workers. Undoubtedly, much remains to be learnt of the genetic organization of *Pseudomonas*, but our current knowledge clearly supports the importance of genetic analysis and techniques as a necessary approach in the biological understanding of this and any bacterium. In this chapter, genetic studies on *P. aeruginosa* will be discussed in detail and comparisons drawn with *P. putida* and other species. With the exception of plasmids affecting metabolic activity, the extent of genetic knowledge in these other species is not as great, but genetic work with *Pseudomonas* is a clear model of how the genetics of individual bacteria can be accomplished, and may be a stimulus to other, possibly younger, workers of the values of using genetically unknown bacteria. Two reviews have previously described the genetics of the genus *Pseudomonas* in some detail (Holloway, 1969; Holloway, Krishnapillai & Stanisich, 1971).

2. THE APPROACH TO GENETIC ANALYSIS

There are three aspects of genetic analysis—the identification of the native genetic variation of the chosen organism, the selection of induced genetic variants—mutants—which can be exploited for a variety of experimental purposes, and some system of genetic recombination by which the genome can be mapped and recombination of components of the genome achieved. For completeness, any genetic investigation should include a study of the physical organization of the genetic material within the cell and the mechanisms by which it undergoes rearrangement and recombination. The two species of *Pseudomonas* which have been studied genetically in detail have been subject to all these techniques although their individual application to each species is somewhat uneven.

Most genetic work with *P. aeruginosa* has been done with two strains, originally described by Holloway (1955). These are strain PAO (originally strain 1) isolated from a patient in Australia in 1954 and strain PAT2 (formerly strain 2) which was originally isolated by Don and van der Ende (1950) in South Africa as strain L-III 3 bi. Both are prototrophs and aeruginocinogenic. Strain PAO is aeruginocinogenic for two aeruginocins, one S type and one R type and strain PAT produces an R-type aeruginocin (Ito, Kageyama & Egami, 1970). (See Chapter 4 for more details of these aeruginocins.) Fortuitously, these two strains have shown a wide range of different genetic properties, and their separate investigation has enabled a broader picture of the genetics of *P. aeruginosa* to be drawn. The study of these two strains has been mainly by means of laboratory produced markers. In more recent years, other strains have been used to study naturally occurring variation, particularly with respect to antibiotic resistance.

With *Pseudomonas putida*, Gunsalus and his associates have used a range of wild-type strains, some isolated from particular environments by selection for particular metabolic properties, and reference should be made to the individual original papers (referred to below) for details of these strains.

It is interesting that with both species these native variants which have particular interest involve plasmids, and it is clear that a complete understanding of the genetic organization of *Pseudomonas* must involve the identification of plasmids, their genetic and physical structure, transmission, phenotypic potential and relationships with the genomes of other bacteria. Genetic studies of plasmids are of lesser significance without a background knowledge of the chromosomal organization of the genome.

There are standard procedures by which genetic analysis can be accomplished. Mutants affecting a desired characteristic an be isolated either by any of the above techniques or by others specifically designed for the purpose. Location of the gene on the chromosome can then be achieved by a process involving recombination—in the case of *P. aeruginosa*, by conjugation or transduction. This is done to distinguish it from other mutants of similar phenotype which may be located at other sites on the chromosome, and hence which are different genetically. Secondly, such procedures may enable recombinant strains to be isolated with combinations of mutant genes unlikely to be obtained by mutation and with properties quite unlike any natural strain (Holloway, 1973). Appropriate genetic analysis can establish whether the gene under consideration is chromosomally located or situated on a plasmid.

2.a. Native Genetic Variation

Taxonomic studies of the genus *Pseudomonas* show that, by any of the criteria used to differentiate species, independent isolates of *P. aeruginosa* have a well-defined group of properties with less variation than many of the other species of this genus (see Chapter 1) (Stanier, Palleroni & Doudoroff, 1966; Lányi, 1969). Altered patterns in pigment production are probably the most common variant (Jessen, 1965) but anyone who has ever cultured *P. aeruginosa* is well aware of the range of variants which occur in artificial culture. References to this dissociation, as it is termed, have been made since the early work on this species. Changes occur in many recognizable properties of the organism, including colonial morphology, pigment production, resistance to deleterious agents, phage and aeruginocin susceptibility and antigenic properties (Hadley, 1927; Zierdt & Schmidt, 1964; Shionoya & Homma, 1968; Homma, 1971). The frequency of these changes, and the aspects of the phenotype involved suggest a genetic mechanism other than point mutation. The fact that 'dissociated' variants often have a variety of phenotypic characteristics altered suggests that either there is a deletion of a short region of the chromosome or that a plasmid is being lost. This latter hypothesis is capable of being tested, although if the plasmids are small and nontransmissible, conclusive evidence may be difficult to obtain. If true, it could point to an important role of plasmids in natural habitats and perhaps a need to develop procedures for maintaining these plasmids during artificial culture.

Strains of *P. putida* with the ability to catabolize unusual substrates for energy can be readily isolated from soil or other habitats by enrichment techniques. The range of substrates involved will be discussed in Chapters 7 and 8, but those which have been genetically analysed include camphor (Rheinwald *et al.*, 1973), salicylate (Chakrabarty, 1972), naphthalene (Dunn & Gunsalus, 1973) and octane (Chakrabarty, Chou & Gunsalus, 1973).

2.b. Isolation of Mutants in the Laboratory

The established range of mutants in *Pseudomonas* is eminently suitable for genetic studies. All species grow well on a chemically defined medium and this aids the isolation of mutants lacking the ability to synthesize a particular amino acid, purine, pyrimidine or vitamin (auxotrophic mutants). Using carbenicillin to contraselect growing cells, enrichment techniques can be used such that following mutagenic treatment, 1–10% of candidate colonies are found to be auxotrophs (Lederberg, 1950; Fargie & Holloway, 1965; J. Watson, unpublished observations). An alternative method of mutation selection applicable for auxotrophs and other types of mutants uses repeated counterselection of non-mutant cells in the presence of D-cycloserine and penicillin (Ornston, Ornston & Chou, 1969). A technique for the selective isolation of auxotrophs of *P. aeruginosa* using dihydrostreptomycin has also been described by Ishida, Seto and Osawa, 1966. Considerable efforts have been made by a number of workers to get a thymidine auxotroph in *P. aeruginosa*, without success. However, this has been achieved in *P. acidovorans* (Kelln & Warren, 1973) and in a marine halophilic *Pseudomonas* (Espejo, Canelo & Sinsheimer, 1971).

Because of the metabolic versatility of this genus, mutants unable to use a particular substrate as an energy source provide a large source of potential genetic markers (Brammer, Clarke & Skinner, 1967; Kemp & Hegeman, 1968). Such mutants may involve either chromosomal changes or be plasmid borne (Chakrabarty, 1972). Care must be used to distinguish mutants of this class with those which are deficient in permeability for a particular component, but which do not lack enzymes to utilize the substance. Substrates studied in this way include amides (Clarke, 1970), β-ketoadipate (Kemp & Hegeman, 1968), mandelate (Rosenberg & Hegeman, 1969; 1971; Wheelis & Stanier,

1970), histidinol (Dhawale, Creaser & Loper, 1972) and nicotinic acid (Leidigh & Wheelis, 1973b). Further exploitation of this class of mutant should be particularly rewarding, especially in *P. aeruginosa* (see Chapter 7).

Mutants resistant to a variety of antibiotics and toxic agents have been isolated and used for a variety of purposes. The well-known native resistance of *P. aeruginosa* to a wide range of inhibitory compounds is well illustrated by its indifference to many analogues of amino acids, purines and pyrimidines (Waltho & Holloway, 1966). One exception has turned out to be the mutants resistant to *p*-fluorophenylalanine (FPA) which can be readily isolated (Waltho & Holloway, 1966; Dunn & Holloway, 1971). These mutants, while not overproducing phenylalanine, as is found with many FPA-r mutants in other bacteria, do affect regulatory mechanisms and have been used in the analysis of the genetic basis of Host Controlled Modification (Dunn & Holloway, 1971). Calhoun and Jensen (1972) showed that increased sensitivity to growth inhibition by amino acid analogues could be obtained using fructose instead of glucose as the source of energy. Mutants were isolated which were resistant to inhibition by *β*-2-thienylalanine and *p*-aminophenylalanine and some of these did produce an excess amount of the amino acid appropriate to the analogue. It is possible that this could be a general method for obtaining regulatory mutants of biosynthetic pathways in *P. aeruginosa* (see Chapter 7, section 4).

The spontaneous frequency of mutants such as are described above is usually too low for practical purposes and mutagens are needed to increase the frequency of mutation. Ethyl methane sulphonate (EMS) (Fargie & Holloway, 1965), manganous chloride (Holloway, 1955), N-methyl-N-nitro-N-nitrosoguanidine (NG) (Holloway, 1966) have all been successfully used. One practical difficulty with NG is its ability to induce aeruginocins, particularly at low concentrations (20 μg/ml), many strains of *P. aeruginosa* being aeruginocinogenic. This results in the death of the majority of cells and leaves a selected fraction of cells which cannot be induced for aeruginocins. For reasons which are not yet clear, this fraction apparently includes many membrane mutants (Holloway, unpublished observations). Although this selection of membrane mutants does not occur at higher concentrations (100 μg/ml) of NG, experience over some years from the author's laboratory would select EMS as the mutagen of choice.

2.c. Mutants Affecting Structural Components

While there is not yet a complete understanding of the biosynthesis of small molecular weight compounds and macromolecules, nevertheless much is understood of the mechanism and control of these processes, particularly in bacteria. The same cannot be said for morphogenesis and the assembly of cellular and sub-cellular structures. A genetic approach to the solution of this problem would involve the development of techniques for the isolation of mutants which have altered structural components or lack the ability to assemble structural components, an approach which has been eminently successful in the analysis of morphogenesis in bacteriophage (Levine, 1969).

Work on this aspect has started in *P. aeruginosa*. One structure of the bacterial cell which attracts considerable attention is the cell membrane, with its relationship to the DNA of the cell, its oxidative-phosphorylation system, permeability functions and its structural relationship to the cell wall. Potentially, a variety of techniques can be used to isolate membrane mutants (Holloway, 1971). For example, mutants which can survive the lethal action of aeruginocins, have been shown to have an altered structure of the membrane, as revealed by polyacrylamide gel electrophoresis of the solubilized membrane. Such mutants show pleiotropy for such characteristics as stability of plasmids, levels of membrane-bound enzymes, permeability, lytic and lysogenic responses to bacteriophages and flagella formation (Holloway, Rossiter, Burgess & Dodge (1974), and are essentially similar to mutants of *Escherichia coli* tolerant to the action of colicins (Reeves, 1972).

Pili (fimbria) have been observed on a number of *Pseudomonas* species and studied particularly in *P. aeruginosa* (Fuerst & Hayward, 1969; Weiss, 1971; Weiss & Raj, 1972; Bradley, 1972). There are apparently different types of pili which can be distinguished by the adsorption or non-adsorption of different phages (Bradley, 1972; 1973). There is no correlation between the occurrence of these pili on the surface of *P. aeruginosa* and the possession by the cell of any known plasmid including sex factors, and the role of pili in conjugation has not been established. The FP pili referred to by Weppelman and Brinton (1970; 1971) are now known not to be coded for by FP sex factors. It is possible to obtain mutants of *P. aeruginosa* which have either mutant pili or lack pili by the selection of clones resistant to various bacteriophages such as PP7 or Pf (Bradley, 1972). Further work may enable the correlation of the properties of such mutants with those of mutants which have structural alterations to the membrane and this could provide an approach for the structural relationships and biogenesis of pili.

Mutants presumably affecting cell wall structure can be readily isolated in *P. aeruginosa* by screening for variations in colonial morphology (V. Krishnapillai, unpublished observation). Van de Putte (unpublished observations) was able readily to isolate non-motile variants of *P. aeruginosa* following mutagen treatment, although it was not established whether flagella were present or not. Temperature-sensitive mutants of *P. aeruginosa*, with a permissive temperature of 37 °C and a restrictive temperature of 43 °C have been isolated (Holloway, unpublished observations) and as with *E. coli*, some of these show alterations in cell morphology even at the permissive temperature.

An understanding of cell wall and membrane structure may help in the understanding of native genetic variation in particular environments. Tseng, Bryan and van der Elzen (1972) have studied the various types of streptomycin resistance found amongst 200 isolates of *P. aeruginosa* from clinical sources. Only about 10 % of strains had resistance attributable to R factors and these strains had high levels of resistance. Other strains had low level resistance, associated with a reduced permeability to streptomycin. Perhaps variation in permeability may be an important but as yet underestimated selective characteristic for this species and, although interesting studies have been made on permeability mutants (Kay & Gronlund, 1969), they have not yet been related to structural alteration in the cell or its occurrence in nature. The variety of available approaches makes the genetic study of morphogenesis in *Pseudomonas* a fertile and attractive venture to undertake.

3. STRUCTURE OF THE BACTERIAL GENOME

Information from a variety of sources indicates that the bacterial genome consists of a chromosome, attached to the bacterial membrane, generally $2–3 \times 10^9$ daltons in size and one or more plasmids, each with a molecular weight of $10–70 \times 10^6$ daltons, also presumably attached to the membrane. The chromosome is a stable structure, but the plasmids have varying degrees of stability. Presumably there must be some control to ensure closely synchronized replication such that the ratio of plasmid numbers to chromosome number remains reasonably constant, in most cases one or two. A variety of physical techniques are available for identification and recognition of the plasmid component. Identification of the plasmid can be based on its ability to be transmitted, its ability to be lost either spontaneously or following treatment with 'curing' agents without causing death of the cell, its contribution to the bacterial phenotype and if possible its lack of genetic linkage to the chromosome (Novick, 1969).

Estimates of the size of the genome in *Pseudomonas* species have been made by Bak, Christiansen and Stenderup (1970) using the technique of renaturation of single-stranded DNA, a method which does not distinguish between chromosomal and plasmid DNA, but estimates the total DNA content of the cell. The four pseudomonads examined appeared to have large genome sizes compared with other bacteria (*P. aeruginosa* 7×10^9 daltons, *P. fluorescens* 5×10^9 daltons, *P. stutzeri* 4×10^9 daltons, *P. oleovorans* 4×10^9 daltons), although this technique gave a value for the molecular weight of the *Escherichia coli* chromosome close to that found with a variety of techniques ($2·3 \times 10^9$ daltons). Recently, Dr. J. Pemberton (personal communication) using electron microscopic examination and velocity sedimentation of the DNA released from spheroplasts, has estimated that the majority of the genome of *P. aeruginosa* is contained in a single duplex DNA molecule with a molecular weight of $2.4 \pm 0.3 \times 10^9$ daltons. The discrepancy between this result and the earlier, larger value recorded by Bak, Christiansen and Stenderup needs reconciliation.

Pemberton and Clark (1973) have examined *P. aeruginosa* strain PAO for physical evidence of plasmids. Strain PAO acts as a female with all known *Pseudomonas* sex factors and shows no donor ability. However, it was shown to harbour a number of cryptic plasmids occurring as covalent circular molecules and amounting to about 3 % of the total DNA of the cell. They were each smaller than the minimal size suggested by Clowes (1972) as necessary for a plasmid to have sex factor activity.

Physical examinations have also been made of several R factors of *P. aeruginosa* and these will be referred to in detail later in this chapter. The R factors have the advantage of being readily transmissible into a variety of genetic backgrounds, clearly distinguishable by physical techniques from the normal genome of the cell and with a definite contribution to the bacterial phenotype. These R factors appear to be excellent material for combined genetic and physical studies.

As yet there is no evidence as to the spatial distribution of the DNA material within the cell of *P. aeruginosa*. By analogy to *E. coli* (Ryter, 1968) it is reasonable to assume that the chromosomal DNA is attached to the bacterial membrane. Likewise there is no direct evidence which proves that plasmids in any bacteria are attached to the membrane. One approach to this problem is the isolation of membrane mutants and an examination of their ability to acquire and maintain plasmids. Such membrane mutants of *P. aeruginosa* have been isolated and characterized in part (Holloway, Rossiter, Burgess & Dodge, 1974).

4. RECOMBINATION ANALYSIS

In *P. aeruginosa* there are processes of reassortment of genetic material which can be used profitably in the genetic investigation of this organism. These are conjugation (promoted by a variety of plasmids), transduction and transformation.

4.a. Transformation

Kahn and Sen (1967) described genetic recombination by transformation in various species of *Pseudomonas* including intra- and interspecific combinations of *P. aeruginosa*. The general features of the transformation system did not differ significantly from those described in other bacteria such as *Bacillus subtilis* or *Haemophilus influenzae*. The transformation frequencies observed were high, up to 5 % of the treated cells in some cases, and a range of markers was transformable.

In view of the ease with which these transformations were achieved, it is a pity that this work has not been followed up either by the original authors or by other workers. An interspecific recombination technique of this type would be of great value for a variety of biochemical, taxonomic and genetic purposes. An earlier report of transformation in *P. fluorescens* (Lambina & Mikhilova, 1964) likewise remains to be consolidated.

4.b. Transduction

The frequency of occurrence of lysogeny in *P. aeruginosa* has been referred to earlier (Chapter 4) and the frequency of occurrence of transducing phages is also high. Those characterized include B3 and F116 (Holloway, Egan & Monk, 1960), G101 (Holloway & van de Putte, 1968) and B (Loutit, 1960). All these phages display general transduction and no example of prophage-linked specialized transduction has been reported. The molecular weights of DNA of phage F116 and G101 are such that they are capable of transducing 1–2 % of the bacterial chromosome, and hence are particularly useful for linkage studies by cotransduction. A variant of F116, called F116L, has been isolated (Krishnapillai, 1971) which shows higher cotransduction frequency of linked markers than F116 but the molecular basis of this effect is unknown.

Recently Moillo (1973) has isolated a temperate phage, M6 from a lysogenic strain of *P. maltophilia*. An extended host range mutant not only plates on *P. aeruginosa*, but can carry out general transduction in that species. M6 has been found to have immunity and physical similarities to the *P. aeruginosa* transducing phage G101 (Holloway & van de Putte, 1968). This phage should be of particular use for linkage comparisons of the genomes of these two species.

In general the properties of these transducing phages of *P. aeruginosa* and the techniques used to obtain and analyse transductants are the same as in *S. typhimurium* or *E. coli* (Hayes, 1968). Their chief value for recombination analysis lies in their distinguishing genetic differences between markers with the same phenotype, in demonstrating the close linkage of markers and in determining the order on the chromosome of closely linked markers.

4.c. Phage Conversion

We have already referred to the ability of some phages to contribute to the bacterial phenotype (Chapter 4). Phage or lysogenic conversion is best viewed as similar to the contributions to the bacterial phenotype made by R factors and other plasmids and it reemphasises the importance of the extra-chromosomal component to the bacterial genome. In view of the frequent occurrence of lysogeny and polylysogeny in *P. aeruginosa*, phage conversion could be an important and underestimated aspect of the bacterial genome.

4.d. Conjugation

This is a process of direct contact of bacterial cells with transfer of genetic material from a donor parent to a recipient parent. It is mediated by plasmids which determine the occurrence of cellular conjugation, the mobilization of DNA in the donor, plasmid-containing cell and its transfer to the recipient cell. For mapping, conjugation is undoubtedly the most useful of the various gene transfer processes in that a large fraction of the chromosome may be transferred in a single conjugal act, and recombination demonstrated between genes situated some distance apart on the chromosome, and not as with transduction and transformation, only between closely linked genes.

A number of plasmids have been shown to promote conjugation in *P. aeruginosa*. In this context we are primarily concerned with those plasmids which promote transfer of chromosome. Some plasmids, for example, certain R factors, enable conjugation to take place, and can promote the transfer of their own genetic determinants, but cannot promote the transfer of chromosomal DNA. Such plasmids and their role in the genetic organization of *Pseudomonas* will be discussed later. The techniques by which conjugation may be detected in *P. aeruginosa*, the recovery and analysis of recombinants and the derivation of linkage relationships are essentially the same as those used in *E. coli* K12 (for details see Hayes, 1968; Clowes & Hayes, 1968), and they are discussed in Chapter 6.

As previously indicated, we can distinguish two aspects of the bacterial genome—a chromosomal component and a plasmid component. Analysis of the former is achieved by recombination analysis using mapping techniques which in essence are the same for bacteria as for Eukaryotes. Two strains with complementary marker composition are crossed and bacterial recombinants selected by the use of appropriate markers. The distribution of non-selected markers in these recombinants is examined by means of the relative proportion of parental and non-parental classes and hence the relative order of genes on the chromosome can be deduced. Techniques singularly applicable to bacteria have been developed, such as interrupted mating, in which the sequence of genes relative to the proximal end of the chromosome is measured by means of the time taken from the beginning of conjugation for each marker to appear amongst recombinant progeny.

Most mapping of *P. aeruginosa* has been carried out using the sex factor FP2 (Loutit, Pearce & Marinus, 1968; Loutit & Marinus, 1968; Loutit, 1969; Stanisich & Holloway, 1969; Pemberton & Holloway, 1972). The author and his associates do not have any evidence to suggest that any markers mapped by them do not fall on to one linkage group and the current map from their data, plus that of several other workers is shown in Figure 5.1. This map has been obtained using both conjugation and transduction techniques. Two markers (*his*-2 and *his*-3) previously thought by Stanisich and Holloway

(1969) to be on a separate linkage group have now been shown to be located on this map. Physical evidence to support this view has recently been obtained by Pemberton (personal communication) who has shown that there appears to be only one major piece of DNA in the genome of *P. aeruginosa.*

By contrast, Loutit and his associates have presented data indicating that there are two linkage groups, the main evidence for this view being the lack of recovery of some markers when selection is made for other markers with about the same time of entry (Figure 5.2). At the time of writing there are some indications that it may be possible to reconcile these findings. Different procedures have been used in each laboratory to obtain interrupted mating data and what is needed to resolve the areas of difference is a marker by marker comparison of the two maps (Figures 5.1 and 5.2), by both trans-duction and conjugation. Where such comparison has been made, there is considerable agreement in relative gene order between the maps published from the two laboratories (Loutit, 1969; Pemberton and Holloway, 1972). Booker and Loutit (1974) have recently studied patterns of chromosome replication in *P. aeruginosa* and obtained evidence which supports the notion that the two linkage groups may belong to a single chromosome. This con-clusion is supported by current physico-chemical studies on the DNA of the genome of strain PAO (Pemberton and Clark, 1973; Pemberton, personal communication). The genetic evidence for more than one linkage group may reflect the particular characteristics of the sex factors used rather than differences between mutants and the techniques of interrupted mating.

The earlier evidence from combined interrupted mating and linkage analysis suggested that FP2 promoted chromosome entry from only one site of origin as marked in Figure 5.1. Recently, Holloway (unpublished observa-tions) has obtained evidence to suggest that FP2 may promote transfer at a lower frequency from a second site. Limited mapping with another sex factor FP39 (Pemberton & Holloway, 1973) is consistent with the view that this sex factor promotes entry of chromosome from a site about ten minutes proximal to the FP2 origin (see Chapter 6) and markers to the right of that site have the same order when crosses with FP39 are made, but have a time of entry about ten minutes than that found with FP2. There is an alternative hypo-thesis. Perhaps FP39 promotes chromosome transfer from the same site of origin as FP2 but there is a delay of ten minutes in the initiation of chromo-some transfer. The data from interrupted mating experiments are consistent with both interpretations and the failure to find any markers between the site or origin of FP2 and the separate site postulated for the origin of FP39 (see Figure 5.1) currently prevents a decision as to precise origin of chromosome promoted by FP39.

No genetic determination of chromosomal circularity has yet been made in *P. aeruginosa.* What is needed is for one sex factor to promote chromosome entry from a range of sites so that what are early and late markers for one site

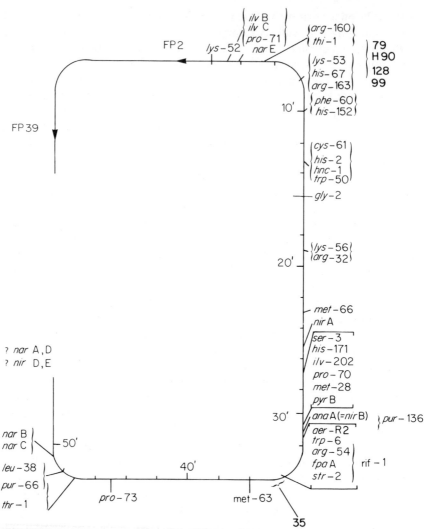

Figure 5.1. Linkage map of *P. aeruginosa* strain PAO derived from data of Holloway, Krishnapillai and Stanisich (1971), van Hartingsveldt and Stouthamer (1973), Krishnapillai and Carey (1972), J. Dodge (personal communication) and K. Carey (personal communication). Marker position is based on times of entry in interrupted matings ±2 min) using FP2 and FP39 donor strains, combined with linkage and transduction data. Origin and direction of chromosome transfer of FP2⁺ and FP39⁺ donor strains is indicated. Allele numbers are provided to aid identification of particular loci. Cotransducible loci are given in brackets; square brackets indicate known order, curved brackets indicate that the precise order remains to be determined. Gene symbols:

aer-R2—aeruginocinogenic determinant (originally designated *pyr*R2 Kageyama (1970))

ana—affected in anaerobic growth

arg—arginine biosynthesis

cys—cysteine biosynthesis

*fpa*A—resistance to *p*-fluorophenyl-alanine

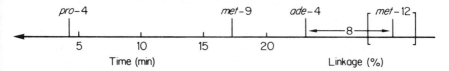

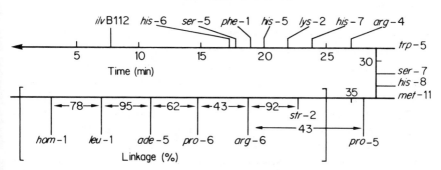

Figure 5.2. Linkage map derived from data of Loutit (1969). Pemberton (1972) using transduction with phage F116 has shown that the following pairs of markers are mutants in the same or closely linked genes and can be assumed to be at equivalent locations on this map and that shown in Figure 5.1.

Markers shown in Figure 5.2	Equivalent marker in Figure 5.1
*ilv*B112	*ilv*-261
lys-2	*lys*-56
arg-6	*arg*-54
met-11	*met*-28
ade-5	*pur*-66

leu-1 of Figure 5.2 is the same allele as *leu*-38 of Figure 5.1. Gene symbols: As for Figure 5.1, with the following additions:

ade—adenine biosynthesis
hom—homoserine biosynthesis

Figure 5.1. *Continued*

gly—glycine biosynthesis
his—histidine biosynthesis
hnc—control of histidine synthesis
ilv—isoleucine and valine biosynthesis
leu—leucine biosynthesis
lys—lysine biosynthesis
met—methionine biosynthesis
nar—dissimilatory nitrate reductase
   activity
nir—dissimilatory nitrite reductase
   activity

phe—phenylalanine biosynthesis
pro—proline biosynthesis
pyr—pyrimidine biosynthesis
rif—rifampicin resistance
ser—serine biosynthesis
str—resistance to streptomycin
thi—thiamin biosynthesis
thr—threonine biosynthesis
trp—tryptophan biosynthesis
pur—purine biosynthesis
35, 79, H90, 99, 128—prophage location

of origin will become contiguous markers for another site of entry. FP2 and FP39 seem to be unsuited for this type of experiment. A search was made for sex factors with other sites of entry without success (Pemberton & Holloway, 1973) but preliminary results with various RP factors (Stanisich & Holloway, 1971) indicate that RP factors do promote conjugation and chromosome transfer. Continued use of this approach and the use of a variety of plasmids will undoubtedly aid the establishment of genetic circularity of the *P. aeruginosa* chromosome.

A remarkable feature of recent genetic research of *P. aeruginosa* has been the increase in the numbers of identified plasmids which has occurred since 1969. This is true for both *P. aeruginosa* and *P. putida*, and it is difficult to avoid the conclusion that plasmids contribute much to the biological success of this genus in a variety of environments. Any classification of plasmids is liable to be ephemeral as more of their properties are identified and understood. At present it is convenient to group them according to their effect on the bacterial phenotype. Undoubtedly, molecular studies, combined with genetic analysis, will ultimately provide a more reliable classification.

The components of the bacterial genome which are determined by plasmids are difficult to determine (Novick, 1969). An appreciation that some phenotypic characteristic is plasmid determined may involve showing that the phenotype can be transferred at conjugation but that transfer of chromosomal genes is not responsible, that the genetic determinant can be 'cured' by such agents as acridines, surface-active agents or mitomycin C, or may even show spontaneous loss at detectable frequencies (*ca.* 1 %). The basic criterion of a plasmid is the demonstration that it lacks genetic linkage to the chromosome—but this may not always be possible, particularly in those species which lack well-mapped chromosomes. A plasmid may be demonstrable in cells by physical techniques such as ultracentrifugation. All or any of these techniques may not be applicable to a given plasmid component and it may require the art of the microbial geneticist to rule out, by Occam's Razor, the various alternative hypotheses to inheritance by plasmid.

5. PLASMIDS IDENTIFIED IN *PSEUDOMONAS AERUGINOSA*

5.a. Sex Factors

Historically these were the first type of plasmid to be identified in this species with the description of FP2 by Holloway and Jennings (1958). The term sex factor refers to the ability of these plasmids to promote the transfer of bacterial chromosome in conjugation. Sex factors can be isolated with relative ease from wild-type strains of *P. aeruginosa* (Pemberton & Holloway,

1973). Further studies have revealed phenotypic characteristics other than the promotion of chromosome transfer. For example, FP2, can confer resistance to mercuric salts on bacterial strains (Loutit, 1971), and other sex factors with this characteristic have been found (Pemberton, 1971; Pemberton & Holloway, 1973). FP39 can confer a Leu$^+$ phenotype (ability to synthesise leucine) on strains of *P. aeruginosa* which are mutant for the *leu*-38 allele situated at about the 48 minute mark on the *P. aeruginosa* chromosome map (see Figure 5.1) (Pemberton, 1971; Pemberton & Holloway, 1973). Possession of various sex factors by *P. aeruginosa* strain PAO can affect the ability to support multiplication of various bacteriophage strains, and a preliminary attempt to group these sex factors by exclusion and incompatibility tests showed a minimum of five groups (P. Dry & V. Stanisich, unpublished observations). Undoubtedly other such associated phenotypic properties remain to be discovered amongst *P. aeruginosa* sex factors.

A new sex factor FP5 has been described by Matsumoto and Tazaki (1973) which promotes chromosome transfer at frequencies comparable to those found with FP2. As FP2$^+$ × FP5$^+$ crosses are fertile, and FP2$^+$ × FP2$^+$ crosses are of very low fertility, it is likely that FP2 and FP5 are not closely related. A mercury-resistant determinant is associated with FP5, but whether it is part of the sex factor genome, as with FP2, or an associated plasmid, as occurs in strain PAT, is not clear. Unlike FP2 or FP39, FP5$^+$ strains can be cured of the sex factor with acriflavine and these females retain the mercury resistance phenotype. The R factor R2-72, but not R39-72, interferes with the donor ability of FP5$^+$ strains.

The only molecular study of FP factors is that of Pemberton and Clark (1973). They have shown that the sex factor FP2 has a density of 1·717 g/cc (equivalent to a G + C content of 58 %) and a molecular weight of 59 × 10^6 daltons. FP39 has a density of 1·719 g/cc (equivalent to a G + C content of 60 %) and a molecular weight of 55 × 10^6 daltons. Under the cultural conditions used for these studies, both sex factors occur in the cell as covalent circular molecules, and there are 2–3 autonomous copies of FP per chromosome, supporting the results of Stanisich and Holloway (1972).

5.b. RP Plasmids

The native high resistance of *P. aeruginosa* to a wide variety of antibiotics has been a source of interest to microbiologists and concern to clinicians. With the demonstration that drug resistance in the Enterobacteriaceae was under both chromosomal and plasmid control, the suggestion was made that the origin of R factors was pseudomonads (Watanabe, 1963). There is still no evidence to support this view, and indeed until 1969 there was little substantial evidence (Lebek, 1963; Smith & Armour, 1966) that R factors could even be transferred to _P. aeruginosa_ from enterobacteria or that any natural strains of _P. aeruginosa_ carried R factors which were transmissible.

In 1969 Lowbury and his collaborators (Lowbury _et al_, 1969; Roe, Jones & Lowbury, 1971) described the isolation of strains of _P. aeruginosa_ highly resistant to carbenicillin from patients in a Burns Unit in Birmingham, England. The nature of their appearance suggested that an R factor was responsible and this was subsequently confirmed by the work of Sykes and Richmond (1970) and Fullbrook, Elson and Slocombe (1970).

The R factors isolated by Lowbury's group have now been investigated in a variety of laboratories. Further independent isolations of R plasmids have been made in various parts of the world, and it is obvious that _Pseudomonas_ R factors are now an important aspect of the epidemiological genetics of this bacterium. The various R factors isolated to date are listed in Table 5.1.

A separate group of R factors has been identified from hospital isolates of _P. aeruginosa_ from American sources (Bryan _et al._, 1973). The transferred markers include streptomycin, tetracycline and sulphonamide resistance but not carbenicillin or ampicillin. Some such as R931 could not be transferred to _E. coli_ but showed ready transfer to _P. aeruginosa_.

Clearly the variety of R factors found in _P. aeruginosa_ now rivals that known to occur in the enterobacteria. They not only differ in antibiotic resistance pattern (Table 5.1) but in physical and biological characteristics. In biological properties some can promote the transfer of bacterial chromosome, some are repressible, can confer mercury resistance, can suppress phage multiplication in _P. aeruginosa_, can limit aeruginocin production and can confer tolerance to aeruginocins (Stanisich & Holloway, 1971; Bryan _et al._, 1973; P. Chandler & V. Krishnapillai, unpublished observations; G. Jacoby, personal communication). It is clear that the current environment of mixed bacterial infections of humans, chemotherapeutic patterns and a spectrum of available plasmids is selective for plasmids capable of replication in _P. aeruginosa_. It is likely that there is an evolution of such plasmids by interchange of genes with the chromosomal and plasmid component of the genome of various bacteria and that we can expect other new plasmids to occur.

Examination by analytical caesium chloride gradient centrifugation, and ethidium bromide–caesium chloride gradient centrifugation to detect covalently closed circular (CCC) DNA, has enabled knowledge of the molecular properties of several _Pseudomonas_ R factors to be obtained. Grinsted _et al._ (1972) and Saunders and Grinsted (1972), have studied two R factors, RP1 and RP8 (originally designated RP4, see Holloway & Richmond, 1973) derived from _P. aeruginosa_ strains isolated in the United Kingdom at Birmingham (Lowbury _et al._, 1969) and Glasgow (Black & Girdwood, 1969). Both these R factors confer resistance to carbenicillin, neomycin/kanamycin,

Table 5.1. Plasmids conferring drug resistance in *P. aeruginosa*

Location of Isolation	Plasmid Designation	Alternative Designation	Drug resistance Pattern	Reference
Birmingham	RP1	R1822	C, K/N, T	1–5
Birmingham	RP1-1	R18-1	C	5, 6
Birmingham	R9169	R91	C, K/N, T	1, 4, 5
Birmingham	R6886	R68	C, K/N, T	1, 4, 5
Glasgow	RP8		C, K/N, T	7
Japan	R2-72		C, S, K	8
Japan	R38-72		T, S	8
Japan	R39-72		T, S	8
Canada	R931		S, T	9, 10
Canada	R679		S, su	9, 10
Canada	R1162		S, su	9, 10
Canada	R3108		S, T, su	10
Canada	R209		S, su, G	10
Canada	R130		S, su, G	10
Canada	R716		S	10
Canada	R503		S	10
Canada	R5265		S, su	10
Paris	R64		A, C, su, G, K	11–13
Paris	R40a		A, K/P, su	13
Boston	Unnamed		G, S, su	14

Abbreviations:
C, carbenicillin; K/N, kanamycin and neomycin; T, tetracycline; S, streptomycin; A, ampicillin; K/P, kanamycin and paromomycin; su, sulphonamide; G, gentamicin; GK, gentamicin and kanamycin.

References
1. Lowbury *et al.*, 1969;   2. Sykes & Richmond, 1970;
3. Fullbrook *et al.*, 1970;   4. Stanisich & Holloway, 1971;
5. Holloway & Richmond, 1973;   6. Ingram *et al.*, 1972;
7. Black & Girdwood, 1969;   8. Kawakami *et al.*, 1972;
9. Bryan *et al.*, 1972;   10. Bryan *et al.*, 1973;
11. Witchitz & Chabbert, 1971;   12. Witchitz & Chabbert, 1972;
13. Chabbert *et al.*, 1972;   14. G. Jacoby, personal communication;
15. Olsen & Shipley, 1973.

and tetracyline, and can be detected as a satellite of CCC DNA when present in strains of *P. aeruginosa*, *P. mirabilis* and *E. coli*. The DNA's of RP1 and RP8 have a molecular weight of 40×10^6 and 62×10^6 daltons respectively, and a buoyant density of $1 \cdot 719$ g/cm^3 corresponding to a G + C content of 60 %. The CCC DNA of RP4 comprises approximately 2–3 % of the total cell DNA of these three bacterial species, and although similar values are obtained for RP1. DNA in both *P. aeruginosa* and *E. coli*, this is increased to about 12 % in the case of the *P. mirabilis* host. Assuming that the chromosomal DNA of these three bacterial species is of similar molecular weight, then it appears that in *P. mirabilis*, replication of RP1 is under relaxed control, with more CCC equivalents per chromosome than is found in either *E. coli* or *P. aeruginosa*.

However not all transmissible resistance determinants have been found associated with satellite DNA (Ingram *et al.*, 1972). A cell line displaying resistance only to carbenicillin, was obtained from the wild-type *P. aeruginosa* strain carrying RP1. This resistance phenotype was freely transferable among strains of *P. aeruginosa* but not to *P. mirabilis*, and only at very low frequency to *E. coli*. The element responsible for this transfer was designated RP1-1. Strains of *P. aeruginosa* carrying the transferable RP1-1, and *E. coli* lines harbouring a non-transferable derivative, RP1-1*, fail to show a satellite band of CCC DNA. In the latter instance, this is consistent with data suggesting that insertion, of at least that part of RP1-1 responsible for drug resistance, into the *E. coli* chromosome has occurred (Richmond & Sykes, 1972). Absence of satellite DNA in *P. aeruginosa* suggests that at least in the majority of cells, this element does not occur as CCC DNA, and at present the state of the element as an autonomous or integrated structure has yet to be clearly established. It is possible that RP1-1 is a derivative of RP1 which has lost, or fails to express, the resistance determinants for neomycin/kanamycin and tetracycline, or alternatively that RP1-1 and RP1 represent distinct plasmids which were coexisting in the original *P. aeruginosa* wild-type strain 1822 isolated in Birmingham.

Bryan and his colleagues (1973) have examined the physical characteristics of R931 in strain 931. This was shown to have CCC DNA, there being two major molecular classes with molecular weight estimated at 25×10^6 daltons and 1×10^6 daltons but the precise relationship of these two classes to the drug resistance phenotype is not yet clear. The content of R factor DNA relative to chromosomal DNA varied from $18 \pm 2 \%$ for R931 in logarithmic phase growth to $8 \pm 3 \%$ for R3108 in stationary phase.

5.c. Compatibility

The identification and differentiation of plasmids is a difficult and as yet unsolved problem in genetic taxonomy. Current opinion puts replicative ability as an important taxonomic feature as measured by the ability of two plasmids to exist and replicate in the same bacterial cell. Such plasmids are said to be compatible.

Different sex factors of *P. aeruginosa* have been shown to show some relationship. FP2 and FP39, two sex factors showing quite distinct properties (see above) are possibly related because *P. aeruginosa* PAO strains carrying both FP2 and FP39 are unstable for both sex factors, suggesting some degree of incompatibility. Furthermore, FP2 resident in a PAO strain can exclude

FP39 while strains containing FP39 can accept FP2 with the same frequency as FP⁻ strains.

Datta and her colleagues (1971) tested RP1 against four compatibility groups; F, an F-like R factor, and I-like R factor and an fi⁺ *col* plasmid compatible with an fi⁺ R factor. No similarity with any of these groups was found and thus RP1 constitutes a new compatibility group of R factors, called P, distinct from the I, N and W groups already identified (Datta & Hedges, 1971; Hedges & Datta, 1971). Further classification of R factors has been carried out by Chabbert *et al.* (1972) and two R factors, R64 and R40a isolated from *P. aeruginosa* in Paris were found to have different compatibilities from the P group and from each other. R931, R3108, R209 and R130 belong to a different compatibility group from RP1 and this group has been termed the P-1 compatibility group by Bryan and colleagues (1973). The compatibility relationships between all the various R factors listed in Table 5.1 have not been determined, but it is almost certain that a number of compatibility groups will be found.

Other evidence suggests that some RP plasmids represent a biologically distinct group from the F-like or I-like plasmids. RP1 and RP4 have been shown to have a greater promiscuity than many other enterobacterial RTF plasmids in terms of their ability to be transferred to such a wide range of other species and genera, suggesting that these other organisms do not carry plasmids of the same compatibility type as RP4. The extent of this promiscuity is described in detail in Chapter 6, but the phenomenon is of particular interest to the genetic organization of *Pseudomonas* because it represents the clearest example of genetic interchange between unrelated bacteria and hence could be an important source of genetic variation for this genus. The fact that some RP factors can promote transfer of chromosome at relatively high efficiency in *P. aeruginosa* (Stanisich & Holloway, 1971), raises the possibility that this class of plasmid may promote genetic exchange of chromosomal material across generic boundaries to result in a wide range of recombinant types. Not only does this have a profound significance as a technique in comparative bacterial genetics, but the acquisition of drug resistance by a wide range of bacteria may have wide-reaching effects for chemotherapy. A number of aspects of this promiscuity need further investigation. Does the plasmid code for the cellular structures which enable cell to cell conjugation to occur? Are the RP plasmids necessarily immune to restriction enzymes? In what form does the plasmid persist in the cell it infects? Can all the recipients transfer the plasmid promiscuously as do the donor *Pseudomonas* strains? Answers to all these questions remain to be determined.

However not all *Pseudomonas* R factors are promiscuous. R931 and other members of the P-2 compatibility group cannot be transferred to *E. coli* recipients. This is uncommon among R factors found in enteric bacteria where intergenic transfer is common (Datta & Hedges, 1972). It will be of interest to determine the reasons for such differences in these *Pseudomonas* plasmids.

5.d. Origin of R Factors

The ability of the RP plasmids to become resident in such a wide range of bacterial genera makes their origin of considerable interest. These RP plasmids were first detected in humans in mixed infections of burns in association with enterobacteria. Lowbury, Babb and Roe (1972) consider that these plasmids originate in enterobacteria and that *P. aeruginosa* is a secondary host. Olsen and Shipley (1973) suggest that the 60 % (G + C) content of the DNA of these plasmids is more characteristic of the terrestrial and aquatic saprophytic pseudomonads and hence that this group of bacteria are the natural hosts of these plasmids.

Whatever the origin of RP plasmids, perhaps the properties of this type of plasmid that give it such biological success are first, promiscuity, namely the ability to transfer from genus to genus and secondly, the ability to undergo recombination with either resident plasmids or the host chromosome to provide additional genetic variability and biological flexibility. The latter property has been demonstrated for one R factor, R30, by Stanisich (1972) in *P. aeruginosa* PAT. This strain carries a plasmid conferring the mercury resistance phenotype which is not self-transmissible. When R30 is passaged through this strain, a range of bacterial recombinants are obtained which are Hg^r and which possess varying combinations of carbenicillin resistance, kanamycin/neomycin resistance and tetracycline resistance. Most have lost the ability to transfer themselves in conjugation. Transduction data shows that there has been a recombination process involving a mercury resistance determinant and the genome of the R factor. This demonstrated ability of the R factor to recombine with other parts of the host genome is of profound importance for the amount of genetic variability available to the bacterium and more information of this type combined with hybridization experiments may be of value for determining the origin of the RP plasmids.

5.e. Regulation

Plasmids may not express their phenotype properties in all cells of a host strain (Meynell, Meynell & Datta, 1968). The RP plasmids have not yet been well studied in this respect. Stanisich & Holloway (1971) showed that R91 showed evidence of repression in terms of its ability to mobilize chromosome in *P. aeruginosa*, in contrast to R68 which appeared to be derepressed. Olsen and Shipley (1973) have shown that as strains of various genera carrying R1822 are equally susceptible to the RNA phage PRR1, giving clear plaques, it is likely that R1822 is derepressed in all these different hosts. R931 seems to be derepressed in *P. aeruginosa* strains 931 and 1310 but in *P. aeruginosa* strain 280 it acts as if repressed. Clearly, this aspect of the *Pseudomonas* **R** factors must be investigated in more detail.

The results are in contrast to FP plasmids which appear to be much more repressed than F in *E. coli* (Holloway *et al.*, 1971). The indices of repression available for fi$^+$ and fi$^-$ R factors are not available for RP plasmids and it is difficult at present to assess the extent or role of repression in the biology of these plasmids.

6. GENETIC ORGANIZATION OF OTHER SPECIES OF *PSEUDOMONAS*

Despite the extensive biochemical knowledge which has been obtained with *P. putida* and *P. fluorescens*, genetic analysis of these species has not proceeded to the extent found in *P. aeruginosa*. Undoubtedly some of the reasons for this are the scarcity of lysogenic strains in these species and until recently the absence of native sex factors. The pioneering work of Gunsalus and Chakrabarty (Chakrabarty & Gunsalus, 1969a,b) has overcome these deficiencies in *P. putida* but the genetic analysis of *P. fluorescens* has been neglected.

Gunsalus and his associates first established a transducing system for *P. putida* using the non-lysogenizing phage pf16 which had been isolated from sewage. Transductants can be isolated in sufficient number by preventing lysis of the recipient cells through the use of antiphage serum and by u.v. irradiating of the phage before use. This phage has been useful for the analysis of the arrangement of both chromosomal and plasmid-carried genes in *P. putida* (see below).

Phage pf16 has additional interesting properties. Some recombinant clones derived from transduction with this phage were found to be immune to lysis by pf16. These were found to harbour a defective phage, pfdm which, in a manner similar to that found with λdg or Pldl in *E. coli*, can be induced to produce high-frequency transducing lysates (Chakrabarty & Gunsalus, 1969a,b). Furthermore, pfdm can persist in a plasmid state and can act to promote conjugation and chromosome transfer with cells not possessing pfdm. As yet the operational efficiency of this system in terms of recovery of recombinants is too low for an effective conjugational linkage analysis, but it indicates that conjugation can occur in this species despite the lack of identification of suitable sex factors of the F or FP type and that suitable plasmids are all that are needed (Chakrabarty & Gunsalus, 1969) (see also Chapter 6).

Gunsalus, Chakrabarty and their associates have gone on to characterize such plasmids, not only finding a range of extrachromosomal genetic elements which will promote conjugation in this species, but also providing new genetic evidence to explain the extensive biochemical versatility of this species.

It is relatively easy to find strains of *P. putida* which have acquired the ability to utilize unusual substances for energy, for example, strains have been isolated by enrichment from soil which can use D-camphor as a sole carbon source (Bradshaw *et al.,* 1959). Rheinwald, Chakrabarty and Gunsalus (1973) have shown that the genetic information required for the catabolism of camphor by such strains is carried by a plasmid. Either spontaneously or at high frequency following mitomycin C treatment, strains may lose this catabolic ability and by transduction analysis it can be shown that a cluster of genes concerned with the utilization of D-camphor has been lost. Transduction studies (Gunsalus & Marshall, 1971) show that there are two clusters of genes specifying the reactions of early and mid-pathway enzymes in the degradation of camphor. One group is plasmid borne, the other chromosomal. The evidence supports the notion of a CAM plasmid, which possesses a cluster of genes for camphor dissimilation and which is able to promote conjugation with CAM⁻ cells, with ready transfer of the CAM plasmid and a low level of transfer of chromosomal genes.

This does not appear to be an isolated example. Chakrabarty (1972) has shown that one entire salicylate degradation pathway in *P. putida* is plasmid borne. Variants can be obtained which, unlike the wild type, are able to use salicylate as a sole carbon source. By treatment of such strains with mitomycin C, derivative strains can be obtained which have lost this ability and such strains appear now to be without the entire genetic segment specifying the meta pathway of salicylate degradation. It is postulated that there is a plasmid SAL, specifying this pathway. This can be transferred readily to *P. putida* strains, lacking the ability to degrade salicylate.

Dunn and Gunsalus (1973) have identified a transmissible plasmid coding for naphthalene oxidation in *P. putida.* As with the SAL plasmid, the ability of a strain of *P. putida* to grow on naphthalene as sole carbon source may be lost spontaneously, or at much higher frequency following treatment with mitomycin C. The naphthalene-utilizing phenotype can be transferred to cured or other non-utilizing strains both by conjugation with the plasmid containing strain or by transduction with phage pf16. Another strain was studied which had the same ability to grow on naphthalene, but is is not clear if this property is chromosomally determined or due to a non-transmissible plasmid.

A further example of plasmid control of catabolic activity comes from the demonstration by Chakrabarty, Chou and Gunsalus (1973) that if the CAM plasmid is introduced into *P. oleovorans*, there is a loss of ability to grow on n-octane. When such strains lose the CAM plasmid, the ability to grow on n-octane is not regained. This result suggested that the CAM plasmid is displacing a plasmid responsible for n-octane metabolism by some mechanism related to the incompatibility of the two plasmids. The notion of an n-octane plasmid is further supported by the fact that when *P. oleovorans* is treated with mitomycin C, the ability to grow on n-octane is lost at significant frequency, suggesting that the plasmid is being cured. The ability to utilize n-octane could be transferred readily to a strain of *P. putida* able to utilize phenol. Not all the genes on the OCT plasmid have been identified but it carries at least the genes specifying enzymes for the degradation of n-octane to octanoic acid and possibly beyond.

If these examples of the genetic basis of metabolic versatility in *P. putida* are typical then it is clear that plasmids can provide the special genetic information required for the degradation of a wide range of substrates and that it is not necessary to postulate an almost omnipotent genome for *P. putida*. It is clear that should the generality of this phenomenon be established, a very likely possibility, it will be of considerable interest to investigate the derivation and evolution of the genetic material from which these plasmids are derived. An additional range of genetic variation provided by these plasmids can be envisaged with the observation by A. Chakrabarty (personal communication) that genetic fusion of the CAM and OCT plasmids can be achieved.

7. GENE ARRANGEMENT IN *PSEUDOMONAS*

The analysis of gene arrangement in such bacteria as *E. coli* and *S. typhimurium* has shown that associations of genes may have significance in terms of the regulation of gene activity—the best known example being the operon. Although the mapping of the chromosome of *P. aeruginosa* and *P. putida* has not been proceeded to the extent of the above bacteria, it is clear that the arrangement of genes in *Pseudomonas* is different from that of some other bacteria. With biosynthetic functions in *P. aeruginosa*, the degree of clustering of genes controlling related functions is much less than that found in *E. coli* or *S. typhimurium* (Fargie & Holloway, 1965; Holloway, 1969; Holloway *et al.*, 1971; Waltho, 1972). Initially Fargie and Holloway looked at the overall situation with respect to gene arrangement of a range of biosynthetic pathways by means of transduction, but more detailed mapping in the histidine pathway (Mee & Lee, 1967; 1969), the isoleucine–valine pathway (Marinus & Loutit, 1969), the arginine pathway (Feary *et al.*, 1969; Isaac & Holloway, 1972), the pyrimidine pathway (Isaac & Holloway, 1968), the methionine pathway (Calhoun & Feary, 1969) as well as the general mapping studies of Pemberton and Holloway (1972) have clearly demonstrated that those pathways which in the enterobacteria have contiguous gene clusters show a less clustered gene arrangement in *P. aeruginosa* (see Chapter 8, section 4).

In *P. putida*, the only biosynthetic pathway to be mapped is that for trytophan, the six structural genes falling into three linkage groups, all with a different pattern of regulation (Gunsalus *et al.*, 1968). It seems reasonable to suggest that the pattern of distribution of genes coding for sequential steps in biosynthetic pathways of *P. putida* will show the diffuse pattern of distribution established in *P. aeruginosa*.

The genetic and physiological aspects of the selective forces which have resulted in this pattern of gene distribution of biosynthetic markers are not yet understood. Certainly the regulatory mechanisms of the biosynthesis of amino acids and pyrimidines of *P. aeruginosa* seem to be less concerned with repressing than in the equivalent situations in *E. coli*. The lack of a partial diploid structure and the difficulties of obtaining regulatory mutants of these pathways has meant that there is not yet sufficient data on which to base any general conclusions. However this pattern of gene distribution is in marked contrast to that found with genes controlling catabolic functions in both *P. aeruginosa* and *P. putida* and it is clear that the evolutionary forces which have dictated the gene arrangements of each of these two metabolic systems have been quite different.

Genes of catabolic pathways show a high degree of clustering. In *P. aeruginosa*, Kemp and Hegeman (1968), using the transducing phage F116, and in *P. putida*, Wheelis and Stanier (1970) and Leidigh and Wheelis (1973a,b) using pf16, have shown that there is a clustering of genes affecting the mandelate and benzoate pathways, and that the former were linked to the genes governing enzymes concerned with anthranilate and benzoate dissimilation. The observed gene arrangements are significant in terms of the regulatory patterns. In three cases (*mdl*A and *mdl*B, *cat*B and *cat*C, *pca*B, *pca*D, *pca*E), the clustered genes are each members of a physiological regulatory unit and hence probably some type of operonic system is operating. However in other cases, linked genes come from a variety of physiological regulatory units and hence this supraoperonic arrangement must be nonrandom and have selective, physiological importance.

In Figure 5.3, the linkage relationships of genes in *P. putida* concerned in the dissimilation of *p*-hydroxybenzoic, quinic, shikimic, benzoic, mandelic, phenylacetic and nicotinic acids and histidine, tyrosine and phenylalanine are shown (see also Chapter 7). This group of genes can be estimated to be spread over some 10–15 % of the *P. putida* chromosome. The gross clustering of all these structural genes occurs together with smaller arrangements of physiological regulatory units, see for example the *pca*BDE cluster. It is of interest to speculate as to the identity of the genes which remain to be identified between the known markers of this region. Perhaps they are concerned with other degradative functions as well as regions for genetic control. Indeed Wheelis and Ornston (1972) and Wu, Ornston and Ornston (1972) have shown by deletion mapping that there is a genetic region for regulatory function to the left of the *cat*B gene. The evolutionary construction of this gene distribution must obviously be related to the mechanisms of gene transfer in this genus. In view of the clustering of related genes on plasmids

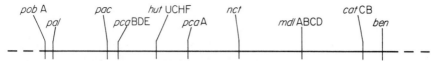

Figure 5.3. Genetic map of chromosomal genes involved in dissimilatory functions in *P. putida* as determined by transduction with phage pf16. (Reproduced with permission from Leidigh & Wheelis (1973a)

Gene symbol	Phenotype or enzyme deficiency
pac	phenylacetate utilization
nct	nicotinic acid utilization
pal	phenylalanine utilization
*pca*A	protocatechuate oxygenase
*pca*B	carboxymuconate lactonizing enzyme
*pca*D	enol lactone hydrolase
*pca*E	transferase
*pob*A	*p*-hydroxybenzoate hydroxylase
ben	benzoate oxidase
*cat*B	muconate lactonizing enzyme
*cat*C	muconolactone isomerase
*hut*U	urocanase
*hut*C	constitutive histidine utilization
*hut*H	histidase
*hut*F	formiminoglutamase
*mdl*A	mandelate racemase
*mdl*B	L-mandelate dehydrogenase
*mdl*C	benzoyl formate decarboxylase
*mdl*D	benzaldehyde dehydrogenase

as seen for the utilization of camphor, octane, salicylate and naphthalene, the clustering of genes for catabolic function on the chromosome could have arisen by integration of plasmids carrying such genes into a particular region of the *P. putida* chromosome. This raises the possibility of selection for some form of regulatory pattern in which different pathways could acquire common regulatory control. As Wheelis and Stanier (1970) have pointed out, the advantages of a clustered gene arrangement of this type are that many pathways are functional only in the entire state, no one part of the pathway being able to function to advantage without the presence of all the rest.

The clustering of catabolic genes in *P. putida* is thus found to occur in two ways—on plasmids, extrachromosomally, and by grouping of related genes on the chromosome. It is to be hoped that future work will be aimed at determining the genetic systems for regulation of these various pathways, and the spatial relationships of control and structural genes. To date, the genetic study of regulatory mechanisms in *Pseudomonas* has not been extensively developed, and it has not been clearly established if the gene arrangements which have a control significance in other bacteria are found in this bacterial group, or whether other arrangements occur. One exception to this situation, outstanding in the extent and precision of the results obtained, has been the genetic and biochemical analysis of the hydrolysis of aliphatic amides by Clarke and her associates (for review, see Clarke, 1970, and Chapter 8). This continuing study represents the most detailed genetic analysis of any individual locus in *Pseudomonas*. The elegant selective procedures developed for the isolation of amidase mutants together with the use of the transducing phage F116 have enabled both biochemical and genetic techniques to be combined with great advantage.

A series of mutations in the structural gene *ami*E have been isolated by selection for altered substrate specificities. These genetic changes have occurred at different sites in the *ami*E gene and selection can thus be made for the wild-type phenotype in crosses between different mutants. By such procedures a fine structure map of the *ami*E gene has been constructed (Betz, Brown, Clarke & Day, 1974). In view of the extensive knowledge of the mutant and wild-type amidases (Brown & Clarke, 1972) it is likely that it will be possible to relate amino acid differences between the different forms of amidase to mutation sites on the chromosome and to construct a functional map of the amidase gene. The way in which the analysis of these mutants with altered substrate specificities can be used for the study of the evolution of enzymes in *Pseudomonas* is described in Chapter 9.

The structural gene *ami*E has been shown by transduction analysis to be contiguous with a regulatory gene *ami*R and it is clear that this gene arrangement has significance for the control of amidase syntheses, although a complete understanding of the mechanisms will require the development and use of a partial diploid structure, comparable to say F' *lac* in *E. coli*.

8. SUMMARY

An understanding of the genetic organization of *Pseudomonas* is clearly necessary for the continued success of biochemical, epidemiological and microbiological research programmes in this genus. The identification and characterization of plasmids, the mapping of chromosomal genes and the development of techniques for the isolation of a wide range of mutants are the major genetic procedures by which this can be accomplished. These techniques are available for a wide range of other problems in *Pseudomonas* biology including morphogenesis and the functional organization of the bacterial cell and their relationship to the biochemical versatility and resistance characteristics of this genus.

9. BIBLIOGRAPHY

Bak, A. L., Christiansen, C. & Stenderup, A. (1970) *Journal of General Microbiology*, **64**, 377.

Betz, J. L., Brown, J. E., Clarke, P. H. & Day, M. (1974) *Genetical Research, Cambridge*, **23**, 222.

Black, W. A. & Girdwood, R. W. A. (1969) *British Medical Journal*, **ii**, 234.

Booker, R. J. & Loutit, J. S. (1974) *Genetical Research, Cambridge*, **23**, 145.

Bradley, D. E. (1972) *Genetic Research, Cambridge*, **19**, 39.

Bradley, D. E. (1973) *Virology*, **51**, 489.

Bradshaw, W. H., Conrad, H. E., Corey, E. J., Gunsalus, I. C. & Lednicer, D. (1959) *Journal of the American Chemical Society*, **81**, 5507.

Brammer, W. J., Clarke, P. H. & Skinner, A. J. (1967) *Journal of General Microbiology*, **47**, 87.

Brown, P. R. & Clarke, P. H. (1972) *Journal of General Microbiology*, **70**, 287.

Bryan, L. E., Semaka, S. D., van Den Elzen, H. M., Kinnear, J. E. & Whitehouse, R. L. S. (1973) *Antimicrobial Agents & Chemotherapy*, **3**, 625.

Bryan, L. E., van den Elzen, H. M. & Jui-Teng Tseng (1972) *Antimicrobial Agents & Chemotherapy*, **1**, 22.

Calhoun, D. H. & Feary, T. W. (1969) *Journal of Bacteriology*, **97**, 210.

Calhoun, D. H. & Jensen, R. A. (1972) *Journal of Bacteriology*, **109**, 365.

Chabbert, Y. A., Scavizzi, M. R., Witchitz, J. L., Gerbaud, G. R. & Bouanchaud, D. H. (1972) *Journal of Bacteriology*, **112**, 666.

Chakrabarty, A. M. (1972) *Journal of Bacteriology*, **112**, 815.

Chakrabarty, A. M., Chou, G. & Gunsalus, I. C. (1973) *Proceedings of the National Academy of Science, U.S.A.*, **70**, 1137.

Chakrabarty, A. M. & Gunsalus, I. C. (1969a) *Virology*, **38**, 92.

Chakrabarty, A. M. & Gunsalus, I. C. (1969b) *Proceedings of the National Academy of Science, U.S.A.*, **64**, 1217.

Clarke, P. H. (1970) *Advances in Microbial Physiology*, **4**, 179.

Clowes, R. C. (1972) *Bacteriological Reviews*, **36**, 361.

Clowes, R. C. & Hayes, W. (1968) *Experiments in Microbial Genetics*, Blackwell Scientific Publications, Oxford.

Datta, N. & Hedges, R. W. (1971) *Nature (London)*, **234**, 222.

Datta, N. & Hedges, R. W. (1972) *Journal of General Microbiology*, **70**, 453.

Datta, N., Hedges, R. W., Shaw, E. J., Sykes, R. B. & Richmond, M. H. (1971) *Journal of Bacteriology*, **108**, 1244.

Dhawale, M. R., Creaser, E. H. & Loper, J. C. (1972) *Journal of General Microbiology*, **73**, 353.

Don, P. A. & van der Ende, M. (1950) *Journal of Hygiene*, **48**, 196.

Dunn, N. W. & Gunsalus, I. C. (1973) *Journal of Bacteriology*, **114**, 974.

Dunn, N. W. & Holloway, B. W. (1971) *Genetical Research, Cambridge*, **18**, 185.

Espejo, R. T., Canelo, E. S. & Sinsheimer, R. L. (1971) *Journal of Molecular Biology*, **56**, 597.

Fargie, B. & Holloway, B. W. (1965) *Genetical Research, Cambridge*, **6**, 284.

Feary, T. W., Williams, B., Calhoun, D. H. & Walker, T. A. (1969) *Genetics*, **62**, 673.

Fuerst, J. A. & Hayward, A. C. (1969) *Journal of General Microbiology*, **58**, 227.

Fullbrook, P. D., Elson, S. W. & Slocombe, B. (1970) *Nature, (London)*, **226**, 1054.

Grinsted, J., Saunders, J. R., Ingram, L. C., Sykes, R. B. & Richmond, M. H. (1972) *Journal of Bacteriology*, **110**, 529.

Gunsalus, I. C., Gunsalus, C. F., Chakrabarty, A. M., Sikes, S. & Crawford, I. P. (1968) *Genetics*, **60**, 419.

Gunsalus, I. C. & Marshall, V. P. (1971) *Critical Reviews in Microbiology*, **1**, 291.

Hadley, P. (1927) *Journal of Infectious Diseases*, **40**, 1–312.

Hayes, W. (1968) *The Genetics of Bacteria and their Viruses*, 2nd ed., Blackwell, Oxford.

Hedges, R. W. & Datta, N. (1971) *Nature, (London)*, **234**, 220.

Holloway, B. W. (1955) *Journal of General Microbiology*, **13**, 572.

Holloway, B. W. (1966) *Mutation Research*, **3**, 167.

Holloway, B. W. (1969) *Bacteriological Reviews*, **33**, 419.

Holloway, B. W. (1971) *Australian Journal of Experimental Biology and Medical Science,* **49**, 429.

Holloway, B. W. (1973) *Conjugation and Episomes,* First International Symposium on the Genetics of Industrial Microorganisms, (Eds. Z. Vanek, Z. Hostalek & J. Cudlin), Academia, Prague, p. 45.

Holloway, B. W., Egan, J. B. & Monk, M. (1960) *Australian Journal of Experimental Biology and Medical Science,* **38**, 321.

Holloway, B. W. & Jennings, P. A. (1958) *Nature (London),* **181**, 855.

Holloway, B. W., Krishnapillai, V. & Stanisich, V. (1971) *Annual Review of Genetics,* **5**, 425.

Holloway, B. W. & Richmond, M. H. (1973) *Genetical Research, Cambridge,* **21**, 103.

Holloway, B. W., Rossiter, H., Burgess, D. & Dodge, J. (1974) *Genetical Research, Cambridge,* **22**, 239.

Holloway, B. W. & van de Putte, P. (1968) 'Lysogeny and bacterial recombination', in *Replication and Recombination of Genetic Material,* (Eds. W. J. Peacock & R. D. Brock), Australian Academy of Science, p. 175.

Homma, J. Y. (1971) *Japanese Journal of Experimental Medicine,* **41**, 387.

Ingram, L., Sykes, R. B., Grinsted, J., Saunders, J. R. & Richmond, M. H. (1972) *Journal of General Microbiology,* **72**, 269.

Isaac, J. H. & Holloway, B. W. (1968) *Journal of Bacteriology,* **96**, 1732.

Isaac, J. H. & Holloway, B. W. (1972) *Journal of General Microbiology,* **73**, 427.

Ishida, T., Seto, S. & Osawa, T. (1966) *Journal of Bacteriology,* **91**, 1387.

Ito, S., Kageyama, M. & Egami, F. (1970) *Journal of General and Applied Microbiology,* **16**, 205.

Jessen, O. (1965) Pseudomonas aeruginosa *and other Green Fluorescent Pseudomonads,* Munksgaard, Copenhagen.

Kageyama, M. (1970) *Journal of General and Applied Microbiology,* **16**, 523.

Kahn, N. C. & Sen, S. P. (1967) *Journal of General Microbiology,* **49**, 201.

Kawakami, Y., Mikoshiba, F., Nagasaki, S., Matsumoto, H. & Tazaki, T. (1972) *The Journal of Antibiotics,* **25**, 607.

Kay, W. W. & Gronlund, A. F. (1969) *Journal of Bacteriology,* **98**, 116.

Kelln, R. A. & Warren, R. A. J. (1973) *Journal of Bacteriology,* **113**, 510.

Kemp, M. B. & Hegeman, G. D. (1968) *Journal of Bacteriology,* **96**, 1488.

Krishnapillai, V. (1971) *Molecular and General Genetics,* **114**, 134.

Krishnapillai, V. & Carey, K. (1972) *Genetical Research, Cambridge,* **20**, 137.

Lambina, V. A. & Mikhailova, T. N. (1964) *Mikrobiologica,* **33**, 800.

Lányi, B. (1969) *Acta Microbiologica Academiae Scientiarum Hungariace,* **16**, 357.

Lebek, G. (1963) *Zentralblatt für Bakteriologie, Parasitenkunde Infektionskrankheiten und Hygiene. Abt. 1 Orig.,* **188**, 494.

Lederberg, J. (1950) in *Methods in Medical Research,* (Ed. J. H. Corrie Jr.), Year Book Publishers, Chicago, p. 3.

Leidigh, B. J. & Wheelis, M. L. (1973a) *Journal of Molecular Evolution,* **2**, 235.

Leidigh, B. J. & Wheelis, M. L. (1973b) *Molecular and General Genetics,* **120**, 201.

Levine, M. (1969) *Annual Review of Genetics,* **3**, 323.

Loutit, J. S. (1960) *Proceedings University Otago Medical School,* **38**, 4.

Loutit, J. S. (1969) *Genetical Research, Cambridge,* **13**, 91.

Loutit, J. S. (1971) *Genetical Research, Cambridge,* **16**, 179.

Loutit, J. S. & Marinus, M. G. (1968) *Genetical Research, Cambridge,* **12**, 37.

Loutit, J. S., Pearce, L. E. & Marinus, M. G. (1968) *Genetical Research, Cambridge,* **12**, 29.

Lowbury, E. J. L., Babb, J. R. & Roe, E. (1972) *Lancet,* **ii**, 944.

Lowbury, E. J. L., Kidson, A., Lilly, H. A., Ayliffe, G. A. J. & Jones, R. J. (1969) *Lancet,* **ii**, 448.

Marinus, M. G. & Loutit, J. S. (1969) *Genetics*, **63**, 547.
Matsumoto, H. & Tazaki, T. (1973) *Japanese Journal of Microbiology*, **17**, 399.
Mee, B. J. & Lee, B. T. O. (1967) *Genetics*, **55**, 709.
Mee, B. J. & Lee, B. T. O. (1969) *Genetics*, **62**, 687.
Meynell, E., Meynell, G. C. & Datta, N. (1968) *Bacteriological Reviews*, **32**, 55.
Moillo, A. M. (1973) *Genetical Research, Cambridge*, **21**, 287.
Novick, R. P. (1969) *Bacteriological Reviews*, **33**, 210.
Olsen, R. H. & Shipley, P. (1973) *Journal of Bacteriology*, **113**, 772.
Ornston, L. N., Ornston, M. K. & Chou, G. (1969) *Biochemical and Biophysical Research Communications*, **36**, 179.
Pemberton, J. M. (1971) 'Conjugation in *Pseudomonas aeruginosa*', *Ph.D. Thesis, Monash University.*
Pemberton, J. M. & Clark, A. J. (1973) *Journal of Bacteriology*, **114**, 424.
Pemberton, J. M. & Holloway, B. W. (1972) *Genetical Research, Cambridge*, **19**, 251.
Pemberton, J. M. & Holloway, B. W. (1973) *Genetical Research, Cambridge*, **21**, 263.
Reeves, P. (1972) *The Bacteriocins*, Springer-Verlag, Berlin.
Rheinwald, J. G., Chakrabarty, A. M. & Gunsalus, I. C. (1973) *Proceedings of the National Academy of Science, U.S.A.*, **70**, 885.
Richmond, M. H. & Sykes, R. B. (1972) *Journal of General Microbiology*, **73**, xv.
Roe, E., Jones, R. J. & Lowbury, E. J. L. (1971) *Lancet*, **i**, 149.
Rosenberg, S. L. & Hegeman, G. D. (1969) *Journal of Bacteriology*, **99**, 353.
Rosenberg, S. L. & Hegeman, G. D. (1971) *Journal of Bacteriology*, **108**, 1270.
Ryter, A. (1968) *Bacteriological Reviews*, **32**, 39.
Saunders, J. R. & Grinsted, J. (1972) *Journal of Bacteriology*, **112**, 690.
Shionoya, H. & Homma, J. Y. (1968) *Japanese Journal of Experimental Medicine*, **38**, 81.
Smith, D. H. & Armour, S. E. (1966) *Lancet*, **ii**, 15.
Stanier, R. Y., Palleroni, N. J. & Doudoroff, M. (1966) *Journal of General Microbiology*, **43**, 159.
Stanisich, V. A. (1972) *Journal of General Microbiology*, **73**, xi.
Stanisich, V. A. & Holloway, B. W. (1969) *Genetics*, **61**, 327.
Stanisich, V. A. & Holloway, B. W. (1971) *Genetical Research, Cambridge*, **17**, 169.
Stanisich, V. A. & Holloway, B. W. (1972) *Genetical Research, Cambridge*, **19**, 91.
Sykes, R. B. & Richmond, M. H. (1970) *Nature, (London)*, **226**, 952.
Tseng, J. T., Bryan, L. E. & van der Elzen, H. M. (1972) *Antimicrobial Agents and Chemotherapy*, **2**, 136.
van Hartingsveldt, J. & Stouthamer, A. H. (1973) *Journal of General Microbiology*, **74**, 97.
Waltho, J. A. (1972) *Journal of Bacteriology*, **112**, 1070.
Waltho, J. M. & Holloway, B. W. (1966) *Journal of Bacteriology*, **92**, 35.
Watanabe, T. (1963) *Bacteriological Reviews*, **27**, 87.
Weiss, R. L. (1971) *Journal of General Microbiology*, **67**, 135.
Weiss, R. L. & Raj, H. D. (1972) *Australian Journal of Experimental Biology and Medical Science*, **50**, 559.
Weppelman, R. M. & Brinton, C. C. (1970) *Virology*, **41**, 116.
Weppelman, R. M. & Brinton, C. C. (1971) *Virology*, **44**, 1.
Wheelis, M. L. & Ornston, L. M. (1972) *Journal of Bacteriology*, **109**, 790.
Wheelis, M. L. & Stanier, R. Y. (1970) *Genetics*, **66**, 245.
Witchitz, J. L. & Chabbert, Y. A. (1971) *Annales de L'Institut Pasteur*, **121**, 733.
Witchitz, J. L. & Chabbert, Y. A. (1972) *Annales de L'Institut Pasteur*, **122**, 367.
Wu, C-H., Ornston, M. K. & Ornston, L. N. (1972) *Journal of Bacteriology*, **109**, 796.
Zierdt, C. H. & Schmidt, P. J. (1964) *Journal of Bacteriology*, **87**, 1003.

CHAPTER 6

Gene Transfer in the Genus
Pseudomonas

VILMA A. STANISICH and MARK H. RICHMOND

1. INTRODUCTION

All the three main types of gene transfer described so far in bacteria have been
shown to take place either between strains of *Pseudomonas aeruginosa* or
other species of the genus *Pseudomonas*. However, to date, only transduction
and conjugation have been used extensively to study this group of organisms.
Transformation, which has been reported both inter- and intraspecifically
in a variety of soil and human and plant pathogens (Lambina & Mikhailova,
1964; Khan & Sen, 1967) has yet to be widely exploited. Transfection—an
entirely artificial process whereby phage infection is used to aid the entry of
naked phage DNA into intact cells—has not yet been demonstrated among
pseudomonads, although infection of spheroplasts of a marine pseudomonad
has been reported (van der Schans, Weyermans & Bleichrodt, 1971).

2. TRANSDUCTION AND CONJUGATION

General Points

This is not the place to give a detailed account of what is known of transduction and conjugation as a means of gene transfer in bacteria. An enormous
amount of work has been done on both topics in enteric bacteria, but
admirable general reviews are available in Adams (1959), Hayes (1968) and
Stent (1971), while a more detailed account of the molecular reactions
involved in transduction are to be found in articles by Ozeki and Ikeda (1968)
and Ikeda and Tomizawa (1965). Conjugation has been even more widely
reviewed. Two useful Symposia are those of the CIBA Foundation (CIBA,
1968) and the New York Academy of Sciences (New York Symposium, 1971)
but other sources of information are the monograph by Meynell (1973) and
reviews by Watanabe (1963), Meynell, Meynell and Datta (1968), Curtiss and
his colleagues (1968; 1969), Novick (1969), Brinton (1971), Clowes (1972) and
Richmond (1973).

As far as the work to be discussed in later sections of this chapter are
concerned, the salient aspects of transduction are as follows. Propagation of
a phage on donor bacteria, or induction of a prophage in them, gives rise to
a preparation that can transfer bacterial DNA when used to infect an appropriate recipient. This DNA may be derived either from the chromosome or
a plasmid in the donor, but is limited in quantity to the amount that can fit
in the head of the phage particle—normally about $30–50 \times 10^6$ daltons of
DNA. Furthermore, because the amount of DNA that can be accommodated
is limited, it is unusual for the phage particle to contain both bacterial and
and phage DNA, and consequently transductants are rarely lysogenized by
the phage that transfers the genes. A second type of transduction found among
enteric bacteria, in which the transferred DNA becomes associated with part
of the phage genome before transfer—so-called specialized transduction—
has not been found among *Pseudomonas* species. In *Pseudomonas putida*
however, the generalized transducing phage pf16 can, under certain conditions, yield high-frequency lysates for genes associated with the metabolism
of mandelate (see this chapter, section 3).

It is still unclear exactly how the DNA from the donor becomes incorporated into a transducing phage particle. Most authors suggest that the
process is random, the phage coat wrapping up the first piece of DNA of
appropriate size that is encountered regardless of its origin (Ikeda & Tomizawa, 1965). However, although the origin of the DNA has little influence on
its chance of being picked up by a generalized transducing phage, the fate of
the DNA once transferred to a recipient depends vitally on its origin. If it
originated in a large replicon it can only be incorporated into the phage as a
fragment and therefore will require to recombine with a resident replicon in
order to replicate and survive over long periods in the recipient population.

If on the other hand, the DNA is present as a plasmid in the donor, the chances are much higher that it will be incorporated in the phage as an intact replicon since the size of many plasmids is just about that needed to make a 'head-full' of DNA for most phages. Under these circumstances, the DNA transferred by the phage is likely to have the necessary genetic information to survive as an independently replicating genetic unit in the recipient, and no recombination with a resident replicon will be necessary.

This difference in behaviour is important since it allows the source of a gene to be inferred from its behaviour following transfer by transduction. If chromosomal in origin, irradiation of the transducing phage will damage the DNA present in the phage head and consequently enhance the chance of survival of the transferred piece by inducing repair and recombination enzymes in the recipient. Thus, ultraviolet irradiation of a phage preparation capable of transferring chromosomal genes will increase the frequency of transduction at low irradiation doses while leading to eventual inhibition after prolonged doses (Arber, 1960; Holloway *et al.*, 1962) (Figure 6.1). If, on the other hand, the DNA being transferred is an intact replicon, irradiation only decreases the probability of transfer, since recombination with a host replicon is unnecessary and the irradiation only serves to inactivate the plasmid-carried genes that specify the plasmid's independent survival mechanism in the recipient. This difference in behaviour on irradiation of transducing phage (the so-called Arber test; Arber, 1960) has been applied to an analysis of the origin of transferred DNA in both Gram-positive and Gram-negative bacteria, and should also be applicable to pseudomonads.

Despite the high potential of bacteriophages as gene vectors in bacteria, their efficacy is limited by two important factors. First, the range of potential recipients is limited by the host range of the phages used to make the transfer. Although this does not confine transfer to within a single species, only narrow taxonomic gaps can usually be crossed because strains susceptible to a given phage are usually members of the same, or at least a closely related species. Thus phage P22 will infect both *Escherichia coli* and *Salmonella typhimurium*, while phage pf20 will plaque both on *Pseudomonas aeruginosa* and on *Pseudomonas putida* (see section 3.b). Nevertheless, transfer across wider taxonomic gaps has not been reported.

Restriction is the other factor that limits the range of accessible hosts with phage as vector. Restricting hosts will reduce the frequency of gene transfer by a phage by a factor of several thousand, or even block it completely when compared with the rate of transfer to restriction-defective variants of the same host. There is certainly more than a single molecular basis for restriction in enteric bacteria (see reviews by Arber & Linn, 1969; Boyer, 1971) and this probably holds for pseudomonads as well (Krishnapillai, 1971).

There are a number of important differences which distinguish conjugation from transduction even if the overall result of the transfer is largely the same.

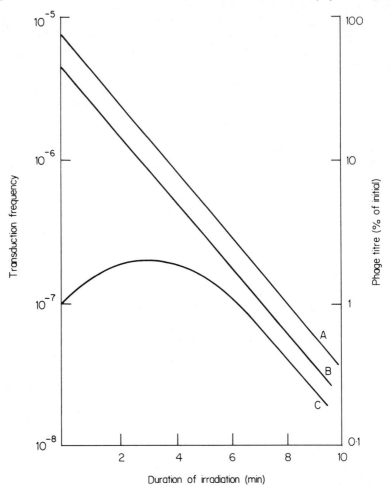

Figure 6.1. An idealized graph of the effect of irradiation of transducing phage on the transfer of chromosomal (C) and plasmid-borne genes (B). The effect of irradiation on the efficiency of plating of the phage is also shown (A). (Ordinates: Transduction frequency phage titre (% of initia); abcissa: Duration of irradiation (min))

Perhaps the most important in practical terms is that cell-to-cell contact is necessary for gene transfer by conjugation and, in order to act as a donor, a cell must carry the necessary genetic information to specify the conjugation apparatus that facilitates this contact. This group of genes is usually called a Transfer Factor: and when it is linked genetically to antibiotic resistance determinants the term is expanded to Resistance Transfer Factor, or RTF (see, for example, Meynell, 1973).

In some instances—notably when the donor cells have transfer factors of the F or I types—the conjugational donor carries a specific surface appendage, the sex pilus, whose formation is specified by genes that comprise part of the transfer factor, and whose presence greatly stimulates the frequency of transfer (Meynell, 1973). However, pili have not been detected on all types of conjugational donors, even after prolonged searches. There is therefore no reason to believe that pili are essential for transfer between all fertile pairs of bacteria. Certainly sex pili have not yet been detected on pseudomonads capable of acting as conjugational donors even though surface appendages of various kinds are common on these organisms (Fuerst & Hayward, 1969; Bradley, 1972). Nevertheless, there seems to be a general belief that there are likely to be some differences between the surfaces of all donor and recipient bacteria since some differences would seem to be inescapable if the mating process is to occur satisfactorily.

Since no phage particle is involved in gene transfer by conjugation this removes one of the limitations found in transducing systems: there is no theoretical limit to the length of DNA that can be transferred. Conjugation may therefore allow substantial amounts of information to pass from donor to recipient provided enough time is allowed for transfer to take place. Furthermore, DNA transfer by conjugation is a much less haphazard process than transduction. The process is confined to the replicon of which the transfer factor is part, so if this region is chromosomal it is the whole chromosome that it transferred linearly starting with the point of insertion of the transfer factor, while if the factor is plasmid borne it is only the element of which the factor is part that passes to the recipient.

The fact that a transfer factor integrated into the chromosome will transfer that replicon has an important practical consequence since it allows the chromosome to be mapped (Hayes, 1968). However, what concerns us primarily here is that conjugation can lead to the transfer of substantial pieces of chromosomal material from one *Pseudomonas* strain to another, and this in turn has important implications for evolution and variation within this group of organisms (see Chapter 9).

Two factors mitigate against ever being able to order plasmid markers by interrupted mating. First, most plasmids are only about 1–2 % as large as the chromosome and transfer of the whole structure takes only a short time— possibly less than 1 minute. Secondly, since it is unusual for the recipient to carry a plasmid similar to the donor, it is unlikely that there will be a replicon for the incoming plasmid to recombine with. Interrupted mating experiments are therefore only suitable for ordering genes on large replicons already represented in the recipient. As with chromosomal transfer, however, our main preoccupation in this chapter is with the positive consequences of transfer. Plasmid transfer between strains and species of the genus *Pseudomonas*, and even outside this range to other members of Pseudomonadales and Eubacteriales, has been shown to occur widely with consequent potential advantage to the strains concerned. It is these aspects of gene transfer that will now be considered in detail.

3. GENE TRANSFER BY TRANSDUCTION

3.a. Transduction in *Pseudomonas aeruginosa*

In this section we are concerned primarily with transduction as a potential means of gene transfer. The use of this procedure for mapping, and the genetic maps that have been deduced by the use of transductional and conjugational crosses are described in detail in Chapter 5.

Transducing phages for *P. aeruginosa* can readily be isolated from wild-type lysogenic strains (Holloway, Egan & Monk, 1960; Patterson, 1965), and a number of these have now been characterized and used for genetic analysis. The generalized transducing phages of this species include F116 (Holloway *et al.*, 1960) and the homoimmune phages F126 and F130 (Krishnapillai, 1971), G101 (Holloway & van de Putte, 1968a), B (Loutit, 1960; Pearce & Loutit, 1965) and B3 (Holloway *et al.*, 1960), and all transfer chromosomal genes at frequencies in the order of 10^{-6} to 10^{-8}/p.f.u. The efficiency of cotransfer of closely linked chromosomal markers varies from phage to phage, and is seen to reflect the size of the DNA which can be carried by them (Table 6.1). A variant of F116, called F116L, has been isolated from the wild-type lysogenic *P. aeruginosa* strain 116, and shows a higher co-transfer frequency than does F116 itself (Krishnapillai, 1971). Whether this phage is, in fact, any larger than the original isolate of F116 has yet to be determined. No case of specialized transduction has been reported in this species although the chromosomal location of the prophage of phage 90, and a number of phages showing homoimmunity to 90, is known (Krishnapillai & Carey, 1972; K. E. Carey, unpublished data). Another temperate phage, 35, having a prophage location considerably distal to 90, is currently being investigated (K. E. Carey, unpublished data) (see Figure 5.1).

Table 6.1. Cotransduction frequency of linked markers in *Pseudomonas aeruginosa* PAO by different transducing phages

Transducing phage	M. wt. of phage DNA (daltons)	Cotransfer (%) *met*-28 and *ilv*-202
B3	20–25×10^6	1
F116	38×10^6	11
F116L	Not known	25
G101	38–41×10^6	34

Phage F116 has been perhaps the most commonly employed of the *P. aeruginosa* temperate phages since it plates on the *P. aeruginosa* strains PAO, PAT, PTS and PAC, all of which have been used extensively for genetic and biochemical studies (Holloway, 1969; Clarke, 1970; Holloway, Krishnapillai & Stanisich, 1971). An additional advantage of F116 is that it is little affected by DNA restriction when passaged through strains of different host specificity. Thus in experiments involving *P. aeruginosa* strains PAO, PAT and PTS, neither phage DNA (as measured by efficiency of plating) nor bacterial DNA (as measured by transduction frequency) are susceptible to DNA degradation under restrictive conditions (Dunn & Holloway, 1970; Krishnapillai, 1971). This property of F116 makes it a particularly useful phage for any studies directed towards a comparative analysis of gene distribution in a variety of strains of *P. aeruginosa*.

As is commonly the case with generalized transducing phages among the enteric bacteria, the phages F116 and G101 can mediate transfer of genes from both chromosomal and plasmid replicons. As far as has been determined at present, all chromosomal genes seem to be equally accessible to transduction but the situation is somewhat less clear cut with genes derived from a plasmid.

A number of *Pseudomonas* R factors which have been transferred from wild-type strains to *P. aeruginosa* strains PAO and PAT, are transducible by phage F116L at frequencies of about 10^{-8}/p.f.u. (Stanisich, 1972, and unpublished data). These include RP1 (Grinsted, *et al.*, 1971) and R30 (Stanisich, 1972) which confer resistance to carbenicillin, neomycin/kanamycin and tetracycline, and R91 (Stanisich & Holloway, 1971) which confers resistance only to carbenicillin. Each transduction event results in the inheritance of all the resistance determinants carried by that R factor as well as of those genes necessary for its transfer by conjugation. That is, transduction of the entire plasmid seems to occur and this would in fact be predicted in the case of RP1 whose molecular weight of 38–40 $\times$ 10^6 daltons (Grinsted, *et al.*, 1971; P. M. Bennett, unpublished data) approximates to that carried by the phage (Table 6.1). Similar transduction frequencies are obtained for the β-lactamase gene of RP1-1 (Ingram *et al.*, 1972), and again those genes for conjugal transfer together with those which prevent plating of the female specific phage ϕ33 (T. Morgan, unpublished data), are also inherited by the transductants.

In addition to these *Pseudomonas* R factors, F116 can transduce the non-self-transmissible plasmid of *P. aeruginosa* PAT which confers resistance to meruric ions (Stanisich, 1972), and also the SAL plasmid derived from *P. putida* (unpublished data cited in Chakrabarty, 1972). On the other hand, F116, F116L or G101-mediated transfer of the FP plasmids FP2 and FP39, has never been found (Pemberton, 1971; Pemberton & Holloway, 1973). The molecular weights of FP2 and FP39 are 59 $\times$ 10^6 and 55 $\times$ 10^6 daltons respectively (Pemberton & Clark, 1973) which are considerably greater than the normal DNA carried by these phages (i.e. 40–50 × 10^6 daltons). Therefore, if transduction of these plasmids occurs at all, it must necessarily involve transfer of only fragments of the DNA. At present, the only known sex factor functions which can conveniently be used in the selection of transductants are mercury resistance (conferred by FP2), and leucine independance (Leu$^+$) (conferred by FP39) (Loutit, 1970; Pemberton & Holloway, 1973). Since transductants displaying these phenotypes are not observed, it suggests that the DNA fragments carrying these genes necessarily lack those functions essential for independent replication in the recipient cell. In addition, the 'rescue' of these markers by their recombination into the host chromosome must occur at too low a frequency to be detectable, if it occurs at all.

Before leaving transductional gene transfer in *P. aeruginosa* it is worth mentioning phage 90 since this is the only phage so far whose prophage location has been determined (Krishnapillai & Carey, 1972). The use of interrupted mating procedures in *P. aeruginosa* PAO have shown that prophage 90 is situated 5–7 minutes from the FP2 origin (see Figure 5.1) (Carey & Krishnapillai, 1974). As yet, no evidence is available to indicate that this phage can produce high-frequency transducing (HFT) lysates similar to those obtained with phage lambda in *Escherichia coli*. Nevertheless it is perhaps worth investigating phage 90 more closely, since the failure to detect HFT lysates may just reflect our ignorance of a marker near enough to the point of integration of this phage to be incorporated in developing phage 90 particles. Furthermore, although phage 90 is only weakly inducible by ultraviolet light it is zygotically inducible as seen from the reduction in recombinant recovery for markers situated distal to the prophage. It is therefore possible that the correct irradiation conditions for ultraviolet induction, and hence perhaps for the production of HFT lysates, have yet to be found.

3.b. Transduction in *Pseudomonas putida*

Phages active against strains of *Pseudomonas putida* have been reported less commonly than for *Pseudomonas aeruginosa*, and to date only three generalized transducing phages have been characterized. Phage PP1 plaques on the *P. putida* strain PMBL-4B1 but not on 13 other *P. putida* strains nor on 11 isolates of *P. fluorescens* tested (Holloway & van de Putte, 1968b). A clear plaque derivative of this phage, PP1c, was found to yield higher titres than the wild-type phage and was consequently used for transduction experiments. Transfer of a variety of auxotrophic markers occurred at frequencies of about 10^{-6}/p.f.u. using multiplicities of infection of less than 0·1.

Two further phages, pf16 (Chakrabarty, Gunsalus & Gunsalus, 1968; Gunsalus *et al.*, 1968) and PX4 (Olsen, Metcalf & Todd, 1968) have been used more extensively for transduction within strains of *P. putida* and also for interspecific transduction between *P. putida* and *P. aeruginosa*.

Phage pf16, isolated on the camphor oxidizing *P. putida* strain PpG1, and its host range variant pf16h1, have been used for the genetic analysis of 6 tryptophan synthetic genes in the opaque and translucent cell lines of this strain (Gunsalus *et al.*, 1968). The standard transduction procedure with this phage involves the use of ultraviolet-irradiated transducing lysates in order to diminish the loss of transductants due to phage kill. Transductional analysis was satisfactory not only in determining linkage between the various *trp* loci, but also in determining the order of linked genes and the fine structure mapping of mutations within individual genes.

This phage can also transfer a number of the *P. putida* plasmids so far reported. Thus the plasmids SAL and NAH, when transferred from their native hosts to *P. putida* PpG1, are transducible by pf16 at frequencies of about 10^{-7}/p.f.u. (Chakrabarty, 1972; Dunn & Gunsalus, 1973). CAM, the native plasmid of strain PpG1, is also transducible (Rheinwald, Chakrabarty & Gunsalus, 1973) while transduction of OCT (Chakrabarty, Chou & Gunsalus, 1973) has not been reported. As far as has been determined, all the known plasmid functions are inherited by the transductants suggesting that the sizes of these plasmids are less than 10^8 daltons, or approximately 3–4 % of the *P. putida* genome (unpublished data cited in Gunsalus *et al.*, 1968; Chakrabarty, 1972; Dunn & Gunsalus, 1973).

Another host range variant of pf16, pf16h2, has been used for interstrain transduction in *P. putida* between the camphor degrading strain PpG2 and the mandelate oxidizing strain PRS1 (Chakrabarty *et al.*, 1968; Chakrabarty & Gunsalus, 1969a). Transduction of the genes associated with the oxidation of mandelate to benzoate (*mdl* genes) occurs from PRS1 to PpG2 at frequencies of about 3×10^{-8}/p.f.u., the cluster of four genes being transferred as a linked unit. The analysis of the *mdl*$^+$ transductants obtained, showed that many of them were immune to pf16 and also unstable for the *mdl* character. Superinfection of these immune *mdl*$^+$ transductants during ultraviolet irradiation gave rise to high-frequency transducing (HFT) phage lysates in that the frequency of *mdl*$^+$ transfer was increased from about 3×10^{-8} p.f.u. by a factor of 10–1000 (Chakrabarty & Gunsalus, 1969a).

Although these are HFT lysates in the sense that the frequency of gene transfer is increased, there is as yet no real evidence that the phage pf16h2 is acting in a manner analogous to the HFT systems that have been studied in *Escherichia coli*. Further analysis indicated that the transductants yielding HFT preparations harboured defective phages (pfdm particles) whose genome consisted of different amounts of phage and bacterial DNA. The properties of these particles are discussed later in this chapter, and also in Chapter 5.

The final *P. putida* phage to be mentioned is pf20, which has been used for a preliminary investigation of genetic homology between strains of *P. putida* and *P. aeruginosa* (Chakrabarty & Gunsalus, 1970). This phage is an EMS-treated derivative of phage PX4 isolated by Olsen and his colleagues (1968) as plating on *P. fluorescens*. The host range of PX4 included *P. aeruginosa* PAT, and ethyl methane sulphonate treatment has extended this to include *P. putida* PRS1. Interspecific transductions between *P. putida* PRS1 and *P. aeruginosa* PAT showed that for certain markers transduction occurred at frequencies of *ca.* 2×10^{-8}/p.f.u., this frequency being only 10-fold lower than when transfer was not across the species barrier. This result implies an insensitivity of the phage to the host-restriction systems of these strains and makes it, as previously discussed for the *P. aeruginosa* phage F116, particularly useful for any studies directed towards an examination of the evolutionary divergence in these strains and species of *Pseudomonas*.

4. GENE TRANSFER BY CONJUGATION

4.a. Conjugation in *Pseudomonas aeruginosa*

A variety of extrachromosomal elements has been identified in strains of *Pseudomonas aeruginosa* and studied to various extents both genetically and biophysically. These include two major classes of transferable elements, the FP and RP (or R) factors, a non-self-transmissible plasmid conferring resistance to mercuric ions, and several cryptic plasmids whose functions have yet to be determined.

The classification of the transferable plasmids as either FP or RP types has been, as with the R and Col factors of the Enterobacteriaceae, based on the procedures or properties used in their detection, and is consequently not intended to imply any major differences between these subgroups. Thus the various FP factors have been identified because of their ability to mediate the transfer of host chromosomal genes at frequencies of at least 10^{-6}/donor cell (Pemberton, 1971; Pemberton & Holloway, 1973). Antibiotic resistance has not been found associated with any of these FP factors although three, including FP2, confer resistance to mercuric ions (Loutit, 1970; Pemberton & Holloway, 1973). Bacteriophage infection and/or multiplication is severely reduced in cells harbouring many of these FP factors (P. J. Dry, unpublished data), while several confer resistance to u.v. radiation (V. Krishnapillai, unpublished data). In contrast, the RP factors have been detected because of the resistance they confer to a variety of antibiotics, in particular carbenicillin, and although they can mediate their own transfer by conjugation (Black & Girdwood, 1969; Lowbury *et al.*, 1969; Datta *et al.*, 1971; Kawakami *et al.*, 1972) their ability to mobilize chromosomal genes is rather more variable (Stanisich & Holloway, 1971; Kawakami *et al.*, 1972).

a.i. Sex Factor (FP) Transfer

The sex factor FP2 was originally detected in *P. aeruginosa* PAT because of the ability of this strain to yield recombinants when mixed with a second *Pseudomonas* strain (PAO) under conditions where transduction and transformation could be ruled out as the means of gene transfer (Holloway, 1955; 1956). Cell-to-cell contact was shown to be necessary for gene transfer to take place and the FP2 factor, as it become known, clearly had much in common with the fertility factor F in enteric bacteria (for reviews see Holloway, 1969; Holloway *et al.*, 1971). The plasmid nature and molecular properties of this sex factor have recently been studied in *P. aeruginosa* PAO (Pemberton & Clark, 1973), and are described in detail in Chapter 5.

Before discussing the ability of FP2 to promote the transfer of chromosomal genes, it is important to discuss the transfer of the plasmid itself. *P. aeruginosa* PAT and derivatives of PAO harbouring the FP2 factor, show greater resistance to mercuric ions than does the uninfected PAO strain (Loutit, 1970; Stanisich & Holloway, 1972). That mercury resistance is a determinant (*mer-r*) carried by the FP2 factor can be inferred from the inability to separate donor properties from those of mercury resistance in conjugational crosses. Thus in PAO intraspecific matings, selection either for *mer-r* transcipients or the screening of the recipient population for cells newly acquiring donor ability, is invariably associated with coinheritance of the alternate property. Admittedly linked transfer in conjugational crosses need not necessarily imply that the two markers form part of the same replicon, however this would seem a reasonable conclusion for FP2 since the biophysical studies of this plasmid in PAO have identified only a single molecular species rather than two or three as might be expected for a situation involving a transfer factor with a dissociable resistance determinant (Pemberton & Clark, 1973).

Several studies on the kinetics of transfer of FP2 have been reported in *P. aeruginosa* PAO (Loutit, Marinus & Pearce, 1968; Loutit, 1970; Pemberton & Holloway, 1973). Selection for mercury ion resistant transcipients provides a convenient index for the inheritance of FP2 among a recipient population, and on this basis transfer can be detected within 5 minutes of mating, increasing to a maximum after 20–30 minutes when approximately 20 % of the recipient population have been infected (Loutit, 1970). Very similar results are obtained in analogous experiments where chromosome transferability rather than mercury ion resistance is used as the criterion of sex factor transfer (Loutit *et al.*, 1968; Loutit, 1970). The results of Pemberton and Holloway (1973) are similar, in that transfer is detectable shortly after mating is initiated, but no plateau is obtained: rather a steady increase with time until about 40 % of the recipient population have inherited the sex factor.

Since neither the conditions of mating nor the male strains used in these experiments were identical, a direct comparison of the results cannot be made.

However from these data it can be estimated that 1 in 10^2 to 10^3 of the males in these FP2$^+$ populations are capable of transferring sex factor at the time of mating. This suggests that in FP2, the genes associated with transfer function are probably in a repressed state, but if this is so, the degree of repression is rather less marked than that observed among many R factors of the enteric bacteria (Meynell *et al.*, 1968) or those of *Pseudomonas* (Stanisich & Holloway, 1971; P. M. Chandler & V. Krishnapillai, unpublished data). What is required to clarify this point is the demonstration either of transient derepression of FP2 transfer in cells newly inheriting this sex factor, or the isolation of derepressed mutants. Both of these lines of investigation should now be simplified using the selective procedure made available by the FP2-linked *mer-r* determinant.

Recently a second sex factor—apart from the *Pseudomonas* R factors— has been detected and christened FP39 (Pemberton & Holloway, 1973). This fertility factor is also a plasmid (Pemberton & Clark, 1973) (see Chapter 5), but does not confer mercury resistance in *P. aeruginosa* PAO. On the other hand, the presence of FP39 in cells carrying certain *leu* mutations will result in restoration of the Leu$^+$ phenotype. Thus, in *P. aeruginosa* PAO, matings of the type FP39$^+$ *leu*$^+$ × FP39$^-$ *leu* always give linked transfer of the sex factor with the Leu$^+$ phenotype. This transfer therefore has the characteristics of an F-prime transfer as found in enteric bacteria, and Hfr transfer can be excluded since all the Leu$^+$ transcipients also express sex factor activity. Whether FP39 carries one or more linked leucine biosynthetic genes, or alternatively a suppressor gene for a particular gene locus or mutation, has yet to be determined. As far as has been examined, the transfer kinetics of FP39 and FP2 are similar except that FP2 transfer is somewhat more efficient.

a.ii. Sex Factor Mediated Chromosome Transfer

FP2 will mobilize chromosomal markers from *P. aeruginosa* strains PAO and PAT in both inter- and intrastrain matings (Holloway, 1955; 1956; Stanisich & Holloway, 1972; J. M. Watson unpublished data). In general, the overall frequency of recombinant formation in the two strains is similar (10^{-3}–10^{-6}/donor cell), although for particular markers it may vary by 50–100-fold or more. As yet, transfer of FP2 or of chromosomal genes by conjugation to other strains or species of *Pseudomonas* has not been reported, although the former should be feasible using mercury resistance as selection.

In *P. aeruginosa* PAO, the distribution of chromosomal markers has been investigated primarily using FP2 donor strains and applying an interrupted mating technique essentially analogous to that employed for enteric bacteria (Loutit, Marinus & Pearce, 1968; Loutit, Pearce & Marinus, 1968; Loutit,

1969b; Stanisich & Holloway, 1969; Pemberton & Holloway, 1972b). Transfer of donor markers is a time-dependent event, the transfer gradients obtained being characteristic of the particular markers studied. These data, combined with transductional analysis and unselected-marker linkage analysis has led to the construction of chromosome maps for *P. aeruginosa* PAO as shown in Chapter 5 (Loutit, 1969b; Pemberton & Holloway, 1972b). The marker distributions obtained span 1–50 minutes, although time of entry data as such is probably only accurate for markers transferred within 30 minutes. For more distal markers, recombinant yields are generally too low for accurate ordering, but this has to some extent been overcome by the application of a mathematical procedure devised by de Haan and his colleagues (de Haan *et al.*, 1969; Pemberton & Holloway, 1972b). At present there is some uncertainty as to whether all the markers studied can be assigned to a single linkage group—but this topic is discussed in detail in Chapter 5.

The mechanism by which FP2 mediates chromosome transfer in *P. aeruginosa* is not known. However, the observation that transfer gradients for donor markers can be obtained, and that only one or a few linkage groups are demonstrable, suggests that the number of transfer origins that can occur on the chromosome is limited. These may be initiated by recombination between FP2 and the chromosome, an event which, irrespective of its stability results in chromosome mobilization, or alternatively FP2 may act simply to provide contact with a recipient cell, chromosome transfer then being initiated by some other process not directly involving the sex factor. Two other observations are also of importance. Firstly, 30–80 % of all chromosomal transcipients inherit FP2 (Loutit, Marinus & Pearce, 1968; Stanisich & Holloway, 1969; Pemberton, 1971), and secondly, genetic evidence in *P. aeruginosa* PAT (Stanisich & Holloway, 1972; Stanisich, 1973) and biophysical evidence in *P. aeruginosa* PAO (Pemberton & Clark, 1973) suggests that there may be more than a single copy of FP2 in donor cells.

This information can be interpreted in several ways. If an interaction between FP2 and chromosome is not a prerequisite for chromosome mobilization, then the recovery of FP2 among a significant proportion of the recombinant population is simply explained as the independent transfer of plasmid and chromosome during a single mating event. On the other hand, a more active involvement of FP2 may require stable integration of the sex factor into the chromosome as is seen with Hfr donors of *Escherichia coli*, or only a transient association sufficient to provide a transfer origin. In either instance it is likely that the sex factor which initiates chromosome transfer remains in the donor cell; that is, it is transferred as a terminal marker in the manner of Hfr donors, while another autonomously co-existing sex factor is transferred independently but during the same mating event.

As far as has been ascertained, chromosome transfer mediated by the sex factor FP39 is very similar to that found with FP2 (Pemberton & Holloway, 1973). The map order obtained with both of these sex factors is identical, suggesting that they transfer chromosome with the same polarity. Sex factor FP39 however, seems to initiate transfer at an origin 10 minutes proximal to that of FP2, although to date no genetic markers have been found to lie in this region. It is therefore possible that both these sex factors initiate transfer from similar sites but that in the case of FP39 a delay occurs either in mobilization in the donor, or in establishment of markers in the recipient.

a.iii. R Factor (RP) Transfer

There have been a number of reports that antibiotic resistance markers could be transferred between strains of *P. aeruginosa*, and even between pseudomonads and enteric bacteria (Lebek, 1963; Smith & Armour, 1966). More recent investigations have shown that such transfer events do occur and are mediated by extrachromosomal elements essentially analogous to the Resistance Factors (R factors) encountered among the enteric bacteria.

Table 5.1 lists the geographical origin of the various R factors derived from strains of *P. aeruginosa*, together with the resistance determinants carried by them. In the main, the genetic and biophysical characterization of these R factors has been confined to those isolated in Britain at Birmingham (Lowbury *et al.*, 1969) and Glasgow (Black & Girdwood, 1969)—primarily the former. These include the R factors designated RP1, R18-1, RP1-1, and R1822 all derived from *P. aeruginosa* 1822 (Sykes & Richmond, 1970; Datta *et al.*, 1971; Grinsted *et al.*, 1971; Stanisich & Holloway, 1971; Ingram *et al.*, 1972; Olsen & Shipley, 1973), R30, R68 and R91 from *P. aeruginosa* 3068, 6886 and 9169 respectively (Stanisich & Holloway, 1971; Stanisich, 1972), and RP8 from *P. aeruginosa* S8 (Saunders & Grinsted, 1972). Their detailed molecular properties are discussed in Chapter 5.

The *P. aeruginosa* R factors have been transferred in both inter- and intrastrain matings involving a variety of different *P. aeruginosa* strains (Stanisich & Holloway, 1971; Grinsted *et al.*, 1971; Bryan, van den Elzen & Tseng, 1972; Ingram *et al.*, 1972; Kawakami *et al.*, 1972; Olsen & Shipley, 1973), as well as in interspecies matings between *P. aeruginosa* and *P. fluorescens*, and *P. fluorescens* and *P. putida* (Olsen & Shipley, 1973) (see Table 6.2). The transfer frequencies obtained range from about 10^{-2} to 10^{-6}/donor cell to some examples of intrastrain matings where the frequency may reach about 5 % in a mating experiment lasting several hours. Because of the different donor–recipient combinations and mating conditions used in these various experiments, it is difficult to make meaningful interpretations as to whether the different transfer frequencies reflect actual differences between these R factors. This is better ascertained in intrastrain matings and this topic will be discussed later in this section.

Table 6.2. Transfer of *Pseudomonas aeruginosa* R factors between various bacterial species

(a) *R factor transfer in interstrain matings of* P. aeruginosa

Plasmid	Donor		Recipient		Plasmid transfer per donor cell
RP1	*P. aeruginosa*	18	[a] *P. aeruginosa*	19	5×10^{-4}
RP1-1	*P. aeruginosa*	18	[a] *P. aeruginosa*	19	6×10^{-4}
R91	*P. aeruginosa*	9169	[c] *P. aeruginosa*	PAT	10^{-2}
R68	*P. aeruginosa*	6886	[c] *P. aeruginosa*	6886	10^{-6}
RP8	*P. aeruginosa*	58	[d] *P. aeruginosa*	19	2×10^{-5}
R1822	*P. aeruginosa*	1822	[e] *P. aeruginosa*	PAT	$5 \cdot 6 \times 10^{-3}$

(b) *R factor transfer in interspecific and intergeneric matings*

R1822	*P. aeruginosa*	PAT	[e] *P. fluorescens*		$1 \cdot 2 \times 10^{-3}$
R1822	*P. fluorescens*		[e] *P. putida*		$1 \cdot 4 \times 10^{-6}$
R1822	*P. fluorescens*		[e] *P. aeruginosa*	PAT	$1 \cdot 6 \times 10^{-2}$
RP1	*P. aeruginosa*	18	[b] *E. coli* K12		7×10^{-4}
RP1	*P. aeruginosa*	18	[b] *Proteus mirabilis*		10^{-5}
RP1-1	*P. aeruginosa*	18	[b] *E. coli* K12		10^{-8}
RP1-1	*P. aeruginosa*	18	[b] *Proteus mirabilis*		$<10^{-10}$
RP8	*P. aeruginosa*	S8	[d] *E. coli*		4×10^{-8}
RP8	*P. aeruginosa*	S8	[d] *Proteus mirabilis*		3×10^{-7}
RP4	*E. coli* K12		[f] *P. aeruginosa* PAO		$2 \cdot 3 \times 10^{-3}$
RP4	*E. coli* K12		[f] *Chromobacterium violaceum*		10^{-6}
RP4	*E. coli* K12		[f] *Rhizobium* spp.		10^{-6}
RP4	*E. coli* K12		[f] *Agrobacterium tumefaciens*		10^{-8}
R1822	*P. fluorescens*		[e] *Neisseria perflava*		$4 \cdot 2 \times 10^{-4}$
R1822	*P. fluorescens*		[e] *S. typhimurium*		$1 \cdot 7 \times 10^{-4}$
R1822	*P. putida*		[e] *Proteus mirabilis*		$2 \cdot 2 \times 10^{-5}$
R1822	*P. aeruginosa*	1822	[e] *Vibrio cholerae*		$2 \cdot 6 \times 10^{-4}$

References: a. Grinsted *et al.*, 1971; b. Ingram *et al.*, 1972; c. Stanisich & Holloway, 1971; d. Saunders & Grinsted, 1972; e. Olsen & Shipley, 1973; f. Datta *et al.*, 1972.

When considering the significance of transfer frequencies, it should be borne in mind that several factors will tend to reduce R factor transfer in certain mating combinations. These are degradation of the transferred DNA by the recipient strain (restriction), and the production of phage or bacteriocin by one, the other or both of the parent strains. Various restriction (Res) and modification (Mod) mutants of *P. aeruginosa* PAO have been used by P. M. Chandler and V. Krishnapillai (unpublished experiments) to demonstrate that the transfer frequency of eight R factors derived from the Birmingham isolates are similarly reduced in matings involving parents with different DNA specificities. The decrease in transfer frequency observed is of the order of 10^3 to 10^4. Such a decrease is to be expected in most instances of inter-strain and interspecies transfer, and where restriction is compounded by bacteriocin or phage kill the transfer frequency may be so diminished as to prevent, or seriously to hamper, the isolation of particular R factor carrying derivatives.

Fortunately, these effects are to some extent counterbalanced by certain properties of the R factors themselves. For example, provided that the transfer frequency of the R factor is fairly high under the mating conditions used (say 10^{-4}/donor cell/30 minute mating), or that the R factor can undergo 'epidemic spread' as is found with R91, then neither restriction nor bacterio-cinogeny presents a formidable problem. In addition, where Res$^-$ mutants are not available growth of the recipient strain at 43 °c prior to mating may be a sufficient measure to produce a transient restriction-deficient phenotype. This phenomenon of a temperature-induced change in restriction phenotype has been observed in several unrelated strains of *P. aeruginosa* (Holloway, 1965; Rolfe & Holloway, 1966; 1969), although whether it is common among isolates of this species, or among pseudomonads in general, is not known.

Differences in the transfer efficiencies of various R factors are more readily seen when they are compared in intrastrain matings. The actual frequencies obtained however, although characteristics of that R factor, will be markedly dependent on the host strain selected for study. This applies also to other characteristics of the R factor such as its stability, its ability to mediate chromosome transfer and the level of expression of its resistance deter-minants (Stanisich & Holloway, 1971; Ingram *et al.*, 1972). Figure 6.2 shows a comparison of the transfer kinetics of two R factors RP1 and RP1-1 in *P. aeruginosa* PAO (T. Morgan, unpublished data). Carbenicillin-resistant transcipients are detectable within 5 minutes of the onset of mating and reach a maximum after about 15 minutes. Similar results are obtained in *P. aeruginosa* PAT although in strain 18 transfer of RP1 is 10 to 100 times more efficient than that of RP1-1.

From such experiments, the number of active donors in established cultures of RP1 and RP1-1 carrying cells is found to be of the order of 10^{-3} to 10^{-4} and 10^{-1} to 10^{-2}/donor cell respectively. Thus, RP1 like its wild-type counterparts among the Enterobacteriaceae is apparently in a repressed state. This also holds true for the R factors R68 and R91, the latter of which is readily derepressed on transfer to recipient lines (Stanisich & Holloway, 1971). The factor RP1-1 on the other hand appears to be naturally dere-pressed as indicated by its high transfer efficiency in both *P. aeruginosa* strains and PAO and PAT.

a.iv. R Factor Mediated Chromosome Transfer

It is difficult to assess the ability of the *Pseudomonas* R factors to mediate the transfer of chromosomal genes. The first difficulty is that, in general, these

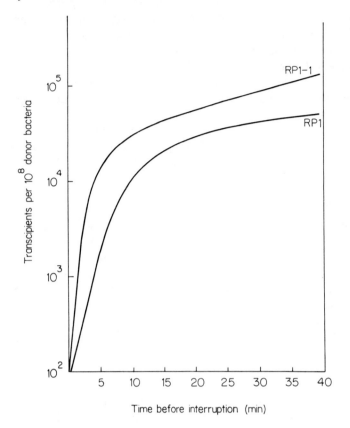

Figure 6.2. Kinetics of transfer of RP1 and RP1-1 in
P. aeruginosa strain PAO

R factors appear to be in a repressed state, so that at any given time only a fraction of the donor population can participate in mating events. This tends markedly to reduce, or even completely to abolish transfer. In addition, since chromosome maps of strains other than *P. aeruginosa* PAO are not available, the study of chromosome transferability will necessarily involve the use of randomly selected genetic markers whose map locations are unknown. Should the R factor under study mobilize chromosome transfer from one predominant site, as is apparently the case with FP2 and FP39, then this will only be detected if an appropriate proximal marker has been included among the test strains. Even with these difficulties, however, some major differences have been shown to exist between the various R factors studied so far, and to some extent these differences are strain specific.

Factors RP1-1 and R18-1, although able to mediate their own transfer with high efficiency within *P. aeruginosa* strains PAO and PAT, can mediate chromosome transfer only at low frequency (10^{-6} to 10^{-8}/donor cell) (Stanisich & Holloway, 1971; T. Morgan, unpublished data). Transfer is however detectable for most of the markers studied and as far as can be ascertained, no preference exists for the transfer of particular chromosomal regions. On the other hand, the R factors belonging to Class II of those studied by P. M. Chandler and V. Krishnapillai (unpublished experiments) (which includes R68, R30 and RP1) can mediate chromosome transfer with rather high frequency in strain PAT (10^{-4}/donor cell) but only at low frequency in strain PAO (10^{-7} to 10^{-9}/donor cell) (Stanisich & Holloway, 1971; Stanisich, 1972 and unpublished data).

The strain specificity shown by these R factors is somewhat unexpected, as FP2-mediated chromosome transfer between strains PAT and PAO invariably yields recombinants irrespective of the chromosomal markers selected, thereby suggesting a gross genetic homology between these strains. This is clearly not sufficient to allow R-mediated chromosome transfer in PAO, and the existence of some further intrinsic difference between these strains is perhaps reflected in the preferential order of marker transfer found in R-mediated matings when compared with those mediated by FP2. The Class II R factors behave similarly in matings with the same auxotrophic recipients, but those markers that are recovered at high frequency in R matings are recovered at low frequency in FP2 matings. The reverse pattern also holds true. This difference may represent a difference in the origins of chromosome transfer initiated by the FP2 and R factors, or at least a difference in their orientation of transfer, but this cannot be established with any certainty until the distribution of chromosomal markers in this strain is known.

One peculiarity of the Class II R factors is their instability in strain PAT (Stanisich, 1972; P. M. Chandler & V. Krishnapillai, unpublished data). It is therefore likely that in any growing culture, loss of the R factor at cell division is followed by reinfection of R^- cells as the cell density increases, so that a significant proportion of the R^+ population is always transiently derepressed and hence efficient in transferring both R factor and chromosomal genes. R91 is an example of an R factor whose transfer can be derepressed in mixed culture with an R^- recipient, and this, as is also found with the enteric R factors, leads to a concomitant increase in its ability to mediate chromosome transfer (Stanisich & Holloway, 1971). Derepression can be effected in both strains PAO and PAT, and in the latter the pattern of marker transfer is similar to that observed in FP2 matings. Of the other R factors studied, chromosome transfer occurs only at low frequency if at all, even under conditions where R factor transfer has been increased some 100-fold by derepression.

a.v. R Factor Transfer from Pseudomonas aeruginosa *to Other Species and Genera*

One particularly interesting property of the *Pseudomonas* R factors which may not be common to other transmissible plasmids of this genus, is the ability of certain of these to transfer across wide taxonomic boundaries. Thus the R factors RP1 (Grinsted *et al.*, 1971), RP4 (Datta *et al.*, 1971) and R1822 (Olsen & Shipley, 1973) can transfer between *P. aeruginosa* and members of the Enterobacteriaceae such as *Escherichia coli* and *Proteus mirabilis* at frequencies of about 10^{-5}/donor cell. Furthermore transfer within the Pseudomonadales occurs widely. Pühler, Burkhardt and Heumann (1972) have used RP4 in their genetic analysis of *Rhizobium lupini* and Datta and her collaborators (Datta *et al.*, 1972) mention that the same plasmid may transfer to other species of *Rhizobium* and to *Chromobacterium*. In addition, the study by Olsen and Shipley (1973) shows that the host range of R1822 extends to such diverse species as *Neisseria* and *Vibrio*; and also to various of the Enterobacteria and to a number of soil saprophytes and photosynthetic species. Examples of various of these interspecific and intergeneric transfers are given in Table 6.2 together with the transfer frequencies obtained. In most instances these figures are not directly comparable because of the different mating conditions which have been used for their investigation, but they nevertheless serve as a guide to the range of frequencies that might be expected.

In the two instances where a detailed investigation has been carried out, these *Pseudomonas* R factors once transferred to *Escherichia coli* behave as though they were 'typical *E. coli* plasmids'. Thus transfer of RP1 and RP8 across the taxonomic gap from *P. aeruginosa* to *E. coli* involves transfer of the entire plasmid, the molecular properties of the elements being indistinguishable in the two species (Grinsted *et al.*, 1971; Saunders & Grinsted, 1972). As mentioned above not all R factors found in *P. aeruginosa* can transfer to members of the Enterobacteria. RP1-1 is a plasmid originally isolated from the same *Pseudomonas* strain that also harboured RP1, and although it will transfer readily between strains of *P. aeruginosa* it will not infect *E. coli* to produce a stable line of RP1-1 carrying *E. coli* cells (Ingram *et al.*, 1972). Why it should be that some of these plasmids will cross the wide taxonomic gap between pseudomonads and members of the enteric bacteria while others will not is quite unknown at present.

The ability to pass from Pseudomonads to enteric bacteria such as *E. coli* is not, however, determined by their compatibility group. Thus RP4 and R40a which belong to compatibility groups P (Datta *et al.*, 1972) and *com*6 (Chabbert *et al.*, 1972) respectively, are both freely transferable from *Pseudomonas aeruginosa* to *E. coli*. It may be that a specialized form of sex pilus, or surface modification, is needed for such a wide dissemination of

R factors and that not all plasmids have the necessary adaptations. Alternatively the low rate of transfer found in some instances may mean that the R factor is not transferred by conjugation but that some other process such as transformation or transduction, might be involved.

Attempts to transfer the R factor RP1-1 to *E. coli* resulted in the isolation of a drug-resistant derivative in which the R factor determinant (carbenicillin resistance) had become integrated into the *E. coli* chromosome (Ingram *et al.*, 1972; Richmond & Sykes, 1973). This stable drug-resistant transcipient, said to carry RP1-1* was no longer capable of promoting transfer either to other *E. coli* strains or back to the parent *P. aeruginosa*. So the R factor RP1-1 may not be as much defective in transfer properties to *E. coli* as in lacking the necessary information to allow the R factor to survive as a plasmid in the recipient; and RP1-1* may have arisen from the abortively transferred RP1-1, perhaps by illegitimate recombination with the *E. coli* chromosome (Richmond & Sykes, 1973).

R factor transfer in strains of *Pseudomonas aeruginosa* was first clearly demonstrated about 3 years ago, but since then there has been an enormous increase in interest in these elements as their properties have been elucidated. In general, they seem to provide one of the most useful tools yet available to us for the genetic analysis of Gram-negative bacteria and particularly strains of *P. aeruginosa*, quite apart from the applied potential of some of them because they can transfer genetic material over such a wide range of bacterial species.

4.b. Conjugation in *Pseudomonas putida* and Other Pseudomonads

b.i. The 'Degradative' Plasmids of Pseudomonas putida *and* Pseudomonas oleovorans

The detection of FP and RP plasmids in strains of *Pseudomonas aeruginosa* has led microbiologists to look for other types of plasmid-linked gene in this and related species. The metabolic diversity of this group of organisms and the unstable nature of some of the degradative properties of the cells suggested that plasmid-borne genes might be involved; and indeed this has been the case. Three of these plasmids, designated CAM (Rheinwald, Chakrabarty & Gunsalus, 1973), SAL (Chakrabarty, 1972) and NAH (Dunn & Gunsalus, 1973), are derived from strains of *P. putida*, and carry the genes necessary for the degradation of camphor, salicylate and naphthalene respectively. A fourth, OCT (Chakrabarty, Chou & Gunsalus, 1973), occurs in a strain of *P. oleovorans*, and specifies the degradation of n-octane (see Table 6.3 and also Chapter 5).

Mitomycin C treatment of these various *P. putida* and *P. oleovorans* strains results in the formation of cured (or deleted) derivatives lacking the

Table 6.3. Plasmid transfer in intrastrain matings of *Pseudomonas putida* and *Pseudomonas oleovorans*

Plasmid	Catabolic processes for which genes are plasmid linked	Native strain	Frequency of intrastrain plasmid transfer (per donor cell)
SAL	salicylate $\rightarrow$ pyruvate	*P. putida* R1	[a] 10^{-6}
CAM	*d*-, *l*-camphor $\rightarrow$ isobutyrate	*P. putida* PpG1	[b] 10^{-5}
OCT	n-octane $\rightarrow$ octanoic acid	*P. oleovorans*	[c] 10^{-9}
NAH	naphthalene $\rightarrow$ catechol	*P. putida* PpG7	[d] 10^{-8}

References: a. Chakrabarty, 1972; b. Rheinwald *et al.*, 1973; c. Chakrabarty *et al.*, 1973; d. Dunn & Gunsalus, 1973.

appropriate degradative functions, and these can only be regained by conjugation with plasmid-carrying cells, or, in the cases of CAM, SAL and NAH, by transduction with phage pf16. The frequency of transfer of these plasmids in intrastrain matings of their native hosts are shown in Table 6.3, the highest transfer frequency obtained being 10^{-5}/donor cell for CAM. The frequency of transfer of SAL is to some extent host specific, as increases of about 10-fold are found for SAL$^+$ derivatives of *P. putida* PpG1 compared to those from the native strain *P. putida* R1 (Chakrabarty, 1972). Nitroso-guanidine mutagenesis has been effective in producing 'high-frequency donor' strains in *P. putida* PpG7 (NAH$^+$), and these can transfer NAH at frequencies some 10^5-fold greater than that obtained from the wild type (Dunn & Gunsalus, 1973). Of these plasmids, only CAM has been shown to mediate transfer of chromosomal genes (cited in Rheinwald *et al.*, 1973).

Table 6.4 summarizes some of the available data on the transfer of these various plasmids in both interstrain and interspecies matings in *Pseudomonas*. Transfer to species outside this genus has not been reported. In general, the transfer frequencies obtained are very low, and this is in part due to DNA restriction. For example, *P. aeruginosa* PAO behaves as a poor recipient of these plasmids, yet Res$^-$ derivatives can accept SAL at frequencies 100- to 1000-fold greater than that shown by the wild type (Chakrabarty, 1972). Nevertheless, restriction alone does not entirely account for the low transfer frequencies and certainly NAH shows little difference in intra- and inter-strain transfer in *P. putida* strains PpG1 and PpG7 (Dunn & Gunsalus, 1973). As yet, no information is available as to the mechanism of transfer of these plasmids, and whether, for example, the 'high-frequency donors' of NAH represent strains derepressed for transfer functions. However, in several instances, interactions between these plasmids have been shown to occur which influence their transferability.

Table 6.4. Plasmid transfer in interspecies and interstrain matings of *Pseudomonas* spp.

Plasmid	Host strain	Recipient	Frequency of plasmid transfer (per donor cell)
SAL	*P. putida* R1	[a] *P. putida* PpG1	10^{-6}
		[a] *P. putida* PRS1	10^{-8}
		[a] *P. aeruginosa* PAO	$<10^{-9}$
		[a] *P. aeruginosa* PAO Res$^-$	10^{-7}
		[a] *P. fluorescens*	10^{-8}
		[a] *P. multivorans*	10^{-7}
OCT	*P. oleovorans*	[b] *P. aeruginosa* PAO	$<10^{-9}$
		[b] *P. multivorans*	$<10^{-9}$
		[b] *P. stutzeri*	$<10^{-9}$
		[b] *P. putida* PpG1	10^{-9}
		[b] *P. putida* U	10^{-7}
NAH	*P. putida* PpG7	[c] *P. putida* PpG1 (CAM$^-$)	10^{-8}
CAM	*P. putida* PpG1	[d] *P. aeruginosa* PAO	$<10^{-9}$
		[d] *P. aeruginosa* PAT	10^{-7}
		[d] *P. fluorescens*	10^{-5}
		[b] *P. oleovorans* (OCT$^+$)	10^{-9}
		[b] *P. olevorans* (OCT$^-$)	10^{-7}–10^{-8}

References: a. Chakrabarty, 1972; b. Chakrabarty *et al.*, 1973; c. Dunn & Gunsalus, 1973; d. Rheinwald *et al.*, 1973.

b.ii.   The Plasmid pfdm

A defective transducing phage which may be detected in cells as CCC DNA gives some idea as to how plasmids like OCT, SAL, NAH and CAM may have arisen. When the host range mutant pf16h2 of phage pf16 is used to transduce the mandelate gene cluster (*mdl*) from *P. putida* PRS1 to a mandelate-deleted derivative of *P. putida* PpG2, a class of transductants can be isolated which harbour defective phages (pfdm particles), in which part of the phage genome is apparently replaced by bacterial *mdl* genes (Chakrabarty & Gunsalus, 1969a). Bacteria carrying pfdm show the presence of covalently closed circular DNA which replicates independently of the bacterial chromosome, and strains carrying these elements will transfer the *mdl$^+$* character to *P. putida* PpG1 in mixed culture (Chakrabarty & Gunsalus, 1969b). Further-more, cell lines carrying pfdm will transfer chromosomal markers to appro-priate PpG1 recipients—albeit at a very low frequency (2–4×10^{-9}/donor cell). It is likely that gene transfer in these experiments is due to conjugation, since cell contact is a prerequisite for transfer and both transformation and phage-mediated transduction can be excluded. However some transient transduction phenomenon cannot be ruled out.

5. FACTORS INFLUENCING THE EFFICIENCY OF TRANSFER

5.a. Mutations that Affect Gene Transfer

Up to this point we have been concerned primarily with the various types of transfer that can occur among pseudomonads and the vectors involved. We now turn to factors that can influence the transfer process itself. Studies on plasmids in enteric bacteria show that the transfer process can be influenced by a large number of factors, one being mutation either in the donor or the recipient of a mating pair. Furthermore, when in the donor, the mutation may be either on the plasmid or on the chromosome.

In *P. aeruginosa* PAO, chromosomal mutations which enhance the frequency of both sex factor and chromosome transfer have been reported (Loutit, 1969a; Pemberton & Holloway, 1972a). One particular mutation designated *fer*-1, increases the recombination frequency obtained for chromosomal markers by 50–100-fold, and also the frequency of transfer of FP2 from certain *P. aeruginosa* donor strains (Pemberton & Holloway, 1972a). The mutation is effective when present in either donor or recipient strains, and apparently affects some component of the cell surface since *fer*-1 strains differ from the wild type in their precipitation characteristics. It may be that the enhanced frequency of gene transfer is due to an increased stability of pair formation as is indeed suggested from interrupted mating experiments involving *fer*-1 donors. Here, entry times for chromosomal markers are only obtained if very vigorous agitation is used to separate mating pairs. However, under these conditions the same map order is obtained from *fer* mutants as with the wild type.

Sex factor mutations which increase donor ability have been reported for FP2 in *P. aeruginosa* strains PAO and PAT (Loutit, 1969a; Stanisich & Holloway, 1972). However, the maximum increase in recombination frequency observed is only about 10-fold, and the nature of the changes involved are not known. Mutants of FP2 and of several R factors which are impaired in their transfer properties have also been isolated. The FP2 mutants have been detected by their reduced ability to transfer chromosomal genes, and the R factor mutants by their reduced ability to transfer R-linked antibiotic-resistance genes. The latter screening procedure is also available for FP2 in *P. aeruginosa* PAO, by virtue of the linked mercury-resistance determinant associated with this sex factor (Loutit, 1970).

The FP2 mutants that have been isolated are of two types (Stanisich, 1973). One class shows only a slight reduction in conjugal fertility which is apparently associated with a change in the cell surface. The precipitation characteristics of these males are identical to those found for recipient strains, and differ from those of wild-type males. The second class of mutants show wild-type precipitation characteristics but a marked reduction in both sex factor and chromosome transferability. Although little is known at this stage of the biochemical changes which have taken place in these mutants, they may well be analogous to the transfer defective (*tra*) F factor mutants isolated in *Escherichia coli* (Willetts, 1972). Transfer defective mutants of the R factors RP1-1 and R91 have also been isolated, but these have yet to be fully characterized (T. Morgan & V. Stanisich, unpublished experiments).

5.b. Plasmid–plasmid Interactions and Their Effect on Plasmid Transfer

Another factor that influences the transfer process is whether there is an interfering plasmid already present in the potential recipient. The investigation of the effects of plasmid interaction on transfer frequencies in *P. aeruginosa* has only recently begun and our understanding of the phenomenon is still very limited. However, in those cases where it has been tested, mutual exclusion occurs between strains of *P. aeruginosa* harbouring homologous (that is, identical) plasmids. This property of plasmid-carrying cells can be identified in either of two ways: the transfer frequency of the plasmid itself will be reduced in matings with strains carrying the same plasmid, as will also the recombination frequency for chromosomal markers. The latter, of course, can only be used as an index of exclusion in those instances where sex factor mediated chromosome transfer occurs at appreciable frequencies. The occurrence of exclusion is therefore taken as evidence of the similarity and perhaps evolutionary relationship of heterologous (that is, independently isolated) plasmids.

Thus, in matings where both parents carry either the FP2, FP39, FP8 and RP1 sex factors, the recombination frequency for chromosomal genes is reduced by a factor of at least 10^3 from that obtained with these donors and plasmid-less recipients (Holloway, 1956; Pemberton, 1971; P. J. Dry & V. Stanisich, unpublished experiments). In the case of FP2, sex factor mutations can result in the formation of donor strains no longer exhibiting entry exclusion. Thus males carrying the mutant sex factor FP* are equally fertile in matings with recipient strains as with strains carrying the wild-type (FP2) or transfer defective (FPd) sex factors. In those matings where two donor strains are involved, i.e. FP* × FP* or FP* × FP⁺, chromosome transfer can occur to either of the parent strains involved (Stanisich & Holloway, 1972; Stanisich, 1973).

Exclusion between *Pseudomonas* strains carrying heterologous sex factors has been observed in several instances, and this may be interpreted as indicating some relationship between these elements. Thus in *P. aeruginosa*, a marked reduction in recombination frequency is observed in matings between FP2 donors and those harbouring the wild-type sex factor FP8 (P. J. Dry, unpublished data), while the presence of FP2 reduces the frequency of inheritance of FP39 by a factor of 10^3 (Pemberton & Holloway, 1973). This latter case of exclusion is however non-reciprocal, since FP2 is transferable to FP39-carrying cells as easily as to those lacking this sex factor. Inheritance of the R factor R1822 is also reduced to strains carrying the wild-type FP2 factor whereas no effect is observed with the mutant FP* (Olsen & Shipley, 1973). This indicates that the mutation in FP* is effective not only in overcoming the barrier to gene transfer between strains harbouring homologous sex factors, but can extend also to include heterologous plasmids. As all the interactions described here involve FP2, it would be of some interest to pursue further an investigation of the exclusion and compatibility properties of these various plasmids as a means of defining more precisely what their relationships might be.

Of the other plasmids which have been studied in *P. aeruginosa*, no detectable effect on transfer frequency has been observed. For example RP1-1 in a recipient line does not affect its inheritance of RP1 (Sykes, 1972), and similarly neither FP2 nor FP39 impair the inheritance of R18-1, R91 or R68 (Stanisich, 1972). In *P. putida*, exclusion seems to be operative in matings between strains harbouring CAM and those harbouring either OCT or NAH (Chakrabarty *et al.*, 1973; Dunn & Gunsalus, 1973). Thus, the presence of either of the latter two plasmids reduces by 10 to 10^3 their frequency of inheritance of CAM. In that instance where the reciprocal mating was also studied, the presence of CAM increased rather than decreased the frequency of inheritance of OCT. However the OCT^+CAM^+ derivatives obtained from either mating were unstable and segregated one or other of the two plasmids (Chakrabarty *et al.*, 1973).

The presence of two self-transmissible plasmids in a donor may influence the frequency of transfer of one of the plasmids to a plasmid-less recipient, and several such examples have been reported in *P. aeruginosa*. Thus males carrying both FP2 and FP39 (Pemberton & Holloway, 1973) or FP2 and R1822 (Olsen & Shipley, 1973) show a lower transfer of FP39 and R1822 respectively than is seen from strains carrying these plasmids alone. On the other hand, strains carrying both FP2 and R91 give lower transfer frequencies for the FP2 component whether this transfer is measured in terms of sex factor transfer or the transfer of chromosomal genes (Stanisich, 1972).

The last type of interaction between two plasmids that has been documented reasonably well is the one where a transmissible element mobilizes a plasmid that is itself not self-transmissible. Thus of eight R factors that have been studied, all will mobilize the mercury-resistance plasmid present in *P. aeruginosa* PAT (Stanisich, 1972), while the sex factor FP* can mobilize mutants of FP2 that have lost their transfer properties (Stanisich, 1973). As is the case with many of the experiments involving plasmid interactions in *P. aeruginosa*, it is difficult yet to be sure what the biochemical processes at work may be. Nevertheless plasmid interactions in *P. aeruginosa*, and particularly the mobilization of non-self-transmissible plasmids by other elements, has great potential importance for those interested in basic studies and those studying the interactions of plasmids in this group of organisms at an environmental level.

6. BIBLIOGRAPHY

Adams, M. H. (1959) in *Bacteriophages*, Interscience, New York.

Arber, W. (1960) *Virology*, 11, 273.

Arber, W. & Linn, S. (1969) *Annual Reviews of Biochemistry*, 38, 467.

Black, W. A. & Girdwood, R. W. A. (1969) *British Medical Journal*, iv, 234.

Boyer, H. W. (1971) *Annual Reviews of Microbiology*, 25, 153.

Bradley, D. E. *Genetical Research, Cambridge*, 19, 39.

Brinton, C. C. (1971) *Critical Reviews in Microbiology*, 1, 105.

Bryan, L. E., van den Elzen, H. M. & Tseng, J. T. (1972) *Antimicrobial Agents and Chemotherapy*, 1, 22.

Carey, K. E. & Krishnapillai, V. (1974) *Genetical Research, Cambridge*, 23, 155.

Chabbert, Y. A., Scavizzi, M. R., Witchitz, J. L., Gerbaud, G. R. & Bouanchaud, D. H. (1972) *Journal of Bacteriology*, 112, 666.

Chakrabarty, A. M. (1972) *Journal of Bacteriology*, 112, 815.

Chakrabarty, A. M., Chou, G. & Gunsalus, I. C. (1973) *Proceedings of the National Academy of Sciences, U.S.A.*, 70, 1137.

Chakrabarty, A. M. & Gunsalus, I. C. (1969a) *Virology*, 38, 92.

Chakrabarty, A. M. & Gunsalus, I. C. (1969b) *Proceedings of the National Academy of Sciences, U.S.A.*, 64, 1217.

Chakrabarty, A. M. & Gunsalus, I. C. (1970) *Journal of Bacteriology*, 103, 830.

Chakrabarty, A. M., Gunsalus, C. F. & Gunsalus, I. C. (1968) *Proceedings of the National Academy of Sciences, U.S.A.*, 60, 168.

CIBA Symposium (1968) *Bacterial Episomes and Plasmids*, (Eds. G. E. W. Wolstenholme & M. O'Connor), J. and A. Churchill, London.

Clarke, P. H. (1970) *Advances in Microbial Physiology*, 4, 179.

Clowes, R. C. (1972) *Bacteriological Reviews*, 36, 361.

Curtiss, R. (1969) *Annual Reviews of Microbiology*, 23, 69.

Curtiss, R., Charamella, L. J., Stallions, D. R. & Mays, J. A. (1968) *Bacteriological Reviews*, 32, 320.

Datta, N., Hedges, R. W., Shaw, E. J., Sykes, R. B. & Richmond, M. H. (1972) *Journal of Bacteriology*, 108, 1244.

de Haan, P. G., Hoekstra, W. P. M., Verhoef, C. & Felix, H. S. (1969) *Mutation Research,* 8, 505.

Dunn, N. W. & Gunsalus, I. C. (1973) *Journal of Bacteriology*, 114, 974.

Dunn, N. W. & Holloway, B. W. (1970) Symposium on *Uptake of Informative Molecules in Biological Systems*, Brussels.

Fuerst, J. A. & Hayward, A. C. (1969) *Journal of General Microbiology*, 58, 227.

Grinsted, J., Saunders, J. R., Ingram, L. C., Sykes, R. B. & Richmond, M. H. (1971) *Journal of Bacteriology*, 110, 529.

Gunsalus, I. C., Gunsalus, C. F., Chakrabarty, A. M., Sikes, S. & Crawford, I. P. (1968) *Genetics, Princeton*, 60, 419.

Hayes, W. (1968) in *The Genetics of Bacteria and Their Viruses*, Blackwell Scientific Publications, Oxford and Edinburgh.

Holloway, B. W. (1955) *Journal of General Microbiology*, 13, 572.

Holloway, B. W. (1956) *Journal of General Microbiology*, 15, 221.

Holloway, B. W. (1965) *Virology*, 25, 634.

Holloway, B. W. (1969) *Bacteriological Reviews*, 33, 419.

Holloway, B. W., Egan, J. B. & Monk, M. (1960) *Australian Journal of Experimental Biology and Medical Sciences*, **38**, 321.
Holloway, B. W., Krishnapillai, V. & Stanisich, V. (1971) *Annual Reviews of Genetics*, **5**, 425.
Holloway, B. W., Monk, M., Hodgins, L. & Fargie, B. (1962) *Virology*, **18**, 89.
Holloway, B. W. & van de Putte, P. (1968b) *Nature, London*, **271**, 459.
  Material, (Eds. W. J. Peacock & R. D. Brock), Australian Academy of Science, Canberra, p. 175.
Holloway, B. W. & van de Putte, P. (1968b) *Nature, London*, 459.
Ikeda, H. & Tomizawa, J. (1965) *Journal of Molecular Biology*, **14**, 110.
Ingram, L. C., Sykes, R. B., Grinsted, J., Saunders, J. R. & Richmond, M. H. (1972) *Journal of General Microbiology*, **72**, 269.
Kawakami, Y., Mikoshiba, F., Nagasaki, S., Matsumoto, H. & Tazaki, T. (1972) *Journal of Antibiotics*, **25**, 607.
Khan, N. C. & Sen, S. P. (1967) *Journal of General Microbiology*, **49**, 201.
Krishnapillai, V. (1971) *Molecular and General Genetics*, **114**, 134.
Krishnapillai, V. & Carey, K. E. (1972) *Genetical Research, Cambridge*, **20**, 137.
Lambina, V. A. & Mikhailova, T. N. (1964) *Mikrobiologica*, **33**, 800.
Lebek, G. (1963) *Zentrallblatt fur Bakteriologie*, **189**, 213.
Loutit, J. S. (1960) *Proceedings of the University of Otago Medical School*, **38**, 4.
Loutit, J. S. (1969a) *Genetical Research, Cambridge*, **14**, 103.
Loutit, J. S. (1969b) *Genetical Research, Cambridge*, **13**, 91.
Loutit, J. S. (1970) *Genetical Research, Cambridge*, **16**, 179.
Loutit, J. S., Marinus, M. G. & Pearce, L. E. (1968) *Genetical Research, Cambridge*, **12**, 139.
Loutit, J. S., Pearce, L. E. & Marinus, M. G. (1968) *Genetical Research, Cambridge*, **12**, 29.
Lowbury, E. J. L., Kidson, A., Lilly, H. A., Ayliffe, G. A. J. & Jones, R. J. (1969) *Lancet*, **ii**, 448.
Meynell, G. G. (1973) in *Bacterial Plasmids*, Macmillan, London.
Meynell, E., Meynell, G. G. & Datta, N. (1968) *Bacteriological Reviews*, **32**, 55.
New York Symposium (1971) *The Problems of Drug-Resistant Pathogenic Bacteria. Annals of the N.Y. Academy of Sciences*, Vol. 182.
Novick, R. P. (1969) *Bacteriological Reviews*, **33**, 210.
Olsen, R. H., Metcalf, E. S. & Todd, J. K. (1968) *Journal of Virology*, **2**, 357.
Olsen, R. H. & Shipley, P. (1973) *Journal of Bacteriology*, **113**, 772.
Ozeki, H. & Ikeda, H. (1968) *Annual Reviews of Genetics*, **2**, 245.
Patterson, A. C. (1965) *Journal of General Microbiology*, **39**, 295.
Pearce, L. E. & Loutit, J. S. (1965) *Journal of Bacteriology*, **89**, 58.
Pemberton, J. M. (1971) *Ph.D Thesis*, Monash University, Australia.
Pemberton, J. M. & Clark, A. J. (1973) *Journal of Bacteriology*, **114**, 424.
Pemberton, J. M. & Holloway, B. W. (1972a) *Australian Journal of Experimental Biology and Medical Sciences*, **50**, 577.
Pemberton, J. M. & Holloway, B. W. (1972b) *Genetical Research, Cambridge*, **19**, 251.
Pemberton, J. M. & Holloway, B. W. (1973) *Genetical Research, Cambridge*, **21**, 263.
Pühler, A., Burkhardt, H. J. & Heumann, W. (1972) *Journal of General Microbiology*, **73**, xxvi.
Rheinwald, J. G., Chakrabarty, A. M. & Gunsalus, I. C. (1973) *Proceedings of the National Academy of Sciences, U.S.A.*, **70**, 885.
Tichmond, M. H. (1973) *Progress in Nucleic Acid Research and Molecular Biology*, **13**, 191.

Richmond, M. H. & Sykes, R. B. (1973) *Genetical Research, Cambridge*, **20**, 231.
Rolfe, B. & Holloway, B. W. (1966) *Journal of Bacteriology*, **92**, 43.
Rolfe, B. & Holloway, B. W. (1969) *Genetics, Princeton*, **61**, 341.
Saunders, J. R. & Grinsted, J. (1972) *Journal of Bacteriology*, **112**, 690.
Schans, G. P. van der, Weyermans, J. R. & Bleichrodt, J. F. (1971) *Molecular and General Genetics*, **110**, 263.
Smith, D. H. & Armour, S. E. (1966) *Lancet*, **ii**, 15.
Stanisich, V. A. (1972) *Ph.D Thesis*, Monash University, Australia.
Stanisich, V. A. (1973) *Genetical Research, Cambridge*, **22**, 13.
Stanisich, V. A. & Holloway, B. W. (1969) *Genetics, Princeton*, **61**, 327.
Stanisich, V. A. & Holloway, B. W. (1971) *Genetical Research, Cambridge*, **17**, 169.
Stanisich, V. A. & Holloway, B. W. (1972) *Genetical Research, Cambridge*, **19**, 91.
Stent, G. S. (1971) in *Molecular Genetics*, W. H. Freeman & Co., San Francisco.
Sykes, R. B. (1972) *Ph.D. Thesis*, University of Bristol.
Sykes, R. B. & Richmond, M. H. (1970) *Nature, London*, **226**, 952.
Watanabe, T. (1963) *Bacteriological Reviews*, **27**, 87.
Willetts, N. (1972) *Annual Review of Genetics*, **6**, 257.

CHAPTER 7

Metabolic Pathways and Regulation: I

PATRICIA H. CLARKE and L. NICHOLAS ORNSTON

1. INTRODUCTION

The enormous metabolic potential of the *Pseudomonas* species has made them a favourite subject of study for biochemists interested in exploring new pathways. For this purpose they have either used strains previously isolated and identified, or freshly isolated strains from enrichment cultures. There are obvious advantages in working with well-known laboratory strains of bacteria and the recent development of genetic systems in *P. aeruginosa* and *P. putida* has provided more incentive to work with known strains of these species. However, variability between strains and the evidence that the genes for some catabolic enzymes may be carried on easily lost plasmids, suggests that freshly isolated strains from natural sources may be more useful for some investigations. The topics which we discuss in this and the following chapter, have to a large extent been selected for us by the people who have worked on the biochemistry of *Pseudomonas*. The biochemical data available cover a very wide range of catabolic and biosynthetic activities but the coverage is very uneven. Some catabolic pathways, such as the *β*-ketoadipate pathway for the breakdown of aromatic compounds, have been subjected to detailed investigations in several laboratories over a number of years. Other pathways have been examined only in enough detail to demonstrate that they exist, and in some cases it is possible to glimpse fascinating and unexploited riches. The biosynthetic pathways in general have been less extensively investigated than the catabolic pathways, although the biochemical and genetic aspects of a few of the amino acid biosynthetic pathways are now fairly well understood.

We have chosen to concentrate on those aspects of the biochemistry of *Pseudomonas* which have been explored in most detail and where it is possible to draw a fairly coherent picture. In addition, we have been biased by a preference for those systems which offer the most interesting solutions to the problems of regulation of enzyme synthesis and enzyme activity. Most emphasis will be laid on *P. aeruginosa* and *P. putida* but we have also attempted to compare the various biochemical pathways and their regulation in some of the other *Pseudomonas* species discussed in Chapter 1. In stressing the special aspects of *Pseudomonas* biochemistry we risk giving a distorted view of the biochemistry of the genus, but the earlier years of microbial biochemistry were devoted to establishing the universality of the fundamental biochemical activities of living organisms and this need not be repeated. Here we are concerned to understand the diversity of a biochemically enterprising bacterial genus and although there will be some discussion of biosynthetic pathways the major emphasis will be on catabolic pathways. In Chapter 7 we discuss the more general aspects of *Pseudomonas* metabolism and its regulation and some pathways will be considered in more detail in Chapter 8.

2. OXIDATIVE REACTIONS

2.a. Cytochromes

The pseudomonads are aerobic organisms deriving their energy from oxidative reactions, although nitrate can replace oxygen as the terminal electron acceptor for the denitrifying species. The cytochromes present have mainly been recognized by their absorption spectra in whole cells, in extracts or in membrane preparations. It is assumed that they are organized in the bacterial membrane as part of the structure of the electron transport chain and are involved in the reactions of oxidative phosphorylation. Recent general reviews on bacterial cytochromes include those by Bartsch (1968), Horio and Kamen (1970), Kamen and Horio (1970). Stanier and his colleagues (1966) determined the difference spectra of the cytochromes present in heavy suspensions of representative strains of several species (Table 7.1). The bands indicated the presence of *a*, *b*- and *c*-type cytochromes in all the species examined with the exception that *P. maltophilia* gave no absorption in the regions 551–554 and 522–524 characteristic of the α- and β-bands of *c*-type cytochromes.

Table 7.1. Wavelengths (nm) of the main absorption peaks in the cytochrome difference spectra of selected strains of the different species of aerobic pseudomonads

Species	α-Bands. Cytochrome type			β-Bands. Cytochrome type		Principal Soret band
	a	*b*	*c*	*b*	*c*	
P. aeruginosa		c. 560	552	530	523	428
P. fluorescens		c. 560	552	530	523	427
P. putida		c. 560	552	530	523	425
P. acidovorans	600		553	530	524	425
P. testosteroni	600		553	530	523	423
P. alcaligenes	600	c. 560	553	530	523	428
P. pseudoalcaligenes	600	c. 560	551	528	522	423
P. multivorans	628, 597	c. 560	554	527	523	428
P. stutzeri		(c. 560)	552	530	523	423
P. maltophilia	628, 597	558		530		430
P. lemoignei		557	552	530	523	422

Cytochromes *c* are defined as possessing haem groups covalently bound to their polypeptide chains. A number of different types of cytochrome *c* have been isolated from bacterial sources, some of which have been soluble and very readily extracted, while others have been solubilized from particulate fractions of the bacterial membrane, sometimes tightly bound to other cytochromes. A low molecular weight cytochrome *c*-551 was isolated from *P. aeruginosa* by Horio and his colleagues (1960) and the sequence was determined by Ambler (1963). This cytochrome *c* contained 82 amino acids in a single polypeptide chain with a haem group covalently attached to cysteine residues 12 and 14. Similar cytochromes *c*-551 were isolated from *P. stutzeri*, *P. mendocina* and *P. fluorescens* biotype C. These species are all capable of denitrification and the various cytochromes may constitute several per cent of the total bacterial protein (Kodama & Shidara, 1969). Ambler and Wynn (1973) determined the sequences of the cytochromes c-551 from the four denitrifying *Pseudomonas* species and found that although the proteins were clearly homologous the differences in the amino acid sequences between pairs of species ranged from 22 % to 39 % suggesting considerable evolutionary divergence. On the other hand, comparison of the sequences of cytochromes c-551 from different strains of *P. aeruginosa* showed very little interstrain variation (Ambler, 1974).

Another c-type cytochrome was isolated from *P. stutzeri* by Kodama and Shidara (1969) which resembled cytochrome c_4 previously found in *Azotobacter vinelandii* with an α-band maximum at about 552 nm and two haem groups covalently attached to each sub-unit of molecular weight about 20,000 to 25,000 daltons. Ambler and Murray (1973) compared the N-terminal amino acid sequences of cytochromes c_4 from *P. aeruginosa*, *P. mendocina*, *P. stutzeri* and *A. vinelandii* and found that they were very similar. They also resembled cytochrome c_2 isolated from *Rhodospirillum rubrum* (Dus, *et al.*, 1968). Ambler and Taylor (1973) also isolated and sequenced cytochrome c_5 from *P. mendocina*. The cytochromes c_5 have α-band maxima at about 555 nm and c_5 from *P. mendocina* was similar in molecular weight to c-551, about 10,000 daltons, with a single haem group covalently linked. There was however very little homology in the sequences of the polypeptides of cytochromes c_5 and c-551 from this organism.

Two cytochrome oxidases have been identified in *P. aeruginosa*. Horio and his colleagues (1961) purified and crystallized a cytochrome oxidase from cultures grown anaerobically in the presence of nitrate. This cytochrome oxidase was shown to have nitrate reductase activity, to be membrane bound and to contain both a c-type and a d-type cytochrome (Yamanaka & Okunuki, 1963; Yamanaka 1964). Cytochromes d do not contain haem but chlorin derivatives (dihydroporphyrin). The nitrate reductase and cytochrome oxidase activities were both present in the purified preparation and it was designated cytochrome c-(551 *P. aeruginosa*): nitrite O_2 oxidoreductase. Nitrite is reduced to nitric oxide with reduced cytochrome c-551 under anaerobic conditions. It cannot oxidize reduced cytochrome c prepared from beef heart muscle but can oxidize cytochromes c from some other bacteria. Yamanaka (1967) suggested that this might be a very primitive cytochrome oxidase which has originally evolved in the pre-oxygen era. When primitive organisms had acquired the ability to reduce nitrate to nitrite a system would be required to remove the nitrite from the environment. If the *Pseudomonas* nitrite reductase had coincidentally acquired the capacity to reduce oxygen, then it would continue to function after the time at which oxygen appeared, although later is might be superseded by more efficient cytochrome oxidases of the haem a type.

While the nitrite reductase–cytochrome oxidase was found only in cultures grown anaerobically in the presence of nitrate, there was evidence for additional cytochrome oxidase activity in both anaerobic and aerobic cultures. Azoulay and Couchoud-Beaumont (1965) partially purified cytochrome oxidase from the particulate fraction of extracts of *P. aeruginosa* grown aerobically. The activity was located entirely in the small particle fraction obtained by disruption of the bacterial membrane. This cytochrome oxidase appeared to have a type a_1 haem and was associated with a cytochrome c-552. It was shown to be able to oxidize reduced beef heart cytochrome c and the bacterial extract could be used to provide a 'NADH reductase' so that it appeared to be able to function at the end of the chain for oxidation of the reduced coenzymes generated in metabolism. It was of particular interest that the aerobic cytochrome oxidase was severely repressed by glucose while the anerobic nitrite reductase–cytochrome oxidase was unaffected by the presence of glucose in the medium. These two cytochrome oxidases could therefore be clearly distinguished by their different regulatory properties.

Very little is known about the structural arrangement of the cytochromes of *Pseudomonas* in the membranes and the composition and spatial arrangements of the electron transport chains. There are some reports of the isolation from *Pseudomonas* of enzyme complexes associated with various cytochromes. The methylene hydroxylase of *P. putida* which is required for the initial metabolic attack on camphor has been isolated and studied by Gunsalus and colleagues. It contains cytochrome P-450$_{CAM}$ together with a flavoprotein reductase and the iron–sulphur protein, putidaredoxin (Sharrock *et al.*, 1973). However this enzyme complex is found in the soluble portion of the bacterial extract and is therefore more readily isolated than the enzyme complexes located in the membrane. The mechanism of this enzyme system has been studied in detail and its function will be discussed further in a later section.

Jones and Hughes (1972) have isolated and studied the nicotinate oxidase complex from *P. putida* (*P. ovalis* Chester) with a view to understanding the structure and functional organization of the oxidative enzymes and electron carriers of the bacterial membrane. Nicotinate oxidase activity was associated with an electron transport chain comprising *b*- and *c*-type cytochromes. They were also able to identify two possible terminal oxidases, a cytochrome *o* component inhibited by CO with an absorption maximum at 417 nm, and another CO-sensitive component which behaved as a cytochrome *c* peroxidase. They suggest that both of these components can act as terminal oxidases for nicotinate and M. V. Jones (personal communication) found that oxidation of nicotinate with the isolated membrane fractions was accompanied by oxidative phosphorylation producing ATP. Cytochrome *o* has also been detected in *P. oleovorans* and Peterson (1970) found that hexane-grown cultures contained 4·5 times as much cytochrome *o* as glucose-grown cultures. This was accompanied by an increase in the ω-hydroxylase which initiates the initial attack on hydrocarbons and fatty acids in this organism. Cytochrome *c* peroxidase was isolated from *P. fluorescens* by Ellfolk and Soininen (1970) who reported that it was located in the membrane and formed 0·5 % of the protein of the cell-free preparations. Among other proteins associated with cytochromes is azurin, a blue copper-containing protein purified from *P. aeruginosa* by Horio (1958), who suggested that it acted as a carrier between cytochrome *c*-551 and cytochrome oxidase. The sequence of *P. aeruginosa* azurin was determined by Ambler and Brown (1967).

2.b. Nitrate and Nitrite

P. aeruginosa can use nitrate as the sole nitrogen source for growth (nitrate assimilation) and can also grow under anaerobic conditions with certain carbon compounds by using nitrate as the terminal electron acceptor providing a pathway for oxidative phosphorylation (nitrate respiration). For nitrate assimilation, nitrate is reduced to ammonia which is assimilated into organic compounds, but in nitrate respiration it is reduce to nitrite and subsequently to nitrogen or oxides of nitrogen. The loss of fixed nitrogen from the soil and from tropical seas by denitrification has been ascribed to bacterial action and the pseudomonads have been implicated (Nason & Takahashi, 1958). Palleroni points out the unsatisfactory nature of the descriptions of the denitrifying bacteria in the earlier bacteriological literature and lists five well-defined *Pseudomonas* species for which denititrification has been shown to be a characteristic property (Palleroni *et al.*, 1970). These are: *P. aeruginosa*; *P. fluorescens* (biotypes B, C, D only); *P. stutzeri*; *P. mendocina*; *P. pseudomallei*; *P. caryophyllii.*

Nitrate respiration depends on nitrate and nitrite reductase activities which are produced under anaerobic conditions in the presence of nitrate. Nitrate cannot support anaerobic growth on compounds for which O_2 is required for the initial metabolic reaction. Most reports suggest that these nitrate and nitrite reductase activities are repressed by oxygen which would fit the observations that denitrification is less marked in well-aerated soils. However, there may be some variation between strains and Kefauver and Allison (1957) found that nitrite utilization by *P. stutzeri* was completely repressed in cultures aerated with 20 % oxygen in nitrogen, but that when the oxygen concentration was reduced to 6 % the cultures were able to utilize oxygen and nitrite simultaneously.

Fewson and Nicholas (1961) purified from *P. aeruginosa* a membrane-bound nitrate reductase containing molybdenum, FAD and cytochrome *c* which reduced nitrate to nitrite in the presence of NADH. This nitrate reductase (A) appeared to be involved in nitrate respiration and was repressed by oxygen while a second nitrate reductase (B), identified by Pichinoty and his colleagues (1969), was constitutive and appeared to be involved in nitrate assimilation although these activities may not be due to completely separate entities. Nitrite reductase (A) was shown by Yamanaka and Okunuki (1963) to contain *o*- and *d*-type cytochromes and to possess cytochrome oxidase activity.

The nitrate reducing systems of the enterobacteria are known to be complex and determined by a number of different genes. Van Hartingsveldt, Marinus and Stouthamer (1971) showed that this is also the case for *P. aeruginosa*. Mutants defective in anaerobic growth in the presence of nitrate were assigned to three groups; *nar* mutants lacking nitrate reductase; *nir* mutants lacking nitrite reductase; *ana* mutants which were normal in both nitrate and nitrite reductase activities but which were unable to grow anaerobically with nitrate. The *nar* mutants were all chlorate resistant and fell into five separate transduction groups. The *nar*C mutants were able to use nitrate as a nitrogen source for growth and the *nar*D mutants could do so in the presence of molybdenum, while the *nar*A B and E mutants were unable to assimilate nitrate. It was suggested by van Hartingsveldt and Stouthamer (1973) that these findings indicated that there were common components of the nitrate assimilatory and respiratory systems.

The *nir* mutants were similarly assigned to three transduction groups (*nir*A, *nir*D and *nir*E), all lacked nitrite reductase activity and accumulated nitrite when grown semi-anaerobically in the presence of nitrate. The cytochrome spectra of these *nir* mutants lacked absorption bands at 465 and 625 nm which van Hartingsveldt and Stouthamer (1973) suggest may be due to the absence of haem *d*. Cytochrome oxidase activities of the *nir*

Table 7.2. Times of entry of genes determining nitrate and nitrite respiration in *Pseudomonas aeruginosa*

Time in minutes + FP2 donor	Markers
9	*nar*E
29	*nir*A
30	*ana*A
44	*nar*B
45	*nar*C
65	*nar*A, *nar*D
70	*nir*D, *nir*E

Data from van Hartingsveldt & Stouthamer (1973).

mutants grown aerobically were the same as the wild type but when the cultures were grown semi-anaerobically in the presence of nitrate, there was a large increase in the cytochrome oxidase activity of the wild-type strain but no increase for the *nir* mutants. Mapping by interrupted mating gave approximate times of entry for these genes during conjugation with an FP2 donor, as shown in Table 7.2 (van Hartingsveldt & Stouthamer, 1973).

2.c. Oxygenases

The first step in the oxidation of many paraffinic and aromatic compounds is carried out by the action of oxygenases, enzymes which incorporate molecular oxygen directly into the chemical structure of the organic substrate. Oxygenase reactions are highly exergonic and none of the free energy released is conserved by formation of pyrophosphate bonds in ATP. Rather, the *Pseudomonas* species appear to use oxygenases to lower the stabilization energy of growth substrates and to render the catabolism of these compounds thermodynamically irreversible. Oxygen is an absolute requirement for the metabolism of these compounds and cannot be replaced by nitrate.

The oxygenases of *Pseudomonas* have been studied in great detail by Hayaishi and his colleagues and several of them have been crystallized. The first of the oxygenases to be crystallized was metapyrocatechase which incorporates two atoms of oxygen across a double bond between one hydroxylated carbon atom and an unsubstituted carbon atom of catechol with the formation of α-hydroxymuconic semialdehyde after ring fission. This enzyme appeared to be very labile in cell-free preparations, but activity was retained in the cell extracts were kept under nitrogen, or if low concentrations of organic solvents such as acetone and alcohol were added to the preparations. Since then, methods have been devised to purify and crystallize other oxygenases of *Pseudomonas* species. These are discussed in an excellent review by Hayaishi (1966).

With the exception of the common substrate oxygen, the oxygenases form a highly diverse group of enzymes which differ in structure, mechanism and coenzyme requirements. Some resemble the oxygenases occurring in mammalian tissues which are particularly important in detoxification reactions (Hayaishi, 1966), but unlike the mammalian enzymes many of the *Pseudomonas* oxygenases are soluble and can be readily extracted from the bacterial cells. The oxygenases can be classified into two groups; the di-oxygenases which incorporate two atoms of oxygen into one molecule of substrate, and the mono-oxygenases which add only one atom of oxygen. The mono-oxygenases add one atom of oxygen to the substrate and the other atom is reduced to water so that these enzymes can be regarded as part oxygenase and part oxidase and are frequently termed mixed-function oxidases. In order to reduce the second atom of oxygen the system requires an electron donor; these may include pyridine nucleotides, flavine nucleotides, ascorbic acid or reduced pteridine nucleotides. In some cases the substrate itself may act as both electron donor and electron acceptor as in the action of lysine mono-oxygenase from *P. fluorescens* (Yamamoto, *et al.*, 1972).

Dioxygenases are concerned at several of the stages of the pathways for the catabolism of aromatic compounds especially in the reactions which give rise to catechol or its derivatives, or convert these compounds into the next intermediates of the pathways. As shown in Figure 7.1, peroxide derivatives have been postulated as intermediates in the oxygenation of benzene (Gibson *et al.*, 1968), benzoate (Reiner, 1972) and anthranilate (Kobayashi *et al.*, 1964). Supporting evidence has come from the isolation of the closely related metabolites benzene glycol (Gibson *et al.*, 1970) and dihydrodihydroxy-benzoate (Reiner & Hegeman, 1971; Reiner, 1971) as intermediates in the utilization of benzene and catechol, respectively. These mechanisms of hydroxylation differ markedly from those proposed for the mammalian enzymes which are thought to introduce vicinal hydroxyl groups *via* a mono-oxygenase with the formation of an epoxide intermediate (Jerina *et al.*, 1971). Peroxide intermediates may be formed in the cleavage of catechol by both '*ortho*' or '*meta*' ring cleavage (Figure 7.2) (Hayaishi, 1962; 1966). Other evidence for the hypothesis of an oxygenated intermediate in these reactions has come from kinetic studies of the mechanism of protocatechuate 3,4-dioxygenase from *P. aeruginosa* (Fujisawa *et al.*, 1972).

Despite the similarity of the oxygenative sequences depicted in Figure 7.1, the oxygenases themselves may utilize quite different mechanisms. An indication of the possible diversity is given by the observation that the '*ortho*' catechol 1,2-dioxygenase contains non-haem ferric ion and is deep red (Kojima *et al.*, 1967) whereas the '*meta*' catechol 2,3-dioxygenase contains ferrous ion and is colourless (Nozaki *et al.*, 1963).

The substrate specificity of some oxygenases is fairly broad. The '*meta*' catechol dioxygenase from fluorescent pseudomonads cleaves a number of ring-substituted alkyl derivatives of catechol (Ribbons, 1970; Sala-Trepat *et al.*, 1972; Bayly & Wigmore, 1973). Figure 7.2 depicts the action of a *Pseudomonas* dioxygenase upon gentisate and a number of related compounds (Hopper *et al.*, 1971). It is clear that these dioxygenases act upon a number of different diphenols and thus contribute a great deal to the metabolic versatility of *Pseudomonas*. The broad specificity of some of these enzymes may allow aromatic compounds with various ring substituents to be metabolized by the same enzymes. Evans and his colleagues (1972) found that the 3,4-oxygenase from a *Pseudomonas* species grown on 2,4-dichlorophenoxy-acetate (2,4-D) was able to attack 3-chlorocatechol and 5-chlorocatechol as well as 3,5-dichlorocatechol. A further demonstration of the catabolic potential of *Pseudomonas* is given by the large number of dioxygenases that

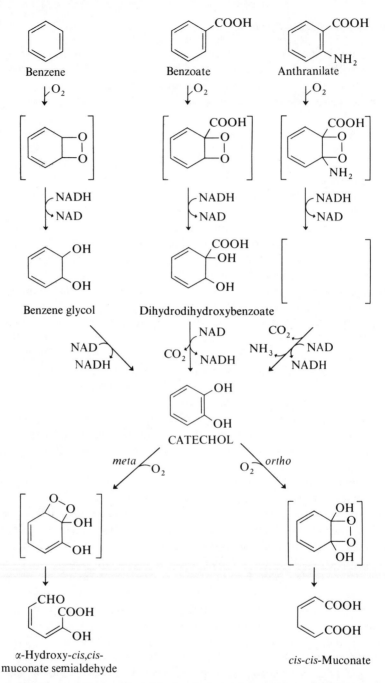

Figure 7.1. The structures enclosed by brackets are postulated peroxide intermediates in the enzymatic formation and cleavage of catechol by *Pseudomonas* species. The intermediates BG (benzene glycol) and DHB (dihydrodihydroxybenzoate) have been isolated (Gibson *et al.*, 1970; Reiner, 1972)

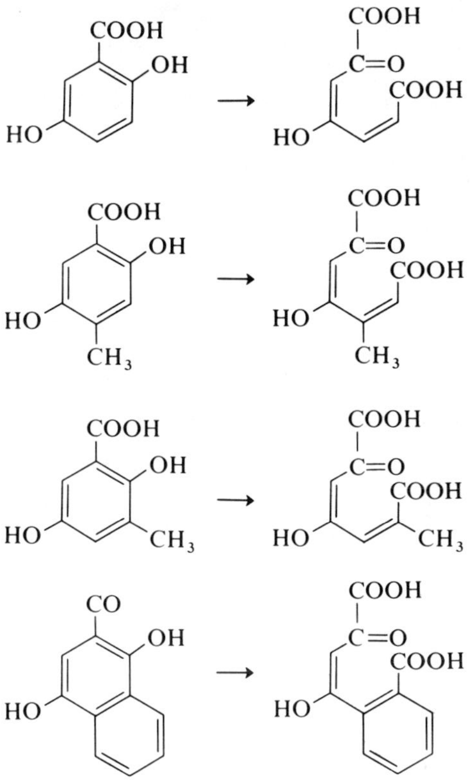

Figure 7.2. Cleavage of substituted gentisates by a *Pseudomonas* dioxygenase (Hopper *et al.*, 1971)

have been characterized. Some examples are described in Figure 7.3 (Chamberlain & Dagley, 1968; Gauthier & Rittenberg, 1971; Jerina *et al.*, 1971; Poillon *et al.*, 1969; Ribbons & Senior, 1970).

Among the mono-oxygenases which have been purified and crystallized are several involved in the metabolism of aromatic compounds (Figure 7.4). These include *p*-hydroxybenzoate hydroxylase from *P. putida*, *P. desmolytica* and *P. fluorescens* (Hosakawa & Stanier, 1966; Yano *et al.*, 1969; Teng *et al.*, 1971; Howell *et al.*, 1972). Crystalline preparations of salicylate hydroxylase and orcinol hydroxylase have also been described (Katagiri *et al.*, 1965; Ohta & Ribbons, 1970). These hydroxylases have molecular weights which range from 60,000 daltons to 93,000 daltons and all have one molecule of FAD bound to each mole of protein. NADH or NADPH act as the electron donors for this group of enzymes and electrons may be transferred from the pyridine nucleotide reductant to the FAD of the enzymes in the absence of

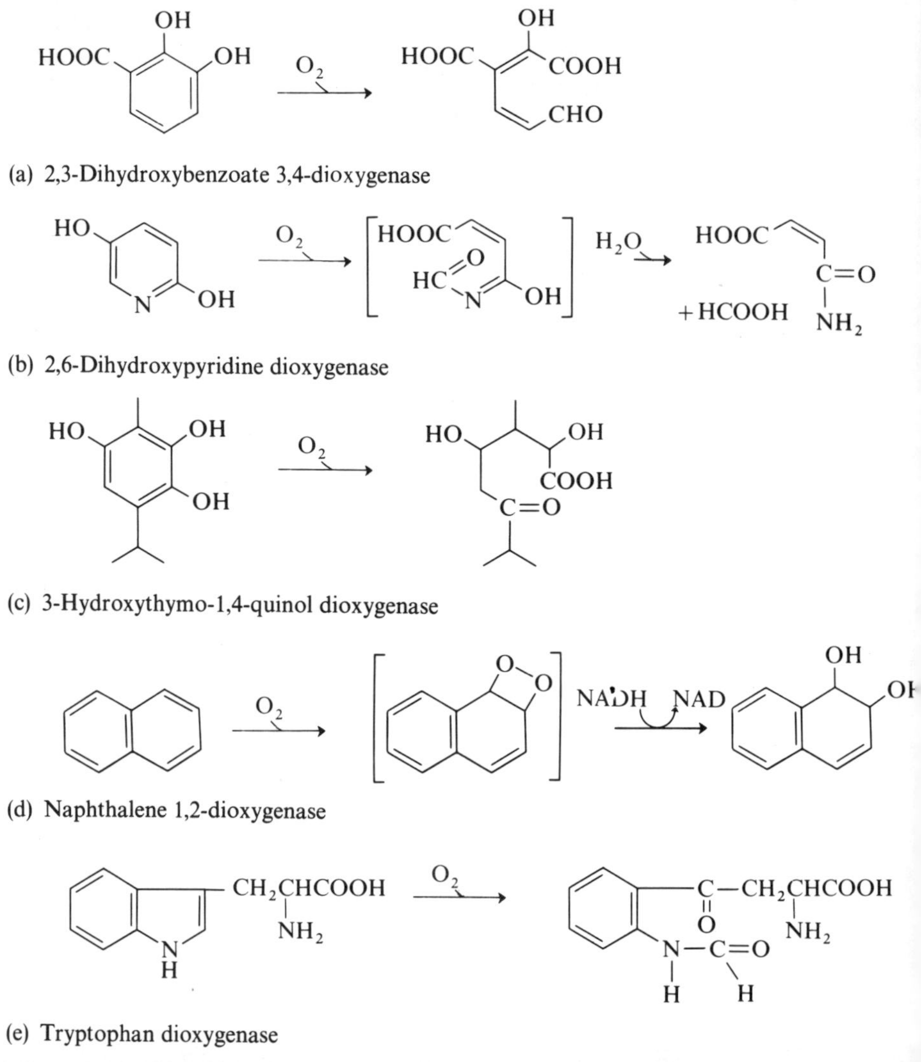

(a) 2,3-Dihydroxybenzoate 3,4-dioxygenase

(b) 2,6-Dihydroxypyridine dioxygenase

(c) 3-Hydroxythymo-1,4-quinol dioxygenase

(d) Naphthalene 1,2-dioxygenase

(e) Tryptophan dioxygenase

Figure 7.3. Different catabolic dioxygenases of *Pseudomonas*

O_2; in each case the binding of the aromatic substrate dramatically increases the rate of this electron transfer.

A complex enzyme system is known to catalyse the mono-oxygenative hydroxylation of methyl groups in *P. oleovorans* which can utilize hexane for growth (Peterson *et al.*, 1966). The ω-hydroxylase system attacks the methyl end of both alkane chains and fatty acids. The enzyme system contains three proteins; a non-haem iron protein rubredoxin (Lode & Coon, 1971),

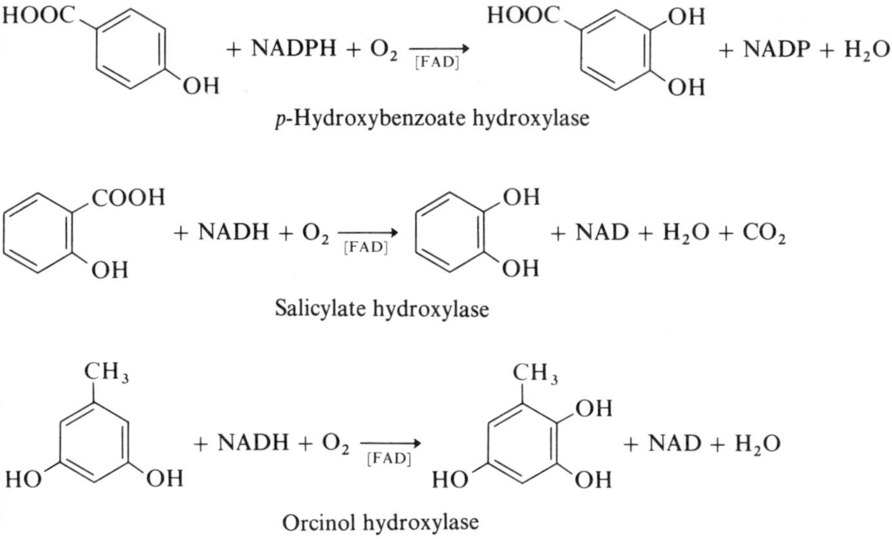

p-Hydroxybenzoate hydroxylase

Salicylate hydroxylase

Orcinol hydroxylase

Figure 7.4. Hydroxylations of aromatic compounds catalysed by mono-oxygenases

a hydroxylase (McKenna & Coon, 1970) and a FAD-containing reductase which uses NADH to reduce the rubredoxin (Ueda *et al.*, 1972). The hydroxylase enzyme is the only component of the system has has not yet been purified to homogeneity. Working with a partially purified hydroxylase system, May and Abbott (1973) observed the epoxidation of alkenes (Figure 7.5), so perhaps the same enzymes may participate in both ω-hydroxylation and epoxidation. Growth on hexane induces the ω-hydroxylase together with cytochrome o (Peterson, 1970).

$$NADH + O_2 + CH_3CH_2(CH_2)_n \cdots \rightarrow NAD + H_2O + HOCH_2CH_2(CH_2)_n \cdots$$

ω-Hydroxylation (Peterson *et al.*, 1966)

Epoxidation (May & Abbott, 1973)

Figure 7.5. The action of complex mono-oxygenases on aliphatic compounds in *Pseudomonas*

Another complex hydroxylase has been characterized by Gunsalus and his associates (1971). The hydroxylation of a methylene carbon in camphor (Figure 7.6a) is catalysed by an enzyme system that contains the three proteins putidaredoxin (an iron–sulphur protein) the soluble cytochrome P-450$_{CAM}$ which binds the camphor substrate and a FAD-containing

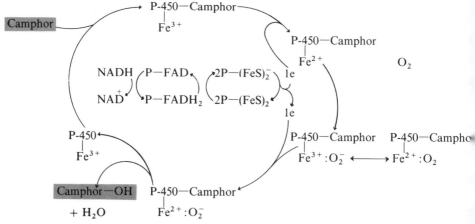

(a) Hydroxylation of methylene carbon of camphor

(b) Proposed mechanism for hydroxylation of camphor (Sharrock *et al.*, 1973)

(c) Ketolactonization of 2,5-diketocamphane

Figure 7.6. Oxygenases of camphor metabolism in *P. putida*

NADH-dependent putidaredoxin reductase. The mechanism of action of the camphor hydroxylase systems has been extensively investigated by Gunsalus and his colleagues by a variety of physicochemical methods. The three proteins are known to function in a tightly coupled electron transport system in a cyclical reaction sequence (Figure 7.6b) and the details are discussed by Gunsalus (1972) and by Sharrock (Sharrock *et al.*, 1973).

Another complex mono-oxygenase which is composed of three different proteins is required for a later step in camphor metabolism in which an oxygen atom is inserted into the ring structure of 2,5-diketocamphane to obtain a lactone (Figure 7.6c). The NADH-dependent reductase has FMN as a prosthetic group (Yu & Gunsalus, 1969).

Mono-oxygenases also catalyse the oxidative cleavage of ether bonds. Cartwright and Smith (1967) demonstrated that vanillate-grown cultures of *P. fluorescens* converted the growth substrate to protocatechuate and formaldehyde. As shown in Figure 7.7, the demethylase consumes stoichiometric amounts of vanillate, NADH and O_2. Like the oxygenative enzymes of the '*meta*' pathways, the demethylases may have fairly broad specificity.

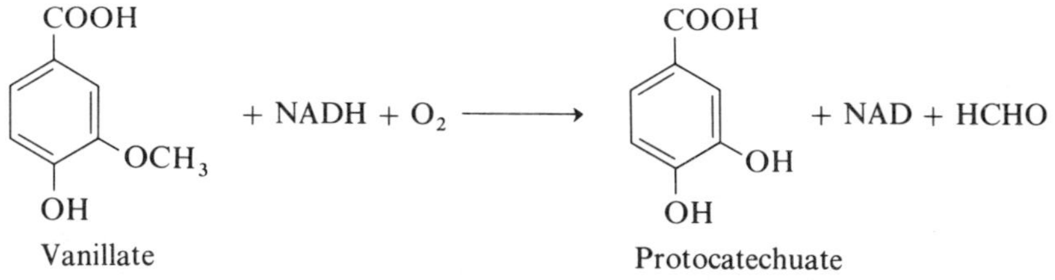

Figure 7.7. Demethylation by mono-oxygenases

Demethylase preparations remove methyl and ethyl groups with equal facility and attack aromatic compounds bearing methoxy substitutions at different sites. These mono-oxygenases have been resolved into two different protein components (Cartwright *et al.*, 1971; Ribbons, 1971) and therefore may possess a complexity comparable to that of the aliphatic mono-oxygenases.

3. CATABOLIC PATHWAYS AND THEIR REGULATION

3.a. Questions Posed by the Biochemical Versatility of *Pseudomonas*

As indicated in Chapter 1, most *Pseudomonas* species are capable of using a wide variety of organic compounds as growth substrates. It is unlikely that any of the compounds occur constantly in the natural environment; it is far more probable that the evolving bacterial species have been exposed to occasional and fluctuating concentrations of the potential growth substrates. The compounds utilized for growth are dissimilated to common intermediary metabolites by inducible enzymes with distinct functions. The nutritional versatility of *Pseudomonas* therefore reflects the presence of a substantial quantity of specialized genetic information that is only occasionally expressed. Consequently, the catabolic pathways of *Pseudomonas* pose three general biochemical questions: (1) How are the catabolic transformations of *Pseudomonas* achieved? (2) How did such marked nutritional diversity evolve? and (3) What mechanisms are used to regulate the synthesis of the catabolic enzymes in response to environmental demand? Efforts to answer the first question include the identification of metabolic intermediates and the elucidation of the chemical mechanisms employed by the enzymes that catalyse the frequently extraordinary reactions of *Pseudomonas* catabolism. The second question, the furthest from resolution, demands a description of the sequence of mutational and selective events that led not only to new structural genes, but also to complex regulatory controls. The third question, the nature of the controls used to govern inducible enzyme synthesis, must be approached by both biochemical and genetic methods and is the main subject of this section.

3.b. Physiological Analysis of Inductive Mechanisms

Enzyme induction is defined as an increase in the differential rate of synthesis of an enzyme (or group of enzymes) in response to a compound that may be present in the growth medium or produced endogenously. Most growth substrates are utilized by a complex series of catabolic reactions in *Pseudomonas*. Correspondingly, addition of a potential growth substrate to a *Pseudomonas* culture usually triggers a sequential series of inductive events as the structural genes for the catabolic enzymes are expressed. Enzyme synthesis is triggered by specific metabolites of the pathway that serve as *inducers*. The inducer of an enzyme is defined as the metabolite that most directly elicits its synthesis. Frequently several enzymes are induced by a single metabolite: such enzymes are under *coincidental* inductive control. Even closer regulatory control is shared by enzymes whose synthesis is governed by *coordinate* induction: the relative rates of synthesis of these enzymes remains constant under all conditions of induction. Thus enzymes that are under coordinate inductive control are governed as a *unit of physiological function*. The extent of induction of any one enzyme in a physiological unit is a precise measure of the extent of induction of the other enzymes in the unit.

Mechanisms of inductive control and the experimental techniques used for their elucidation can best be illustrated by considering a single catabolic sequence, for example the pathway employed for the utilization of L-tryptophan by fluorescent pseudomonads (Figure 7.8). All of the 12 enzymes depicted in Figure 7.8 are induced by growth with L-tryptophan despite the fact that two of them (CMLE and CMD), are not required for its catabolism (Palleroni & Stanier, 1964). Addition of L-tryptophan results in a 50-fold to over 3000-fold increase in the differential rate of synthesis of these enzymes. The inductive role of L-tryptophan is indirect and indeed Palleroni and Stanier (1964) showed that L-tryptophan has no inducer activity. The first clue came from the observation that L-kynurenine, formed by the largely irreversible hydrolysis of *N*-formylkynurenine, triggered the induced synthesis of all of the enzymes required for the oxidation of L-tryptophan. Anthranilate, the fourth metabolite in the sequence, elicited the synthesis of the enzymes

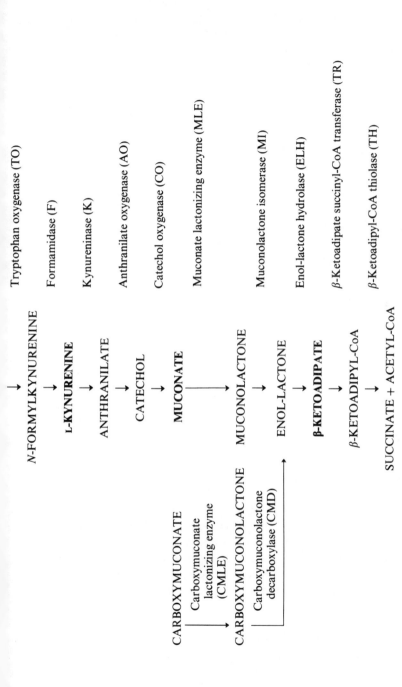

Figure 7.8. The pathway of L-tryptophan utilization in fluorescent pseudomonads

Inducing compounds are printed in bold type. In the text the enzymes are represented by the abbreviations shown in brackets. Two enzymes for the metabolism of carboxymuconate which do not participate in the catabolism of tryptophan are induced gratuitously by growth with this amino acid.

required for its own oxidation, but did not cause an increase in the rate of synthesis of the first three enzymes (tryptophan oxygenase, TO; formamidase, F; kynureninase, K in Figure 7.8). Thus it appeared that L-kynurenine was the inducer of these three enzymes of the pathway. That this was the case was shown by Palleroni and Stanier (1964) who demonstrated that mutant strains that were unable to synthesize formamidase (F) (and consequently were unable to interconvert kynurenine and *N*-formylkynurenine), still formed induced levels of tryptophan oxygenase (TO) and kynureninase (K) when exposed to kynurenine. These results showed that the inductive effect of L-kynurenine could not be the indirect consequence of its conversion to *N*-formylkynurenine. The same mutant strain was used to show that L-tryptophan and *N*-formylkynurenine were not inducers since in the absence of the enzyme required for their conversion to kynurenine these two compounds did not induce enzyme synthesis.

Further evidence for this inductive pattern was provided by the inducer analogue γ-hydroxykynurenine (Figure 7.9), which was not metabolized

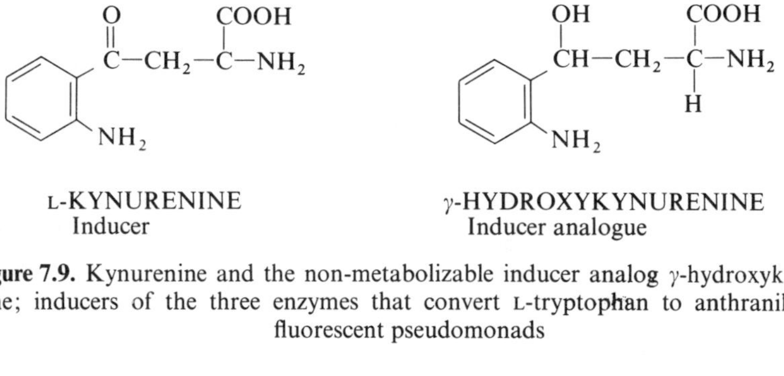

L-KYNURENINE
Inducer

γ-HYDROXYKYNURENINE
Inducer analogue

Figure 7.9. Kynurenine and the non-metabolizable inducer analog γ-hydroxykynurenine; inducers of the three enzymes that convert L-tryptophan to anthranilate in fluorescent pseudomonads

by the bacteria, but nevertheless caused the induced synthesis of the three enzymes that convert L-tryptophan to anthranilate and L-alanine. By varying the concentration of the inducer analogue in the growth medium, Palleroni and Stanier (1964) were able to show that the induction of tryptophan oxygenase (TO) and formamidase (F) was strictly coordinate and hence that these two enzymes were members of a single unit of physiological function (Unit I, Table 7.3). However, higher concentrations of γ-hydroxykynurenine were required for full induction of kynureninase (K). Therefore, this enzyme is a member of a separate unit of physiological function (Unit II, Table 7.3) despite the fact that its induction is coincident with that of tryptophan oxygenase (TO) and formamidase (F).

The studies of Palleroni and Stanier (1964) illustrate the carefully controlled investigations that are required for the unambigous identification of inducers. The physiological studies had indicated that L-kynurenine was the metabolite

Table 7.3. Physiological units of induction in the L-tryptophan pathway of fluorescent pseudomonads

Physio-logical unit	Enzyme	Inducer
I	Tryptophan oxygenase (TO) Formamidase (F)	L-Kynurenine
II	Kynureninase (K)	L-Kynurenine
III	Anthranilate oxygenase (AO)	Anthranilate
IV	Catechol oxygenase (CO)	Muconate
V	Muconate lactonizing enzyme (MLE) Muconolactone isomerase (MI)	Muconate
VI	[a] ⌈Carboxymuconate lactonizing enzyme (CMLE)⌉ ⌊Carboxymuconolactone decarboxylase (CMD)⌋ Enol-lactone hydrolase (ELH) β-Ketoadipate succinyl-CoA transferase (TR)	β-Ketoadipate

[a] Gratuitous induction of these two enzymes of the convergent protocatechuate branch of the β-ketoadipate pathway.

that most directly elicited synthesis of the first three enzymes. This was confirmed with a mutant which lacked formamidase and was unable to convert the precursors to L-kynurenine, and by making use of the non-metabolizable analogue γ-hydroxykynurenine. Anthranilate oxygenase (AO) is induced in wild-type bacteria, or in a mutant lacking kynureninase, by anthranilate. Catechol and its products to not induce the enzyme, so that it can be concluded that anthranilate itself is the inducer of anthranilate oxygenase (Unit III, Table 7.3).

Catechol is produced as an intermediate in the breakdown of many aromatic compounds in addition to tryptophan. This will be discussed in more detail later in this chapter and in Chapter 8, and here we will be concerned mainly with the regulation of the enzymes required for catechol metabolism as part of the tryptophan pathway. Catechol does not appear to induce any of the enzymes that participate in its dissimilation. Convincing evidence in support of this conclusion was provided by Kemp and Hegeman (1968) and by Bird and Cain (1968) who employed a novel method, physiological restriction, to prevent the metabolism of catechol by wild-type strains of *P. aeruginosa*. Catechol oxygenase (CO, Figure 7.8), the enzyme that cleaves catechol to muconate, requires molecular oxygen as a second substrate. Cultures of *P. aeruginosa* were grown with nitrate instead of oxygen as the terminal electron acceptor in respiration so that the bacteria could be exposed to catechol under conditions in which it was non-metabolizable. There was no induction by catechol in the absence of oxygen but admission of oxygen to the cultures permitted the enzymatic cleavage of catechol to occur and resulted in the full induction of the enzymes that mediate its dissimilation.

Wild-type *Pseudomonas* cultures do not grow at the expense of muconate because the cell membrane is not permeable to this dicarboxylic acid. Mutant strains which are permeable to muconate occur with high frequency and these may be selected directly from the wild type using muconate as the sole growth substrate (Ornston, 1966). Exposure of permeable strains to muconate elicits not only the synthesis of all the enzymes required for its catabolism but also catechol oxygenase (Ornston, 1966). Since catechol is not an inducer and because the oxygenative cleavage that gives rise to muconate is irreversible, it is clear that muconate, or a catabolite thereof, is the *product inducer* of catechol oxygenase. The direct inductive effect of muconate was demonstrated with mutant strains that lacked muconate lactonizing enzyme (MLE, Figure 7.8) and consequently accumulated muconate when exposed to catechol or metabolic precursors of this compound. Under conditions leading to the accumulation of muconate, both catechol oxygenase (CO) and muconolactone isomerase (MI) are fully induced. Enol-lactone hydrolase (ELH) and subsequent enzymes of the pathway are not induced by muconate in MLE mutant strains, showing that muconate must be metabolized in order to elicit the synthesis of the later enzymes.

Muconate lactonizing enzyme and muconolactone isomerase are regulated as a single physiological unit (Unit V, Table 7.3) in *Pseudomonas*. Indeed the unified control may be at the level of gene expression because the structural genes for the two enzymes are tightly linked and are governed by a neighbouring regulatory gene in *P. putida* (Wheelis & Ornston, 1972; Wu *et al.*, 1972). Although catechol oxygenase is also induced by muconate, its synthesis is relatively more sensitive to catabolite repression and hence catechol oxygenase comprises a separate regulatory unit (Unit IV, Table 7.3). Mutant strains have also been used to confirm that muconolactone is not an inducer in fluorescent pseudomonads. Strains lacking the muconate lactonizing enzyme (MLE) do not grow at the expense of muconolactone apparently because muconolactone does not induce muconolactone isomerase (MI). Muconolactone has been used as growth substrate to select secondary mutant strains that are constitutive for muconolactone isomerase from strains that are blocked in the synthesis of muconate lactonizing enzyme (MLE) (Parke & Ornston, unpublished).

In *P. putida* the three enzymes (CMLE, CMD and ELH, Figure 7.8) that convert carboxymuconate to β-ketoadipate are induced as a regulatory unit (known as the carboxymuconate unit, Unit VI, Table 7.3). Further evidence of their united regulatory control is provided by the observation that the three enzymes are produced at high levels when constitutive strains are grown in the absence of inducer. Thus, the synthesis of all three enzymes appears to be under the control of a single regulatory gene (Parke & Ornston, unpublished). The studies of Kemp and Hegeman (1968) showed that the co-ordinate regulation extends to the control of the synthesis of β-ketoadipate succinyl-CoA transferase (TR) in *P. aeruginosa*. The regulation of the synthesis of the thiolase that cleaves β-ketoadipyl-CoA to succinyl-CoA and acetyl-CoA has not been studied.

Growth of either *P. putida* or *P. aeruginosa* at the expense of β-ketoadipate elicits fully induced levels of all the enzymes of the carboxymuconate unit. The hydrolytic reaction that gives rise to β-ketoadipate is essentially irreversible so that it appears that β-ketoadipate (or β-ketoadipyl-CoA) serves as an inducer of the three enzymes that mediate its formation from carboxymuconate. This conclusion is strengthened by the observation that β-ketoadipate causes full induction of carboxymuconate lactonizing enzyme in a mutant of *P. putida* lacking enol-lactone hydrolase (ELH) (Ornston, 1966). Dysfunction in the structural gene for β-ketoadipate succinyl-CoA transferase does not impair the ability of β-ketoadipate to serve as an inducer of the other enzymes of the carboxymuconate regulatory unit. Therefore it is evident that β-ketoadipate itself serves as an inducer. This evidence does not of course exclude the possibility that β-ketoadipyl-CoA is also an inducer.

Analysis of the catabolic sequence for L-tryptophan has revealed diverse biochemical intermediates and several different types of regulatory control but relatively similar techniques were used for elucidation of all the inductive mechanisms that govern the pathway. In order to ensure the unambiguous identification of inducers, some compounds thought to be the most likely candidates were converted to non-metabolizable analogues, as in the chemical modification of kynurenine to produce the non-substrate inducer γ-hydroxykynurenine. Another general approach was to introduce mutations which resulted in the absence of certain enzymes, so that the metabolism of those intermediates which were thought to be inducers could be restricted. In one case the metabolism was restricted by depriving the enzyme, catechol oxygenase, of its essential second substrate oxygen.

When an inducer has been identified the extent of its regulatory control may be established. It must be stressed that in such studies the inducer should not be metabolized since coincident induction is established in terms of the enzymes which are coincidently synthesized and not with respect to the rates of synthesis. Inducing metabolites can be formed quite rapidly from their precursors and the addition of a metabolizable primary inducer to a culture could result in the apparently simultaneous synthesis of enzymes that are actually induced sequentially by successive metabolites of the pathway (Hosakawa, 1970). Kinetic experiments may indicate enzymes that are subject to independent inductive control but cannot provide proof of coincident induction. A far more sensitive approach, and one which can indicate the existence of coordinate as well as coincident control of synthesis of groups of enzymes, is to measure the rates of enzyme synthesis under a range of different conditions in which the growth rates or the concentrations of inducer or of repressing metabolites are varied.

As summarized in Table 7.3 the nine enzymes that convert L-tryptophan to β-ketoadipyl-CoA are induced as six separate physiological units and four of the ten metabolites of the pathway serve as inducers. In some cases enzymes are induced by their substrates, occasionally an enzyme is induced by its product and in several instances an inducer is more than one metabolic step away from the enzyme that it induces. Two of the four inducers (kynurenine and muconate) are endowed with dual inductive function: they elicit the synthesis of the regulatory unit of enzymes that gives rise to them as well as the regulatory unit of enzymes that mediate their dissimilation. In order for substrates to be converted to product inducers, several of the enzymes of the tryptophan pathway must be present at significant levels in uninduced cells; there must be sufficient tryptophan oxygenase (TO) and formamidase (F) to form kynurenine from tryptophan, catechol oxygenase (CO) to form muconate and enol-lactone hydrolase (ELH) to form β-ketoadipate. In fact, enzyme activities at 2 % of the fully induced levels appear to be sufficient to permit the indirect mechanism of product induction to be employed efficiently (Palleroni & Stanier, 1964; Ornston, 1966).

3.c. Simultaneous Adaptation and Inducer Specificity

The principle of *simultaneous adaptation* (Stanier, 1947; Karlsson & Barker, 1948; Suda *et al.*, 1950) has been successfully employed for the elucidation of the broad outlines of many catabolic pathways, including some of the early steps of tryptophan catabolism shown in Figure 7.8. The theory rests upon the assumption that catabolic enzymes are induced in the presence of their substrates and consequently that an organism adapted to the oxidation of a given growth substrate will oxidize all of the metabolites formed subsequently by the catabolism of that substrate. As was noted at the time of the formulation of the concept of simultaneous adaptation, the intermediate metabolites that do not permeate the cell membrane would not be oxidized even if the organisms possessed the full complement of enzymes necessary for their catabolism (Stanier, 1951). Thus, muconate might be falsely excluded as an intermediate in the tryptophan pathway unless precautions were made to overcome permeability barriers. The induction of the muconate enzymes could be established by preparation of cell-free extracts or simultaneous adaptation to muconate could be observed with permeable mutant strains. Similar problems of cell permeability were encountered when attempts were made to use the principle of simultaneous adaptation to establish the existence of the tricarboxylic acid cycle in *Pseudomonas* (see section 5 of this chapter).

Another limitation to the principle of simultaneous adaptation is introduced by the observation that a number of catabolic enzymes may be induced in the absence of their substrates. For example, only eight of the twelve enzymes induced by the growth of fluorescent pseudomonads with kynurenine actually play a role in the dissimilation of that compound. Product induction causes the gratuitous induction of the two enzymes required to convert typtophan and *N*-formylkynurenine to kynurenine and consequently endows kynurenine-grown cells with the ability to catabolize tryptophan. At a later stage in the tryptophan pathway two enzymes of the protocatechuate branch of the β-ketoadipate pathway (Figure 7.10) are induced by β-ketoadipate. These two enzymes are necessary only when the bacteria are grown with a substrate such as *p*-hydroxybenzoate which is metabolized by this branch of the pathway, so that growth in the presence of kynurenine results in the gratuitous synthesis of a total of four enzymes. Accordingly the evidence provided by simultaneous adaptation experiments must be regarded as highly tentative.

In the foregoing example the gratuitous synthesis of enzymes is not due to a lack of inducer specificity. Rather, specific metabolic products are endowed with extensive function that includes induction of the enzymes that give rise to them. In other catabolic systems a wide range of metabolites may serve as inducers for a single set of enzymes. Hegeman (1966a) has shown that the five enzymes of the mandelate regulatory unit which convert mandelate to benzoate are induced by a number of different metabolites in *P. putida*. This broad inductive specificity may supplement the broad substrate specificity of the mandelate enzymes that can also catalyse the corresponding transformations of *p*-hydroxymandelate to *p*-hydroxybenzoate (Figure 7.10). A broad specificity of both induction and catalysis leads to a broad metabolic flexibility that permits a wide range of compounds to be dissimilated by the mandelate enzymes. It should be noted that the mandelate enzymes are induced by a quite different mechanism in *P. aeruginosa*. Inductive controls are relatively fragmented in this species. L-mandelate dehydrogenase is induced by L-mandelate, benzoylformate decarboxylase and benzaldehyde dehydrogenase are induced by benzoylformate and most remarkably, benzaldehyde dehydrogenase is also induced by β-ketoadipate (Rosenberg, 1971).

Gunsalus *et al.* (1965) first proposed that broad specificity of both induction and catalysis could provide metabolic flexibility from their analysis of the enzymes of the camphor pathway in *P. putida*. Both D(+)-camphor (2-boranone, compound **1** in Figure 7.11) and 3,4,4-trimethyl-5-carboxymethyl-Δ^2-cyclopentanone (compound **5** in Figure 7.11) appear to serve as inducers of the enzymes that dissimilate camphor to acetate and isobutyrate. Subsequent investigations (Hartline & Gunsalus, 1971) demonstrated that many different chemical derivatives of boranane were inducers of the camphor enzymes. Some of these compounds were metabolizable and some were not, but in general the inductive specificity was broader than the specificity of catalysis. Hartline and Gunsalus (1971) pointed out that there

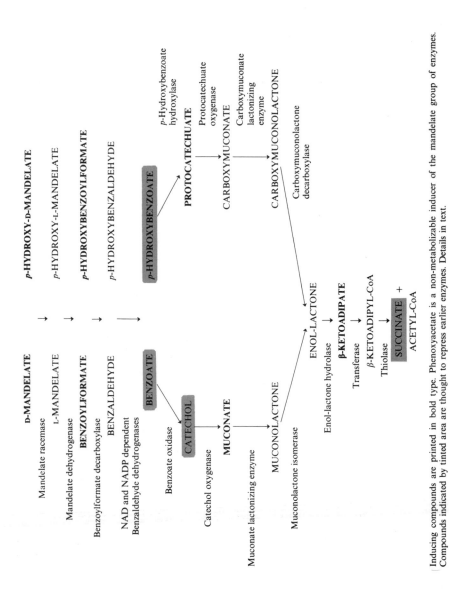

Figure 7.10. Inducers and repressors of the enzymes of the mandelate and *p*-hydroxy-mandelate pathways in *Pseudomonas putida*

could have been no selective pressure against inducer function by chemical analogues that had not been encountered by the bacteria in the natural environment and therefore the control of enzyme synthesis may be more rigorous in nature 'than it appears to be in the laboratory.

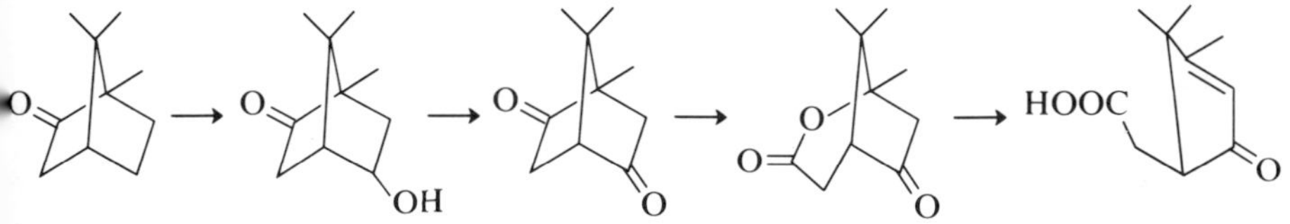

				3,4,4-Trimethyl- 5-carboxymethyl- Δ^2-cyclopentanone
(+)-Camphor	2	3	4	5
1				

Figure 7.11. Early steps in the dissimilation of D(+)-camphor by *P. putida*

3.d. Repression by Alternative Substrates and Metabolic Products

A decrease in the differential rate of enzyme synthesis is termed *repression*. Inducible enzymes may be regulated by at least two different types of repressive control. One of these is *catabolite repression* which has been studied most extensively for the *lac* operon of *Escherichia coli*. It has long been recognized that the synthesis of β-galactosidase in the presence of an inducer is repressed by compounds such as glucose which support a rapid growth rate. Current views on catabolite repression of β-galactosidase are that it acts at the level of gene transcription (Emmer *et al.*, 1970). The transcription of the *lac* operon at the maximum rate both *in vitro* and *in vivo* requires a specific protein variously known as the CRP or CAP protein determined by a gene unlinked to the *lac* operon. For full activity the CRP protein requires 3′,5′-cyclic adenosine monophosphate (c-AMP) whose synthesis is dependent *inter alia* on the activity of adenyl cyclase which catalyses the formation of c-AMP from ATP. The addition of glucose to a culture of *E. coli* growing in a minimal medium with glycerol, reduces the level of c-AMP within the bacterial cells. The addition of c-AMP results in partial relief of glucose repression of β-galactosidase synthesis in both inducible and constitutive strains. This is a general control for inducible catabolic enzymes and mutations in the CRP gene can result in mutants which are pleiotropically negative and are unable to utilize lactose, arabinose maltose or galactose (Emmer *et al.*, 1970; Zubay *et al.*, 1970). Catabolite repression in *E. coli* is not restricted to glucose but is also produced by other carbon compounds. In *Pseudomonas* glucose is a relatively poor growth substrate and the compounds producing the most severe catabolite repression are succinate and other intermediates of the tricarboxylic acid cycle. Succinate represses the synthesis of inducible enzymes required for the catabolism of tryptophan, histidine, acetamide, mandelate and of many other compounds that can be used as growth substrates.

The catabolite repression exerted by substrates that support vigorous growth has been used to vary the differential rate of enzyme synthesis and thus to establish very clearly whether or not enzymes are subject to independent or coordinate regulation. For example, the three enzymes that convert catechol to β-ketoadipate enol-lactone are induced by muconate.

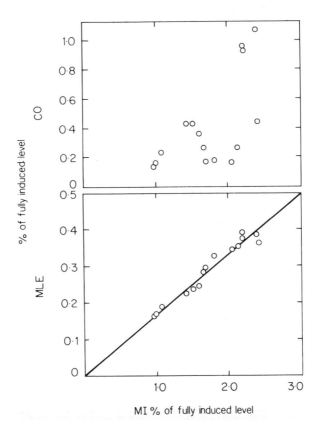

MI % of fully induced level

Figure 7.12. Independent catabolite repression of enzymes induced by muconate in *P. putida*. Enzymes were induced by exposing cells to benzoate, a metabolic precursor of catechol. Catabolite repression was caused by adding either succinate or glucose to the growth medium. Muconate lactonizing enzyme (MLE) and muconolactone isomerase (MI) are repressed coordinately; the synthesis of catechol oxygenase (CO) is relatively more sensitive to catabolite repression. (Reproduced with permission from Ornston (1966) *Journal of Biological Chemistry*, **241**, 3800)

As shown in Figure 7.12, the synthesis of two of the enzymes (MLE and MI) is repressed coordinately; the synthesis of the other enzyme, CO is relatively more sensitive to catabolite repression. Since catechol oxygenase is required for the production of the inducer, muconate, from catechol, the severe catabolite repression of this enzyme may reduce the rate of production of inducer and thus may indirectly reduce the rate of synthesis of all the enzymes that participate in the dissimilation of catechol. A constitutive mutant strain of *P. putida* produces high levels of CO, MLE and MI under all growth conditions (Wu *et al.*, 1972). The synthesis of the MLE–MI regulatory unit is not regulated in this strain; both enzymes appear at 40 % of fully induced levels in the presence or the absence of inducer. The synthesis of CO remains sensitive to catabolite repression in the mutant strain, further indicating the independent repressive control of the synthesis of these enzymes.

The mechanism of catabolite repression in *Pseudomonas* has not yet been completely elucidated but some detailed studies have been carried out on the repression of amidase synthesis in *P. aeruginosa* (Clarke, 1970). The enzyme is induced by its substrate, acetamide, and is very severely repressed by succinate and less severely by lactate and glucose. The rate of enzyme synthesis is strongly influenced by the dilution rate when the wild-type strain is grown in continuous culture with carbon limitation. At low dilution rates the enzyme is synthesized relatively slowly as would be expected if the enzymatic hydrolysis of the inducer maintains it at a low intracellular concentration. As the dilution rate is increased the rate of amidase synthesis increases, but at the highest dilution rates, the rate of amidase synthesis drops to a low level. Clarke *et al.* (1968) concluded that the rapid growth rate that occurs at high dilution rates is directly related to the severity of catabolite repression. Preliminary evidence that catabolite repression of amidase synthesis is dependent on the intracellular levels of c-AMP was obtained by Smyth and Clarke (1972) who showed that catabolite repression of amidase by lactate was partially relieved by c-AMP. Catabolite repression of inducible enzymes in *Pseudomonas* may therefore be similar to that known to occur in *E. coli*.

A more specific regulatory control is end-product or *metabolite repression*, in which products of the pathway directly reduce the rate of synthesis of the enzymes that give rise to them (Magasanik, 1961). The procedures required for the unambiguous identification of product repressors are much like those used for the identification of inducers. In order to establish that the metabolite exerts its effect directly it must be rendered non-metabolizable. As with inducers this goal may be achieved either by the chemical modification of the presumed repressor compound or by the isolation of mutant strains that cannot metabolize the repressing metabolite.

Mandelstam and Jacoby (1965) and Stevenson and Mandelstam (1965) found that metabolites formed from mandelate severely repressed some of the enzymes required for the breakdown of this compound by *P. putida* (Figure 7.10). Mandelate is converted to benzoate by a series of enzymes, which comprise a single regulatory unit, that are also able to convert *p*-hydroxymandelate to *p*-hydroxybenzoate. This group of enzymes is repressed by benzoate, *p*-hydrobenzoate, catechol and succinate. Benzoate is converted to catechol by benzoate oxidase which is repressed by catechol and succinate. It was also found, as was discussed earlier, that the enzymes required for catechol metabolism are repressed by succinate but it is possible that succinate repression is due to general catabolite repression and not to repression by a specific metabolite. There is as yet no evidence for the site of action of metabolite repressors. Higgins and Mandelstam (1972a) showed that *p*-hydroxybenzoate strongly represses the mandelate enzymes which are less severely repressed by benzoate. In mutants that are unable to metabolize *p*-hydroxybenzoate the mandelate group of enzymes are still repressed by *p*-hydroxybenzoate and similarly benzoate represses these enzymes in a benzoate-negative mutant. This suggests that benzoate and *p*-hydroxybenzoate may have a direct effect on the repression of the mandelate enzymes. In the presence of mandelate and either benzoate or *p*-hydroxybenzoate, the mandelate enzymes are repressed almost completely and the alternative substrate is preferentially utilized for growth. Some evidence was obtained for an inducible mandelate transport system and this may have complicated the measurements of the inducer and repressor constants for this system (Higgins & Mandelstam, 1972b).

The amidase of *P. aeruginosa* is subject to repression by various amides which do not support growth. This was termed *amide analogue repression* by Brammar and Clarke (1964) and is an effect of the particular specificity of the amidase regulator protein. Butyramide is an amide analogue repressor for amidase in *P. aeruginosa*, but in *P. putida* it acts as an inducer (Clarke, 1972). In both organisms it is likely that butyramide is bound to a regulator protein but only in *P. putida* is the conformational change produced which allows transcription of the amidase genes.

3.e. Regulation by Enzyme Inhibition

Thus far we have considered only regulation at the level of enzyme synthesis, but considerable evidence has been adduced to support the view that catabolic flow is strongly influenced by regulation of enzyme activity in *Pseudomonas*. One well-characterized catabolic enzyme is the inducible glucose 6-phosphate dehydrogenase that initiates the catabolic Entner–Doudoroff pathway. Lessie and Neidhart (1967a) found that high concentrations of ATP or GTP inhibited this enzyme by decreasing its ability to bind glucose 6-phosphate in *P. aeruginosa*. Thus, a high energy charge decreases the rate of catabolic flow from glucose. Further control of glucose 6-phosphate dehydrogenase was identified by Lessie and Vander Wyk (1972) who found two forms of the enzyme in *P. cepacia* (*multivorans*). One form was active with either NAD or NADP as substrate and was inhibited by ATP and the other glucose 6-phosphate dehydrogenase required NADP and was not inhibited by ATP. *P. cepacia* also synthesizes two forms of 6-phosphogluconate dehydrogenase and the activity of these enzymes was modulated in a corresponding manner. The activity of the tricarboxylic acid cycle enzymes is regulated by inhibition of citrate synthase by NADH and reactivation by AMP (Weitzman & Jones, 1968). Other examples are discussed in Chapter 8.

3.f. Regulation of Uptake Systems

The metabolism of *Pseudomonas* is frequently regulated by control of the uptake of potential growth substrates from the external environment. It is not surprising that many of the transport systems for compounds that are metabolized by inducible enzymes are themselves inducible. The synthesis of inducible uptake systems can often be shown to be inhibited by chloramphenicol, or other inhibitors of protein synthesis, indicating that during the induction period there is synthesis of specific transport proteins.

In a series of investigations, Kay and Gronlund (1969a,b,c; 1971) studied the uptake of amino acids by *P. aeruginosa* and found that while energy-dependent transport systems for 18 of the 20 common amino acids were present in glucose-grown cells, it was possible to increase the activity of transport systems for particular amino acids by growing the bacteria in the presence of the amino acid. Amino acid transport is discussed in more detail in Chapter 8.4.b. The advantage in having low constitutive levels of transport systems for potential growth substrates is that when the compound is present in the environment it can be taken into the cells and then induce synthesis of the metabolic enzymes and of its own transport system.

It has now been shown that the phosphoenolpyruvate phosphorylating system for sugar uptake by Kundig and his colleagues (1964) does not participate in the uptake of glucose by *P. aeruginosa* (Phibbs & Eagon, 1970; Romano *et al.*, 1970). After a general survey of uptake of sugars by bacteria Romano and his coworkers (1970) concluded that the phosphoenolpyruvate system was operative in facultative anaerobes in which substrate level phosphorylation plays a major energetic role, but was not present in the strict aerobes which derive most of their ATP by oxidative phosphorylation. Phibbs and Eagon (1970) showed that the hexose uptake system in *P. aeruginosa* was energy dependent and inducible. They suggested that the transported hexoses were trapped as phosphate derivatives formed by intracellular kinases. A different conclusion was reached by Midgley and Dawes (1973) who found that the analogue, α-methylglucoside was transported, but not phosphorylated, by cultures of *P. aeruginosa*.

An inducible glycerol uptake system in *P. aeruginosa* has been examined and it was found that cold shock released a glycerol binding protein, which was not glycerol kinase (Tsai *et al.*, 1971). A transport deficient mutant did not produce the glycerol binding protein and these observations led to the conclusion that this protein is an essential component of the uptake system for glycerol.

Indirect evidence led Higgins and Mandelstam (1972b) to look for a concentrating uptake system for mandelate in *P. putida*. In this strain the five enzymes that convert D- and L-mandelate to benzoate are induced by mandelate (Hegeman, 1966a,b,c). Higgins and Mandelstam (1972b) found that low inducer concentrations caused only slight induction of the mandelate enzymes when cultures were grown in a chemostat. However, cells that had been preinduced by growth in the presence of a high concentration of mandelate maintained fully induced levels of the enzymes when grown subsequently in the presence of low concentrations of inducer. These observations suggested that the preinduced cells contained a concentrating uptake system that maintained the intracellular inducer at a level high enough to cause full induction even in a medium with a low external concentration of mandelate. Direct measurement of the uptake of radioactive mandelate revealed the presence of an inducible transport system which could concentrate mandelate in the cells. This could be detected only with low mandelate concentrations (5×10^{-5} M) while at higher concentrations the passive diffusion of mandelate overwhelmed the effect of the uptake system. It is worth noting that organisms which live in environments which are likely to contain potential growth substrates at very low concentrations would be at an advantage if they possessed transport systems which could concentrate these compounds.

The ability of a concentrating uptake system to maintain induction during the growth of preinduced cells in the presence of extremely low inducer concentrations was established by Novick and Weiner (1957). These authors pointed out that preinduced cells differed from non-induced cells in a stable character—the synthesis of inducible enzymes—so long as a low inducer concentration was maintained in the growth medium. Under these conditions the preinduced wild type appeared to have acquired constitutivity, but this character was lost by transfer to a medium without inducer.

Another example of the acquisition of pseudo-constitutivity by the activity of a uptake system came from the isolation of mutant strains of *P. putida* that, unlike the wild type, grow at the expense of β-carboxy-*cis*,*cis*-muconate (Meagher *et al.*, 1972). The mutant organisms were enriched by twenty sequential transfers between media containing β-carboxy-*cis*,*cis*-muconate or succinate as sole carbon source. A similar procedure, involving transfer between inducing and non-inducing media was used for the isolation of constitutive mutants of *Escherichia coli* (Cohen-Bazire & Jolit, 1953). The transfer between β-carboxy-*cis,cis*-muconate and succinate did not yield strains of *P. putida* producing constitutive levels of the β-carboxy-*cis,cis*-muconate enzymes. Instead, the mutant strains isolated by this procedure had acquired the ability to form an inducible uptake system for β-carboxy-*cis,cis*-muconate and differed in this respect from the wild-type parent (Meagher *et al.*, 1972). Thus an enrichment procedure that yields constitutive mutant strains can also lead to the isolation of organisms capable of pseudo-constitutive behaviour due to an inducible permease.

Compounds which appear to repress the synthesis of inducible enzymes may act by repressing the synthesis of specific transport systems and also by inhibiting the entry of potential growth substrates into the bacterial cell. This has not been studied in detail in *Pseudomonas* except for a very few systems. The glucose enzyme of *P. aeruginosa* are repressed by succinate (Ng & Dawes, 1973). Midgley and Dawes (1973) reported that the synthesis of the inducible glucose transport system (methyl-α-glucoside–glucose transport system) is repressed by succinate and citrate, and the uptake of the non-metabolizable analogue, α-methylglucoside by induced cells is inhibited by succinate and some other organic acids. Thus, a repression of synthesis of specific transport systems and inhibition of transport activity may regulate metabolic pathways.

3.g. Convergent Pathways

In every round of genetic replication *Pseudomonas* species must devote a significant amount of carbon and energy to the synthesis of DNA bearing the specialized genes that give these bacteria their diverse catabolic potential. It is difficult to imagine how this genetic information could be retained in the absence of favourable selective pressures in the natural environment. Analysis of the pathways and their regulation may permit us to postulate what some of these selective pressures are.

In general we may surmise that a mutation that places a general long-term demand upon the biosynthetic capacity of the cells will not be favoured by evolution unless it introduces a compensating benefit to the metabolism of the organism. Convergent catabolic pathways, that permit a single structural gene to participate in the utilization of a range of different growth substrates, provide an efficient use of genetic information and thus might be expected to be selectively advantageous. Many examples of such convergent pathways are known; camphor (Gunsalus *et al.*, 1967) and valine (Marshall & Sokatch, 1972) are utilized *via* isobutyryl-CoA; four different carbohydrates (Dagley & Trudgill, 1965) and L-hydroxyproline (Gryder & Adams, 1969) are dissimilated *via* α-ketoglutarate semialdehyde and β-ketoadipate enol-lactone (Figure 7.13) is formed in the catabolism of over a dozen different known growth substrates (Ornston & Stanier, 1964). The full unregulated expression

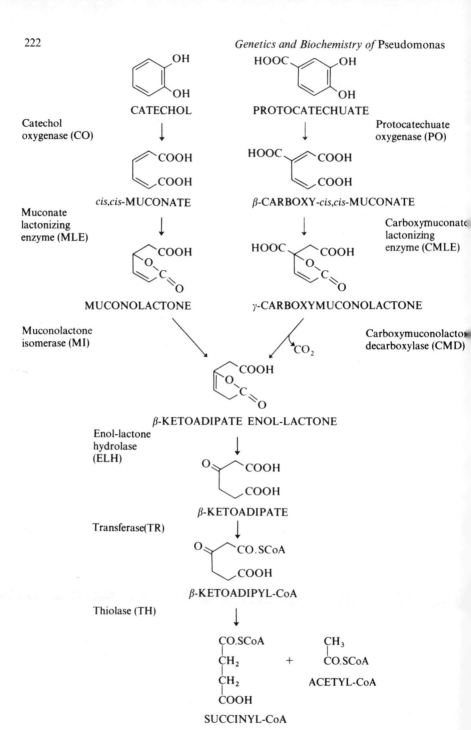

Figure 7.13. The central reactions of the β-ketoadipate pathway

of the structural genes for the catabolic pathways of most *Pseudomonas* species would place an intolerable biosynthetic demand on the organisms. A substantial fraction of gene expression must be devoted to the formation of regulatory mechanisms that restrict the expression of the genes for the catabolic enzymes to circumstances in which the gene products benefit cell growth. Just as metabolic convergence permits a single structural gene to code for an enzyme that participates in numerous catabolic functions, unified inductive control permits a single regulatory gene to govern the synthesis of several enzymes. To the extent that regulatory mechanisms are unified, the metabolic burden placed by their own biosynthesis is reduced. In general, competition between two forces, one the simplicity of regulatory control and the other the efficiency of gene expression, appears to have shaped the induction pattern in catabolic pathways. In some cases the resolution of this conflict has been one sided and can be interpreted readily. In the L-tryptophan pathway, sequential inductive steps occur at the level of anthranilate and catechol (Figure 7.8). The former is a commonly occurring metabolite and the latter is a site of catabolic convergence. Sequential inductive controls at these steps permit utilization of the aromatic pathway intermediates without necessitating the gratuitous synthesis of the catabolic enzymes that give rise to them. In this case relatively complex regulatory controls permit efficient regulation of structural gene expression. In other instances simplified inductive mechanisms impose an occasional burden of gratuitous enzyme synthesis. For example, the three enzymes that convert typtophan to anthranilate (Figure 7.8) are all induced by the intermediate kynurenine. As a consequence of this simplified regulatory mechanism kynurenine can be used as a growth substrate only with the gratuitous synthesis of the two enzymes that mediate its formation from L-tryptophan. Another possible evolutionary solution to the problem—the selection of L-tryptophan rather than kynurenine as the inducer—would preclude the utilization of kynurenine as a growth substrate.

The conflict between selective forces for efficient gene expression and those for unified control has been resolved in a variety of ways as can best be exemplified by a comparison of the mechanisms used to control the synthesis of the enzymes of the β-ketoadipate pathway in *Acinetobacter* (Cánovas & Stanier, 1967) and in *Pseudomonas* (Ornston, 1966). The central enzyme reactions of this pathway are shown in Figure 7.13. Two diphenols, protocatechuate and catechol, represent the final aromatic intermediates in the breakdown of a large number of growth substrates. Thus the diphenols themselves represent major sites of metabolic convergence. The reactions shown in Figure 7.13 convert the two diphenols to common intermediary metabolites (succinate and acetyl-CoA) by metabolic pathways that converge at the level of β-ketoadipate enol-lactone. The inductive controls governing the pathways have been studied in detail in a number of other bacterial genera as well as *Pseudomonas*.

The simplest inductive controls are found in *Acinetobacter*. As shown in Table 7.4, all of the enzymes that convert protocatechuate to β-ketoadipyl-CoA are governed as a single regulatory unit induced by protocatechuate. Muconate serves as a product inducer of catechol oxygenase and also induces the enzymes of another regulatory unit that convert muconate to β-keto-adipyl-CoA. A clear selective advantage is apparent in the simplicity of the highly unified regulatory controls of *Acinetobacter* but this mode of regulation places some metabolic burdens on the members of this genus. It appears that there are duplicate structural genes for enol-lactone hydrolase (ELH 1 and ELH 11) and for the transferase (TR 1 and TR 11) so that protocatechuate and muconate independently induce these enzymes. By the selection of early metabolites as inducers the regulatory mechanisms of *Acinetobacter* preclude the direct utilization of some of the metabolic intermediates (e.g. carboxy-muconate or β-ketoadipate) as growth substrates.

With the exception of the acidovorans group, a common control mechanism is found among *Pseudomonas* species (Ornston, 1966; Kemp & Hegeman, 1968). As shown in Table 7.4 this regulatory mechanism is more fragmented that that of *Acinetobacter*. The inductive role of protocatechuate is restricted to the induction of its oxygenase (PO) and the rest of the enzymes of this branch of the pathway are induced by β-ketoadipate. Muconate induces catechol oxygenase (CO) as one unit and muconate lactonizing enzyme (MLE) and muconolactone isomerase (MI) as another unit. Product induction by β-ketoadipate permits *Pseudomonas* strains to grow at the expense of β-ketoadipate and carboxymuconate although a permeability mutation is required for the expression of the carboxymuconate uptake system (Meagher *et al.*, 1972). The use of β-ketoadipate as an inducer for ELH and TR also avoids the necessity for duplicate structural genes for these enzymes; since β-ketoadipate is formed during growth with either protocatechuate or muconate the single set of structural genes is expressed in either instance. An apparent metabolic burden placed by the simplified regulatory control that unites induction of CMLE, CMD, ELH and TR is that under some growth conditions (e.g. growth with catechol, muconate or β-ketoadipate) the two enzymes of this regulatory unit which belong to the protocatechuate branch of the pathway are induced gratuitously.

Analysis of the regulation of the β-ketoadipate pathway in *Alcaligenes* (Johnson & Stanier, 1971) and in *Nocardia* (Rann & Cain, 1969) has revealed novel control mechanisms that differ in the extent of coordinate control as well as in the metabolites that serve as inducers. Each induction pattern offer potential selective benefits and disadvantages, suggesting that these are effectively counterbalanced in any of the regulatory systems that have been examined. If this is indeed the case, then the primary factors that shaped the evolution of the regulatory mechanisms were not so much the forces of natural selection as the chance mutations that led to the initial evolution of

Table 7.4. Induction patterns in *Acinetobacter* and *Pseudomonas* species. Brackets indicate the enzymes that are subject to coordinate control. The regulation of β-ketoadipyl-CoA thiolase has not been studied

	Acinetobacter				*Pseudomonas*		
Inducer	Enzyme	Inducer	Enzyme	Inducer	Enzyme	Inducer	Enzyme
Protocatechuate	PO	Muconate	CO	Protocatechuate	PO	Muconate	CO
	{ CMLE		{ MLE	β-Ketoadipate	{ CMLE		{ MLE
	{ CMD		{ MI		{ CMD		{ MI
	{ ELH 1		{ ELH 11		{ ELH		
	{ TR 1		{ TR 11		{ TR		

Figure 7.13 gives an outline of the central reactions of the β-ketoadipate pathway and a key to the enzyme abbreviations.

the regulatory systems. Once formed, the regulatory mechanisms appear to have been strictly conserved within well-defined bacterial groups (Cánovas *et al.*, 1967). This raises the possibility that ecological factors may combine to favour retention of specified regulatory mechanisms in an organism with a well-defined ecological niche.

The β-ketoadipate pathway does not appear to play a major role in the dissimilation of aromatic compounds in the *Pseudomonas acidovorans–testosteroni* group. The primary function of the pathway in these organisms appears to be the utilization of muconic and carboxymuconic acids (Robert-Gero, Poiret & Stanier, 1969; Ornston & Ornston, 1972). The enzymes of the pathway are induced by distinctive mechanisms in the acidovorans group. Unlike the other bacteria that have been examined, they employ carboxymuconate (or carboxymuconolactone) as the inducer of CMLE and CMD. β-Ketoadipate induces the synthesis of MI in *P. testosteroni* but not in *P. acidovorans* so that the induction patterns may be used to distinguish between these two species (Ornston & Ornston, 1972).

3.h. Divergent Pathways

Catechol may be metabolized by any one of three metabolic routes in *Pseudomonas*. One of these is initiated by catechol 1,2-oxygenase ('*ortho*' cleavage) and leads via β-ketoadipate to succinate and acetyl-CoA and has been discussed in detail in the previous section. The other two pathways begin with the 2,3-oxygenative fission of catechol ('*meta*' cleavage) and result in the production of acetaldehyde and pyruvate (Dagley *et al.*, 1960; Dagley, 1971). 2-Hydroxymuconic semialdehyde the product of the '*meta*' oxygenase may undergo either of two metabolic fates (Figure 7.14). The direct action of a hydrolase produces 2-ketopent-4-enoic acid and formate (Dagley & Gibson, 1965; Bayly & Dagley, 1969) or, on the other hand, 2-ketopent-4-enoic acid may be formed as the result of three enzymic reactions initiated by the NAD-dependent oxidation of 2-hydroxymuconic semialdehyde (Nishizuka *et al.*, 1962; Sala-Trepat & Evans, 1971).

Several investigators have provided convincing evidence that all three pathways may occur in a single strain of *P. putida*. Davies & Evans (1964) demonstrated that growth of one *Pseudomonas* strain with benzoate elicited the synthesis of the enzymes of the '*ortho*' pathway, but that growth with naphthalene caused induction of the '*meta*' pathway. Similar catabolic divergence was reported by Cain and Farr (1968) who found that a *P. putida* strain utilized catechol derived from benzenesulphonate via the '*meta*' pathway but employed the '*ortho*' pathway for catechol derived from benzoate. Subsequent studies (Sala-Trepat *et al.*, 1972; Bayly & Wigmore, 1973) have shown the coexistence of both the hydrolase and the NAD-dependent dehydrogenase for 2-hydroxymuconic semialdehyde in *Pseudomonas* strains in which the '*meta*' enzymes are induced.

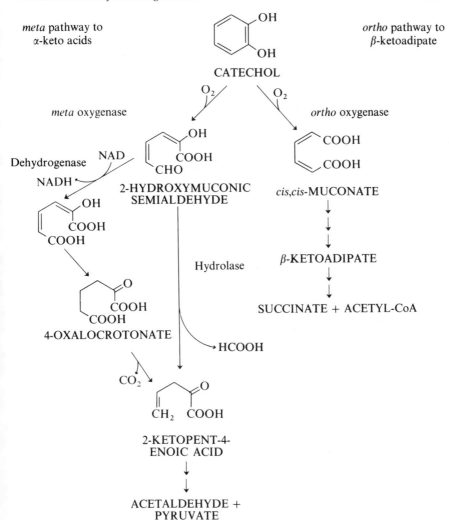

Figure 7.14. Divergent reactions in the catabolism of catechol by *Pseudomonas* species.

In the light of the tendency towards metabolic convergence in catabolic pathways the presence of the striking divergence in the catabolism of catechol is remarkable. Certain *Pseudomonas* strains appear to possess three different sets of genetic information, any one of which would permit the cells to achieve the single metabolic goal of converting catechol to common intermediary metabolites.

Analysis of the regulation of the pathway in *P. putida* U (Dagley's strain) indicates that the roles of the divergent catabolic pathways are largely independent (Feist & Hegeman, 1969a; Sala-Trepat *et al.*, 1972; Bayly & Wigmore, 1973) and therefore that the apparent metabolic redundancy of the divergent pathways is not as great as it might seem. The '*ortho*' pathway is highly specific in both its catalysis and induction. Catechol must be converted to muconate in order to trigger the induction of the '*ortho*' oxygenase. This mechanism appears to heighten the specificity of induction because methyl derivatives of catechol do not elicit synthesis of the '*ortho*' enzymes. Thus the '*ortho*' pathway is restricted to the utilization of catechol under circumstances in which this metabolite accumulates to a degree sufficient to trigger product induction of the 1,2-oxygenase. In contrast, the '*meta*' oxygenase cleaves a range of alkyl derivatives of catechol in addition to catechol itself. The broad catalytic specificity of the '*meta*' oxygenase is mirrored in the range of compounds that mediate its induction. Phenol *o*-, *m*- and *p*-cresol are each potent inducers of all the enzymes of the '*meta*' pathways in *P. putida* strain U (Figure 7.15b, c and d). Thus, catechol produced during the utilization of phenol is metabolized exclusively by the '*meta*' pathway because it never achieves the endogenous concentration necessary to trigger the induction of the specialized '*ortho*'-pathway enzymes. Feist and Hegeman (1969a) and Bayly and Wigmore (1973) demonstrated that phenol (but not the cresols) is utilized by the '*ortho*' pathway if the 1,2-oxygenase is removed by mutation. Benzoate is not an inducer of the '*meta*' enzymes and consequently is converted to catechol which indirectly initiates induction of the '*ortho*' enzymes.

An indication of the role of 2-hydroxymuconic semialdehyde hydrolase was provided by Sala-Trepat and Evans (1971) who showed that this enzyme was induced only to a low level in *Azotobacter*. These bacteria grew well with phenol and with *p*-cresol, but not with *o*- or *m*-cresol. Thus it appeared that the NAD-dependent 2-hydroxymuconic semialdehyde dehydrogenase (which is inducible in *Azotobacter*) permitted growth with compounds converted to catechol of 4-methylcatechol (Figure 7.15b and d) but that the hydrolase was required for the catabolism of compounds converted to 3-methylcatechol (Figure 7.15c). Investigations by Sala-Trepat and colleagues (1972) and by Bayly and Wigmore (1973) confirmed that this is indeed the case in *P. putida*. Although the hydrolase cleaves 2-hydroxymuconic semialdehyde to 2-ketopent-4-enoate and formate, the relative activity of the NAD-dependent dehydrogenase is considerably higher in extracts of cells induced for the '*meta*' enzymes. Conversely the dehydrogenase is inoperative in the 3-methylcatechol pathway (Figure 7.15c) because there is no free aldehyde produced (Sala-Trepat *et al.*, 1972). Bayly and Wigmore (1973) confirmed the essential role of the NAD-dependent dehydrogenase in the catabolism of catechol by showing that a mutant strain lacking the enzyme grew with phenol (using 2-hydroxymuconic semialdehyde hydrolase) at only a fifth of the wild-type rate.

Since both the hydrolase and the dehydrogenase are induced by growth with phenol or the cresols, it appears that the induction of one or the other enzyme must be gratuitous. In fact this may not be the case in the natural environment where the aromatic compounds are likely to occur as mixtures rather than singly.

It is worth noting that in some *P. putida* strains the '*meta*' enzymes are induced by an even wider range of compounds. Benzoate and toluate, trigger the synthesis of the '*meta*' enzymes in one *P. putida* strain (*P. arvilla*) and accordingly these compounds are metabolized by '*meta*' cleavage in this organism (Feist & Hegeman, 1969b; Murray *et al.*, 1972).

3.i. Linkage of Genes for Catabolic Enzymes

The evolution and the maintenance of the varied catabolic sequences of *Pseudomonas* require mechanisms for developing and preserving many different genes in natural populations. A considerable amount of evidence has been adduced to show that much of the genetic information for the vast catabolic potential of *Pseudomonas* populations is possessed by only a fraction of the representative strains and may be transferred from one organism to another by transduction or conjugation. In general these mechanisms of genetic transfer lead to the acquisition of a relatively small segment of genetic information by the recipient organisms. Recombinants can be selected only if the environment offers them a definite nutritional advantage. In our consideration of regulatory patterns, we observed that certain groups of catabolic genes appear to operate as units of physiological function. They are subject to coincident induction and in most natural environments all of the enzymes must function together for growth to occur. Consequently translocations that favour the clustering of structural genes for closely related catabolic functions would be favoured by evolution because they would permit the coincident selection of units of physiological function after partial genetic transfer. The foregoing interpretation was offered by Wheelis and Stanier (1970) to account for the remarkable amount of clustering of catabolic genes observed by them in *P. putida*, and by Kemp and Hegeman (1968) and Rosenberg and Hegeman (1969) in *P. aeruginosa*. Not only are the genes for enzymes within physiological units of function tightly linked, but a number of different physiological units have been shown to be cotransducible. Leidigh and Wheelis (1973) used transduction to demonstrate clustering within 10 to 15 % of the *P. putida* chromosome of genes specifically associated with the following catabolic sequences: benzoate, histidine, *p*-hydroxybenzoate, mandelate, nicotinate, phenylacetate, phenyl-alanine, quinate, shikimate and tyrosine. It appears that a relatively large amount of catabolic information may be cotransferred by a relatively small chromosomal fragment.

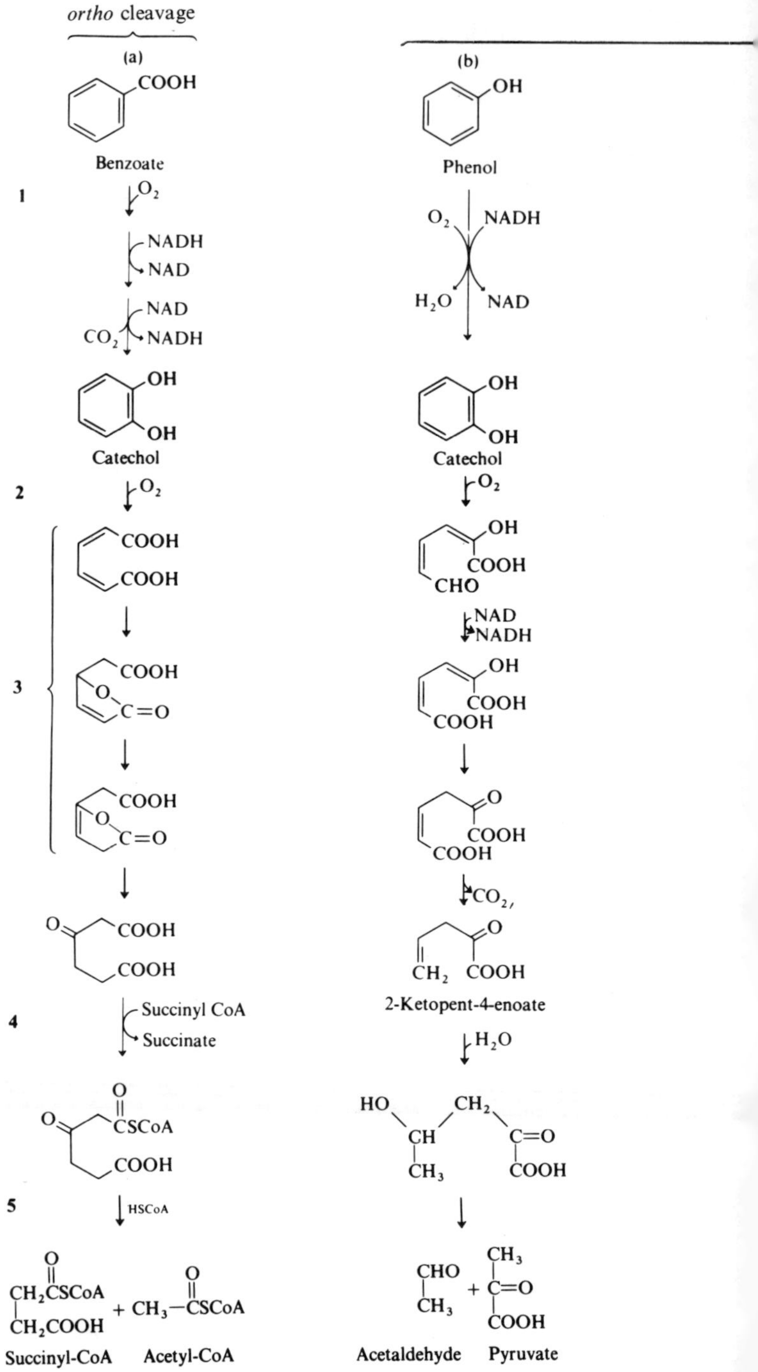
ortho cleavage
(a)
(b)
COOH
Benzoate
OH
Phenol
1
O2
NADH
NAD
NAD
CO2
NADH
O2
NADH
H2O
NAD
OH
OH
Catechol
OH
OH
Catechol
2
O2
O2
3
COOH
COOH
C=O
OH
COOH
CHO
NAD
NADH
OH
COOH
COOH
O
COOH
COOH
CO2
O
CH2 COOH
2-Ketopent-4-enoate
COOH
COOH
4
Succinyl CoA
Succinate
H2O
O
CSCoA
COOH
HO
CH2
CH
C=O
CH3
COOH
5
HSCoA
CH2CSCoA
CH2COOH
+ CH3—CSCoA
Succinyl-CoA
Acetyl-CoA
CHO
CH3
CH3
+ C=O
COOH
Acetaldehyde
Pyruvate

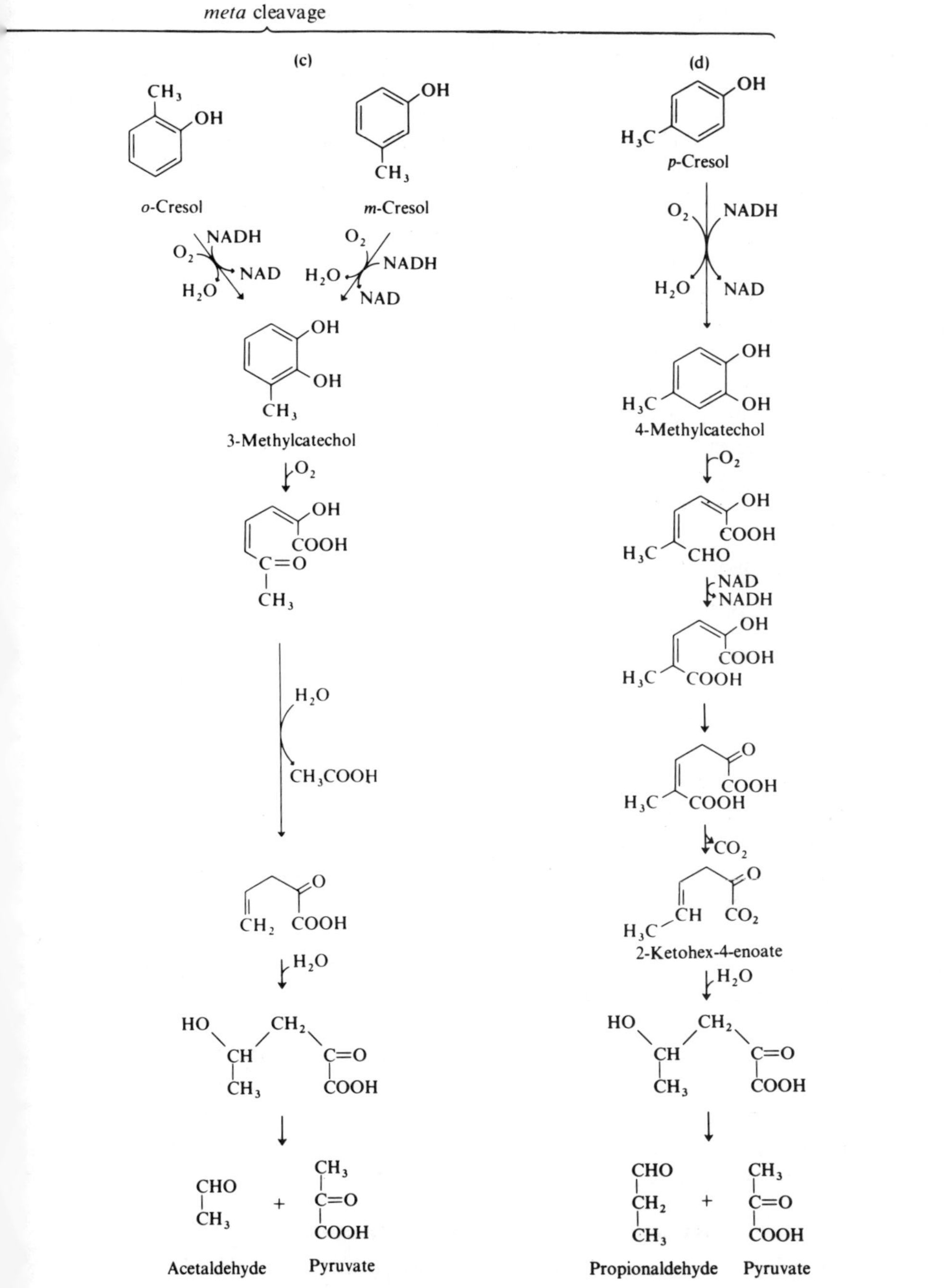

Figure 7.15. Pathways for the dissimilation of catechol and its methyl derivatives in *Pseudomonas*

In several instances the genes for catabolic function are contained on plasmids that may be transferred among different *Pseudomonas* species. Transmissible extrachromosomal replicating units have been described for mandelate (Chakrabarty & Gunsalus, 1969), camphor (Rheinwald *et al.*, 1973), salicylate (Chakrabarty, 1972) and naphthalene (Dunn & Gunsalus, 1973). The latter two functions are metabolically related (naphthalene is metabolized *via* salicylate) and the genes for some of the enzymes of these pathways may be on a single plasmid. The genetic organization within a single *Pseudomonas* species may vary. Thus the genes for naphthalene degradation appear to be associated with the chromosome in some strains of *P. putida* but with a plasmid that can be cured by mitomycin treatment in other strains of the same species.

Rather than presenting a rigid unified structure, genetic explorations into the catabolic pathways of *Pseudomonas* have revealed a range of structures in dynamic equilibrium, opportunities for the development and maintenance of a maximum of genetic variability appear to be exploited by frequent recombinational events within natural populations.

4. BIOSYNTHETIC PATHWAYS AND THEIR REGULATION

4.a. Growth Requirements

Almost all *Pseudomonas* species are non-exacting and are able to grow in simple salt media with ammonium salts providing the nitrogen source and any of a large number of organic compounds providing the carbon source. They are therefore able to synthesize amino acids, purine and pyrimidine nucleotides, and all essential cellular compounds. However, far less attention has been paid to biosynthetic pathways than to catabolic pathways. One reason for this is that the biosynthetic pathways were expected to reveal few novelties. Comparative studies of biosynthetic pathways in a very wide range of species, including bacteria, plants and animals, had shown that, with a few very interesting exceptions, the same sequence of reactions occurs for the synthesis of any particular amino acid. On the other hand, studies on the degradation of organic molecules were expected to elucidate new metabolic pathways, and perhaps new regulatory systems, and these expectations have been amply justified by results. Another reason for the relative dearth of information about the regulation of biosynthetic pathways is that although it is not difficult to obtain auxotrophic mutants of *Pseudomonas* for amino acids, purines and pyrimidines, or vitamins it is only in the last few years that mutants with altered regulatory characteristics have been obtained.

4.b. Arrangements of Genes for Biosynthetic Pathways

In any discussion of the biosynthetic pathways of *Pseudomonas* it is inevitable that comparisons will be made with *Escherichia coli* and *Salmonella typhimurium.* These two latter species have been subjected to very intensive genetic biochemical study and the enzymes of the pathways, the genes determining them, and their regulation are now very largely known. Two types of control can be recognized, regulation of enzyme activity by feedback inhibition and regulation of enzyme synthesis by end-product repression. For a typical biosynthetic pathway of *E. coli* the genes determining the enzymes are clustered together to form an operon of linked genes under the control of a single regulator gene and the rate of synthesis of the enzymes of the pathway is controlled by coordinate repression by the end-product combined with the aporepressor determined by the regulator gene. It should be emphasized that even in *E. coli* this is not the only pattern for the bio-synthetic pathways. The genes for arginine biosynthesis are arranged in six separate groups on the chromosome although all the enzymes are repressed by arginine and an aporepressor determined by a single regulator gene (Gorini, Gundersen & Burger, 1961). The histidine genes form a single operon, but the regulation is complex and can be altered by mutation in any one of five different genes. There is evidence that charged histidinyl transfer RNA may be involved in repression of synthesis of the histidine enzymes rather than histidine itself (Lewis & Ames, 1972). The genes for methionine biosynthesis are scattered on the chromosome in four groups and the regulation of the biosynthesis of this amino acid is also complex (Lawrence, Smith & Rowbury, 1968).

In *P. aeruginosa* and *P. putida* it has been found that the genes for the enzymes of each of the biosynthetic pathways investigated are far more widely scattered than those for the corresponding enzymes of *E. coli* (Fargie & Holloway, 1965). The genes for histidine biosynthesis for example, which form one single giant operon in *E. coli* are found in *P. aeruginosa* in five separate transduction groups (Mee & Lee, 1969). However, when an enzyme from *Pseudomonas* species has been compared in detail with the corresponding enzyme from *E. coli* and *Salmonella typhimurium* it is found that it not only carries out the same reaction in the same way but considerable homology may exist at the molecular level (Crawford & Yanofsky, 1971). This relatively recent approach to comparative biochemistry has underlined the essential similarity of the enzymes of biosynthetic pathways. It is not unreasonable to suppose that the biosynthesis of amino acids was one of the earliest achieve-ments in the evolution of living things and that it became fixed in a pattern which has persisted in the diverse living species of the present day. Within this general pattern, modifications would occur resulting in differences in enzyme properties or complete loss of certain enzymes, from some species.

It is now accepted that the scattered arrangement of biosynthetic genes which is found in *P. aeruginosa* also occurs in other bacteria such as *Acineto-bacter* (Sawula & Crawford, 1972). It has long been known that the bio-synthetic genes of eucaryotes are not tightly clustered and in the case of fungi such as *Neurospora* or *Aspergillus* the genes for a single biosynthetic pathway may be carried on different chromosomes. It appears now that the gene arrangement of the species belonging to the Enterobacteriaceae may be unusual even among bacteria and that the particular arrangement of scattered genes found in *Pseudomonas* species is therefore not to be considered exceptional. The biosynthetic pathways which have been subjected to genetic analysis in detail will be discussed further in Chapter 8.

4.c. Regulatory Mutants and the Use of Metabolic Analogues

Major differences between the *Pseudomonas* species and the Enterobacteriaceae are found in the patterns of regulation of enzyme activity and of enzyme synthesis. The classical methods for studying enzyme regulation depend on the comparison of wild-type strains with regulatory mutants selected for resistance to growth inhibition by appropriate metabolic analogues. This has until recently been remarkably unsuccessful with *Pseudomonas* species which are notoriously resistant to metabolic analogues as well as to many drugs and antibiotics.

Waltho and Holloway (1966) found that of a large number of analogues tested only *p*-fluorophenylalanine gave significant growth inhibition of *P. aeruginosa* and Kay and Gronlund (1969c) in a further survey with 53 amino acid analogues found that only 3,4-dehydroproline, 4-nitropyridine-*N*-oxide, *m*-fluorophenylalanine and *m*-fluorotyrosine gave good growth inhibition on glucose-minimal agar plates, while very weak inhibition was produced by a few other compounds. This resistance to analogues may in some case be ascribed to general impermability but when it is considered that *P. aeruginosa* has constitutive permeases for almost all the amino acids (Kay & Gronlund, 1969a) it is clear that a specificity is an important factor in resistance. The transport systems, the regulatory systems, and the enzymes of *P. aeruginosa* all have their own particular affinities for normal cell metabolites, and also for analogues of these metabolites and these are characteristic and differ from those of other bacterial genera. It is now known that it is possible to isolate both feedback inhibition-resistant mutants, and mutants with altered regulation of synthesis of biosynthetic enzymes, as mutants showing resistance to certain analogues of amino acids. A list of those compounds which have been used successfully in this way will be found in Chapter 8 in the section dealing with the amino acid relevant to the inhibitory compounds.

In some cases differences in sensitivity to analogues may be found between *Pseudomonas* species so that although *P. aeruginosa* is inhibited by 5-fluoro-tryptophan at 10 μg/ml and gives rise to mutants derepressed for some of the tryptophan enzymes (Calhoun, Pierson & Jensen, 1973a), it required 200 μg/ml to inhibit *P. putida* and obtain the same class of mutants (Maurer &

Crawford, 1971). Calhoun and Jensen (1972) found that they could increase the sensitivity of *P. aeruginosa* to growth inhibition by β-thienylalanine by growing the cultures with fructose as a carbon source and were able to isolate a resistant mutant with a prephenate dehydratase which was less sensitive to feedback inhibition by phenylalanine. In this case the authors suggest that with fructose as the carbon source the level of precursors of the aromatic amino acids was reduced and this allowed the phenylalanine analogue to produce sufficient inhibition of an enzyme essential for phenyl-alanine synthesis to prevent growth. Manipulation of the metabolic pool by selection of the appropriate carbon source may make it possible to get growth inhibition by other analogues which are ineffective in glucose minimal medium which was used for most of the earlier studies.

Detailed studies of some of the amino acid biosynthetic pathways have now shown that control of metabolic flow through the pathway by feedback inhibition is general. For a few pathways it has now been confirmed that some of the enzymes are repressed by the end-product. For the tryptophan pathway in both *P. aeruginosa* and *P. putida*, it has been shown that some enzymes are repressible, other are inducible and one appears to be invariable (Gunsalus *et al.*, 1968; Calhoun *et al.*, 1973a).

4.d. Regulation of Enzyme Activity

The maintenance of metabolic balance requires that the flow of metabolites through a biosynthetic pathway should be adjusted to the requirements of the cells for growth and this is normally achieved by feedback inhibition control of one or more enzymes of the pathway. Feedback inhibition may be recognized by the rapid cessation of growth when a non-metabolizable analogue, which mimics the action of the end-product, is added to the culture. There are very specific structural demands which must be met before a compound can act in this way; it must be sufficiently like the end-product to enter the cell via a transport system, to bind at the regulatory site of the feedback inhibition-sensitive enzyme and it must not be able to substitute for the end-product for cellular syntheses. The physiological significance of growth inhibition by an analogue is demonstrated by showing that inhibition by the analogue can be relieved by the end-product itself. Presumptive evidence about the particular enzyme which is regulated by feedback inhibition may be obtained by the use of crude cell-free extracts, but conclusive evidence can only be obtained by the study of purified proteins.

The elegant series of investigations of Jensen and Calhoun and their colleagues on the biosynthesis of the aromatic amino acids of *P. aeruginosa* include studies on the feedback inhibition of this biosynthetic pathway at each of the levels mentioned above. One of the phenylalanine analogues which inhibits the growth of *P. aeruginosa* is β-thienylalanine (TA). As was mentioned earlier, inhibition by this compound is only noticeable when fructose is used as the carbon source and in such a medium growth inhibition by β-thienylalanine is reversed by phenylalanine and also by the intermediate, shikimate. The effect of shikimate is to increase the concentration of precursors for aromatic amino acid synthesis and not to supply the product of an earlier enzyme inhibited by β-thienylalanine. In cell-free extracts the prephenate dehydratase of the wild-type strain was inhibited by a low concentration of phenylalanine while a TA-resistant mutant had similar prephenate dehydratase activity but was only slightly inhibited by the same concentration of phenylalanine. The physiological effect of the mutation conferring TA-resistance is to release the feedback inhibition control and this results in the over-production of phenylalanine which is excreted by these mutants (Calhoun & Jensen, 1972). When prephenate dehydratase was purified from *P. aeruginosa* by Calhoun and his colleagues (1973b) they found that the activity was carried on a bifunctional protein which also had chorismate mutase activity and catalysed the formation of prephenate from chorismate. Both activities were inhibited by phenylalanine, the chorismate mutase partially and the prephenate dehydratase completely; chorismate mutase activity was strongly inhibited by prephenate. These studies illustrate the way in which analogues, resistant mutants and enzyme kinetic studies have been applied to the elucidation of one feedback sensitive enzyme of a biosynthetic pathway.

Effective control of a pathway which leads to a single end-product can be achieved by feedback inhibition of one of the early enzymes of the pathway, usually the first one. In the example discussed previously the enzyme inhibted by phenylalanine is required only for the synthesis of phenylalanine and is therefore part of a linear pathway. However, phenylalanine is one of three end-products of the complex branching pathway which leads to the synthesis of the three aromatic amino acids, phenylalanine, tryptophan and tyrosine. If one of the earlier enzymes before the branch points were to be inhibited by any single product then growth might cease because of depletion of precursors required for the synthesis of the other two end-products (Stadtman, 1963; Datta, 1969). Comparative studies of feedback inhibition controls in a range of bacterial genera have revealed that several different solutions have evolved to overcome the problem of the effective control of branched pathways. Jensen, Nasser and Nester (1967) confirmed that other genera of the Enterobacteriaceae resembled *E. coli* in producing isofunctional enzymes with DAHP synthetase activity for the first step of the aromatic pathway, the conversion of phosphoenolpyruvate and erythrose phosphate to DAPH (3-deoxy-D-arabinoheptulosonate-7-phosphate). All the *Pseudomonas* species produced a single DAHP synthetase which was inhibited in cell extracts by tyrosine, although the acidovorans group also showed some sensitivity to phenylalanine. The production of a single enzyme species at the feedback inhibition-sensitive step by *Pseudomonas* species, is characteristic of this genus. A similar difference between *Pseudomonas* species and entero-bacteria occurs in the regulation of the branched pathway for the synthesis of the amino acids derived from aspartate. In *E. coli* there are three iso-functional aspartokinases which catalyse the first step of the branched pathway leading to threonine, lysine and methionine. Two of these are inhibited by threonine and lysine respectively, and the third is under only repression control by methionine. *Pseudomonas* species produce a single aspartokinase so the regulation of this pathway again differs from that of *E. coli* (Cohen, Stanier & Le Bras, 1969).

The feedback control of branched pathways in some species of bacteria is dependent on inhibition by a single amino acid and in this case when one of several possible end-products is added to a minimal salt medium it may produce growth inhibition (Datta & Gest, 1964). Organisms which behave this way are probably not exposed in their natural environment to high concentrations of only a single amino acid so that this is a laboratory hazard rather than a natural phenomenon. More satisfactory solutions have evolved in which the control point enzyme is completely inhibited only if all the end-products, either directly or indirectly, are involved in the inhibition. In *P. aeruginosa* and *P. fluorescens* it appears from studies with crude cell extracts that DAHP synthetase was sensitive to inhibition by tyrosine alone and that phenylalanine and tryptophan were as ineffective as inhibitors (Jensen *et al.*, 1967). This result had to be reconciled with the observation that the addition of tyrosine did not inhibit bacterial growth. Jensen and his colleagues (1973), in a more detailed study of the allosteric properties of DAHP synthetase from *P. aeruginosa*, found that the regulation of this enzyme was more complex. Tyrosine and phenylpyruvate together produced more than 90 % inhibition at saturating concentrations and partial inhibition was produced by low concentrations of typtophan. It is interesting that regulation by phenyl-alanine is indirect and is effected by its transamination product phenyl-pyruvate, and this provides a novel control mechanism (Figure 7.16). However feedback regulation of DAHP synthetase in *Bacillus* species (Jensen & Nester, 1966) is effected by intermediates and not end-products of the path-way. Jensen and his coworkers (1973) found that inhibition of *P. aeruginosa* DAHP synthetase by tyrosine and tryptophan was competitive with respect to phosphoenolpyruvate and inhibition by phenylpyruvate was competitive with respect to the other substrate erythrose phosphate. They pointed out that the phosphoenolpyruvate level in the metabolic pool is an indication of the energy state of the cell and that inhibition by tyrosine would be most marked at low intracellular levels of phosphoenolpyruvate when the cell was deficient in available energy sources. Phosphoenolpyruvate is an important cell metabolite and is known to regulate the activity of some key

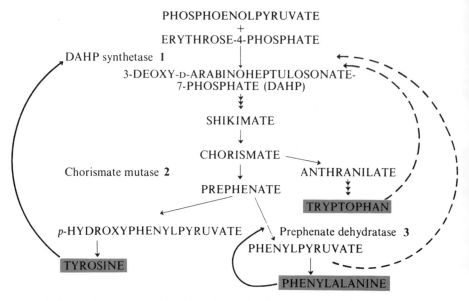

Solid line indicates strong inhibition, dotted line indicates weaker inhibition.

Figure 7.16. Feedback inhibition of the enzymes of the branched pathway for the biosynthesis of the aromatic amino acids in *Pseudomonas aeruginosa* (Calhoun *et al.*, 1973; Jensen *et al.*, 1973)

enzymes including isocitrate lyase which is essential for growth on acetate (Kornberg, 1966b).

Studies so far on *Pseudomonas* indicate that patterns of feedback inhibition of biosynthetic pathways have evolved which are effective in cell economy and characteristic of the genus. The branched pathways do not involve iso-functional enzymes at the branch points but depend on some type of multi-effector inhibition (Jensen *et al.*, 1973). In the few instances where detailed comparisons have been made between *Pseudomonas* species, only minor differences in sensitivities of enzymes to feedback inhibitors have been observed.

4.e. Regulation of Enzyme Synthesis

The synthesis of biosynthetic enzymes in *E. coli* is normally repressed in the presence of the end-product. While feedback inhibition provides a fine control which responds rapidly to changes in the concentration of the end-product, the repression of enzyme synthesis can be regarded as a course control which effects greater cell economy by conserving energy and substrates which would have been needed to make the enzymes. In *E. coli* this type of regulation can be recognized by derepression of enzymes when the bacteria are grown in minimal medium, and repression when the end-product is present; derepression of bradytrophic mutants when the bacteria are grown in a medium in which the end-product limits growth; repression of the wild type when the bacteria are grown in the presence of an appropriate analogue of the end-product which may result in growth inhibition and the isolation of derepressed mutants which produce the enzymes even in the presence of the end-product or its analogue.

Few of these effects have been observed in *Pseudomonas* and this could be (a) because the biosynthetic enzymes of *Pseudomonas* are not regulated by repression or (b) because the regulation by repression is so efficient that it cannot be disturbed by the methods which are applicable to *E. coli*, or (c) because the regulation follows a different pattern.

A difference in the enzyme activities of wild-type strains grown in different media which could be interpreted as repression or derepression of enzyme synthesis in the presence and the absence of the end-product of the pathway has been observed for a few biosynthetic pathways. Crawford and Gunsalus (1966) found in *P. putida* that the *trp*ABD enzymes for steps 1, 2 and 4 of the tryptophan biosynthetic pathway were repressed by tryptophan, while the enzymes of the *trp*C and *trp*EF linkage groups were unaffected (see Chapter 8 and Figure 8.32 for details of the pathway). Similar results were obtained for *P. aeruginosa* where Calhoun and his colleagues (1973a) found that the enzyme activity varied up to 50-fold under repressed and derepressed conditions. This group of tryptophan genes behaves as an operon under typical end-product repression control. Derepressed mutants have been isolated in both *P. putida* (Maurer & Crawford, 1971) and *P. aeruginosa* (Calhoun *et al.*, 1973a) which were selected for resistance to 4- or 5-fluoro-tryptophan. The final enzyme of the pathway, tryptophan synthetase, determined by *trp*EF was found to be induced by its substrate, indoleglycerol phosphate. Each of the three groups of tryptophan genes is therefore under different regulation. The *P. aeruginosa* mutants derepressed for *trp*ABD excrete tryptophan so that when the control is removed from these early enzymes the metabolites flow through the pathway. The *P. putida* mutants excrete anthranilate and Maurer and Crawford (1971) suggested that the rate of operation of the constitutive *trp*C gene was such that it limits the production of indoleglycerol phosphate required for induction of the final enzyme so that the intermediates are excreted and not the final product of the pathway. If this is so, it might be possible to obtain a mutant of *P. putida* in which the operation of *trp*C was 'set high' so that derepression of *trp*ABD could lead to excretion of tryptophan.

Marinus and Loutit (1969a,b) found that two of the enzymes of the iso-leucine–valine pathway were regulated by multivalent repression by isoleucine, valine, leucine and pantothenate, all of which require these enzymes for their biosynthesis. The two enzymes, acetohydroxyacid synthetase and acetohydroxyacid isomeroreductase are determined by the linked genes *ilv*BC. The synthesis of the enzymes was coordinate and certain mutations affected both enzymes so that it was thought that they form an operon under multivalent end-product repression control. The genes for the other enzymes *ilv*A, *ilv*D and *ilv*E are not linked to *ilv*BC and the enzymes are not subject to multivalent repression, whereas in *E. coli* the five genes form a single operon regulated by multivalent repression.

The activity of the first enzyme of the branched biosynthetic pathway starting from aspartate was found to be regulated by feedback inhibition by two of the end-products, lysine and threonine, in all the *Pseudomonas* species (Cohen *et al.*, 1969). In *P. putida* this enzyme was also repressed by lysine, threonine and methionine added singly, and the most severe repression was produced by methionine (Robert-Gero *et al.*, 1971). On the other hand, Jensen and his colleagues (1973) were unable to detect any repression of DAHP synthetase, the first enzyme of the branched pathway for the biosynthesis of the aromatic amino acids.

Mutants requiring purines or pyrimidines for growth have been isolated from *P. aeruginosa*. Isaac and Holloway (1968) showed that the pathway for pyrimidine synthesis is the same as in *E. coli*. Genes for four of the enzymes of the pathway were identified; aspartate transcarbamylase, dihydroorotate dehydrogenase, orotidine monophosphate pyrophosphorylase and orotidine monophosphate decarboxylase. No linkage could be detected between these genes by transduction or conjugation and no repression of these enzymes was observed in cultures grown in the presence of uracil. There was some evidence for feedback inhibition of aspartate transcarbamylase which is known to be subject to allosteric regulation by feedback inhibition in *E. coli* (Gerhart & Pardee, 1962).

Further investigations may reveal repression and induction controls of enzymes of other biosynthetic pathways and meanwhile we might agree with Holloway (1969) that the control mechanisms of *Pseudomonas* are not fundamentally different from those of other organisms but may be thought of as 'variations on a theme'.

5. THE ROLE OF THE TRICARBOXYLIC ACID CYCLE IN METABOLISM

5.a. The Tricarboxylic Acid Cycle Reactions

The tricarboxylic acid (TCA) cycle is of central importance in both catabolism and biosynthesis in *Pseudomonas*. The reactions are shown in Figure 7.17. The first attempts to demonstrate the operation of the cycle were based on applying the principle of sequential induction. Cultures grown on acetate

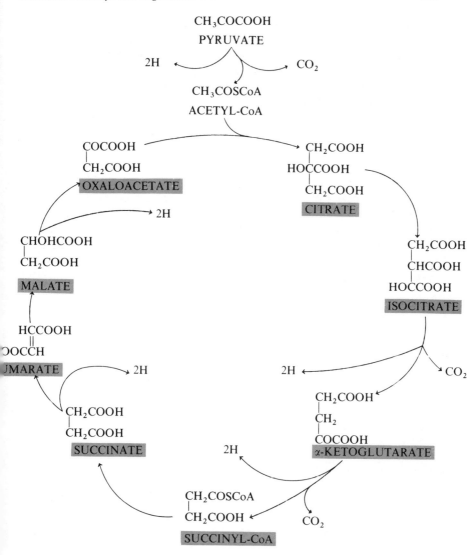

Figure 7.17. Reactions of the tricarboxylic acid cycle

were expected to be able to oxidize all the cycle intermediates if acetate was oxidized by means of the cycle reactions. However, this presented some difficulties and several groups of workers found that washed suspensions of *P. aeruginosa* or *P. fluorescens* which had been grown on acetate, were able to oxidize citrate or succinate only after a lag period. Cultures grown on citrate oxidized citrate without lag, but there was a lag period before they oxidized succinate at a linear rate, and similarly cultures grown on succinate had a lag period before they oxidized citrate. Dried preparations of bacteria, or cell-free extracts, oxidized the TCA cycle intermediates without lag. It was concluded that the TCA cycle enzymes were present in cells grown on acetate, or any of the intermediates of the cycle, and that the differences in the oxidative capacities of the bacteria grown on different substrates were due to differences in cell permeability. It was known that protein synthesis occurred during the lag period before the bacteria reached the linear rate of substrate oxidation, but this was due to the synthesis of specific permeases and not to the synthesis of metabolic enzymes (Campbell & Stokes, 1951; Barrett & Kallio, 1953; Kogut & Podoski, 1953; Clarke & Meadow, 1959).

The precise pathway of metabolic flow is determined largely by the intermediates formed from the primary growth substrate and Figure 7.18 gives an indication of the range of TCA cycle intermediates which may be produced by different metabolic pathways. Some growth substrates are converted into pyruvate or acetyl-CoA but many give rise most directly to the dicarboxylic acids α-ketoglutarate, succinate, fumarate and malate. The complete

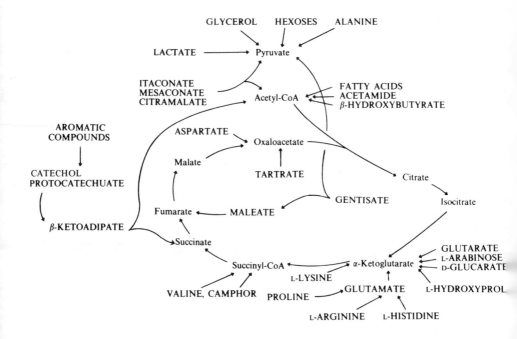

Figure 7.18. Some of the primary growth substrates converted to tricarboxylic acid cycle intermediates by *Pseudomonas*

oxidation of the dicarboxylic acids requires their conversion to acetyl-CoA, a process involving the NADP-dependent oxidation of L-malate to pyruvate by the 'malate enzyme' (malate dehydrogenase decarboxylating).

$$HOOC.CH_2.CHOH.COOH + NADP \rightarrow CH_3.CO.COOH + CO_2$$
$$+ NADPH$$

Jacobson and colleagues (1966) found that the malate enzyme is strongly repressed when cultures of *P. putida* are exposed to acetate. Thus, this repressive control appears to conserve the dicarboxylic acids of the TCA cycle by reducing the rate of decarboxylation.

Pyruvate is converted to acetyl-CoA by the pyruvate dehydrogenase complex, but when it is present as the sole carbon source for growth it is necessary also to synthesize the 4-carbon acids of the cycle. This is necessary to provide catalytic amount of the TCA cycle intermediates for the oxidation of acetyl-CoA and also, as will be discussed in more detail later, to replenish losses due to removal of intermediates for biosynthetic reactions. In most *Pseudomonas* species pyruvate is converted to oxaloacetate by direct carboxylation by the ATP-dependent pyruvate carboxylase.

$$CH_3CO.COOH + CO_2 + ATP \rightarrow HOOC.CH_2.CO.COOH$$
$$+ ADP + Pi$$

Pyruvate carboxylase was purified from *P. citronellolis* by Seubert and Remberger (1961) who showed that, unlike the enzyme isolated from mammalian tissues or from *Arthrobacter*, it does not require acetyl-CoA for activity. A different system for oxaloacetate synthesis from pyruvate is found in *Escherichia coli*. Pyruvate is first converted to phosphoenolpyruvate by phosphoenolpyruvate synthase, then oxaloacetate is formed by carboxylation of phosphoenolpyruvate. It appeared that a similar system might operate in *Pseudomonas* AM1, an organism which contains the specialized enzymatic pathways required for utilization of 1-carbon compounds. However, although *Pseudomonas* AM1 contains phosphoenolpyruvate carboxylase, and not pyruvate carboxylase, it has been shown by Salem and colleagues (1973a) that this enzyme is not required for growth on pyruvate which is metabolized by an alternative route.

Acetyl-CoA enters the TCA cycle by condensing with oxaloacetate to form citrate under the action of citrate synthase and the major control of the rate of operation of the cycle appears to occur at this stage. Weitzman and Jones (1968) showed that the citrate synthases of *Pseudomonas* species were inhibited by NADH and reactivated by AMP. They suggested that the sensitivity to NADH could be considered as feedback inhibition since NADH is one of the products of the action of the cycle. (The citrate synthase of eucaryotic cells is inhibited by ATP which can likewise be considered as a product of the TCA cycle.) The level of AMP relative to ADP and ATP can be taken to represent the energy charge of the cell (Atkinson, 1969) so that the reactivation by AMP represents a second control on the activity of citrate synthase related to energy demand. The *Pseudomonas* citrate synthases are 'large' enzymes with molecular weights of about 250,000 daltons (Weitzman & Dunmore, 1969). Other Gram-negative bacteria such as *Escherichia coli* produce 'large' citrate synthases which are inhibited by NADH but are not reactivated by AMP. Weitzman and Jones (1968) pointed out that AMP is an activator for the citrate synthases of *Pseudomonas* and also for the citrate synthases of other Gram-negative bacteria which are strict aerobes. For *E. coli*, and the facultative anaerobes among the Gram-negative bacteria, AMP has no activating role for citrate synthase and it was suggested that for organisms which employ the Embden–Meyerhof pathway for glucose breakdown, regulation by AMP might be more significant for the glycolytic enzymes than for those of the TCA cycle.

Another difference between the regulation of the citrate synthases of *Pseudomonas* species and *E. coli* lies in the response to α-ketoglutarate. The citrate synthase of *E. coli* is inhibited by α-ketoglutarate but the *Pseudomonas* enzymes are insensitive (Weitzman & Dunmore, 1969). This difference in enzyme regulation can again be related to the different metabolic patterns of the two groups of organisms. The TCA cycle of *Pseudomonas* is almost always necessary for both biosynthesis and terminal respiration whereas the facultative anaerobes can grow even under aerobic conditions with the TCA cycle functioning almost entirely for biosynthesis (Amarasingham & Davis, 1965).

One of the questions to be answered is whether the synthesis of the TCA cycle enzymes of *Pseudomonas* is regulated by induction or repression. In *E. coli* it has been established that the TCA cycle enzymes are repressed by glucose in a minimal salt medium, by growth in a complex medium containing amino acids and by growth in the absence of oxygen (Gray *et al.*, 1966). The activities of several TCA cycle enzymes of *P. aeruginosa* grown in batch culture in glucose or succinate medium were compared by Tiwari and Campbell (1969) who concluded that the cycle enzymes were constitutive and not repressed by glucose. Hamlin and his coworkers (1967) found that the aconitase activity of batch cultures of *P. aeruginosa* grown in glucose medium was only about half that of cultures grown in citrate medium while there was little difference in the isocitrate dehydrogenase activities. In peptone medium the levels of both aconitase and isocitrate dehydrogenase were reduced. Ng and Dawes (1973) measured aconitase and isocitrate dehydrogenase under ammonia-limitation of growth in continuous culture, in steady state and transient conditions, with various proportions of citrate and glucose providing the carbon sources. Some differences were observed in levels of both aconitase and isocitrate dehydrogenase which suggests that at least these two enzymes are under regulatory control although this was clearly not coordinate.

The inducible glucose enzymes were severely repressed by citrate in continuous culture and Hamilton and Dawes (1959) had previously observed a citrate–glucose diauxie in which citrate was used preferentially as the growth substrate. Citrate affects glucose metabolism by repressing the glucose transport system induced by glucose and also by inhibiting glucose transport, but it may also act more directly by repressing the synthesis of the inducible glucose enzymes within the cell (Midgley & Dawes, 1973; Mukkada *et al.*, 1973). The dominant metabolic position of the TCA cycle is also evidenced by repression of many other peripheral catabolic enzymes by succinate and other TCA cycle intermediates. This means that cultures utilize the TCA intermediates in preference to more complex organic compounds which require the induction of one or more enzymes for their metabolism. Succinate represses the synthesis of *P. aeruginosa* amidase and histidase (Brammar & Clarke, 1964; Lessie & Neidhardt, 1967b), the enzymes of the aromatic pathways of *P. aeruginosa* and *P. putida* (Mandelstam & Jacoby, 1965; Ornston, 1966) and the early enzymes of camphor metabolism in *P. putida* (Hartline & Gunsalus, 1971).

5.b. The Glyoxylate Cycle and Growth with 2-Carbon Compounds

The compounds used as growth substrates provide intermediates for biosynthetic pathways as well as acting as energy sources for growth, and the tricarboxylic acid cycle is the starting point for the biosynthesis of very many of the important cell metabolites (Figure 7.19). These biosynthetic pathways place a constant drain on the TCA intermediates and under many conditions of growth it is essential that the pool should be replenished. Kornberg (1966a) suggested that enzyme reactions whose function was to replenish and maintain the pool of biosynthetic intermediates should be termed *anaplerotic sequences*. The function of such enzymes can be seen most clearly by considering the metabolic fate of 2-carbon or 3-carbon compounds under conditions where they provide the sole carbon source for growth. It was mentioned earlier that growth with pyruvate requires the synthesis of 4-carbon compounds and in most *Pseudomonas* species this occurs by the direct carboxylation of pyruvate to form oxaloacetate. However, the most thoroughly investigated anaplerotic sequence is that employed during growth with acetate or acetogenic compounds. Under these circumstances intermediates of the TCA cycle are replenished by the activity of two enzymes, isocitrate lyase (Campbell *et al.*, 1953) and malate synthase (Wong & Ajl,

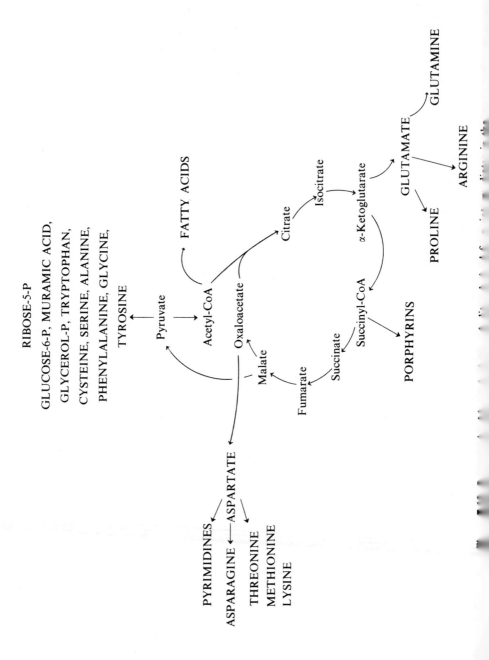

1956). The former enzyme catalyses the cleavage of isocitrate to give glyoxylate and succinate, and the latter enzyme catalyses the condensation of glyoxylate with acetyl-CoA to form malate. Addition of these two enzymes to those of the TCA cycle permits the net synthesis of one molecule of a dicarboxylic acid (succinate) from two molecules of acetyl-CoA (Figure 7.20) by the 'glyoxylate cycle' (Kornberg & Madsen, 1958).

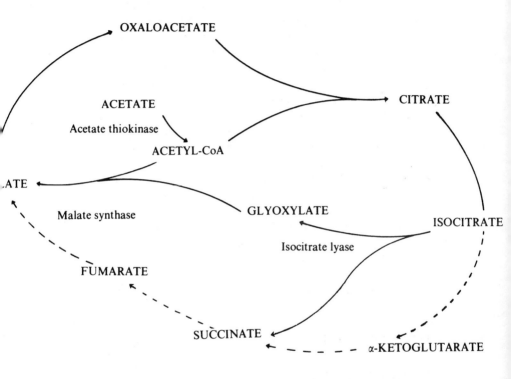

The glyoxylate cycle reactions are indicated by solid lines and the reactions belonging only to the tricarboxylic acid cycle by dotted lines.

Figure 7.20. The glyoxylate cycle reactions during growth with acetogenic compounds

$$HOOC.CHOH.CH.COOH.CH_2COOH \xrightarrow[\text{lyase}]{\text{isocitrate}} CHO.COOH$$
$$+ HOOC.CH_2CH_2COOH$$
$$CHO.COOH + CH_3COSCoA \xrightarrow[\text{synthase}]{\text{malate}} HOOC.CHOH.CH_2COOH$$

Although isocitrate lyase was first demonstrated in *P. aeruginosa* by Campbell (Campbell *et al.*, 1953) the details of the regulation of the synthesis

and activity of this enzyme have been most extensively investigated in
E. coli (Kornberg, 1966b). Cultures of *P. aeruginosa* or *E. coli* grown in
acetate medium have much higher levels of isocitrate lyase than cultures
grown in succinate medium, but Kornberg and his colleagues showed by
using mutants of *E. coli* blocked in various enzymes, that this apparent
induction was due to the removal of oxaloacetate and that isocitrate lyase
synthesis was controlled by repression by phosphoenolpyruvate. The
activity of isocitrate lyase is regulated by inhibition by phosphoenolpyruvate,
and less severely by pyruvate itself. The enzyme is therefore under both
repression and inhibition controls. Mutants lacking isocitrate lyase are
unable to grow on acetate (Kornberg, 1966b) and in *P. aeruginosa* Chapman
and Duggleby (1967) showed that isocitrate-lyase negative mutants were
unable to grow on the carboxylic acids containing even numbers of carbon
atoms (4C–8C) although they could grow on the carboxylic acids containing
odd numbers of carbon atoms (3C–9C), presumably because propionyl-CoA
would be produced as one of the intermediates. Growth on some carbon
compounds allows gratuitous synthesis of isocitrate lyase. Skinner and
Clarke (1968) found that *P. aeruginosa* grown on propionate had high levels
of isocitrate lyase although this enzyme is not necessary for growth on
propionate since isocitrate-lyase negative mutants, which are unable to
grow on acetate, can grow at a normal rate on propionate. They also found
that a mutant which had only about 10 % of the citrate synthase activity of
the wild type had a very low isocitrate lyase activity and this was thought to
be due to accumulation of oxaloacetate and phosphoenolpyruvate by this
mutant. Malate synthase activities of *E. coli* are also higher in acetate-grown
than in succinate-grown cultures but for *P. aeruginosa* there does not appear
to be much difference which suggests that malate synthase is regulated
independently from isocitrate lyase.

5.c. Growth with Glyoxylate

Growth with glyoxylate itself, or compounds which give rise to it in
metabolism, requires a further series of enzymes which comprise the tartronic
semialdehyde pathway (Dagley, *et al.*, 1961; Kornberg & Gotto, 1961;
Hansen & Hayaishi, 1962). Glyoxylate is formed by many *Pseudomonas*
species during growth with glycolate, glycine, allantoate or oxalate (Figure
7.21). A metabolic divergence occurs in the dissimilation of allantoate which
is converted to ureidoglycollate by a single hydrolytic cleavage by *P.
aeruginosa* (reaction 1, Figure 7.21) but requires two successive hydrolytic
steps in *P. putida* (reactions 2 and 3, Figure 7.21) (Trijbels & Vogels, 1967;
Wu *et al.*, 1970).

The first step in the metabolism of glyoxylate is complex and involves
two molecules of glyoxylate which are subjected to a decarboxylative

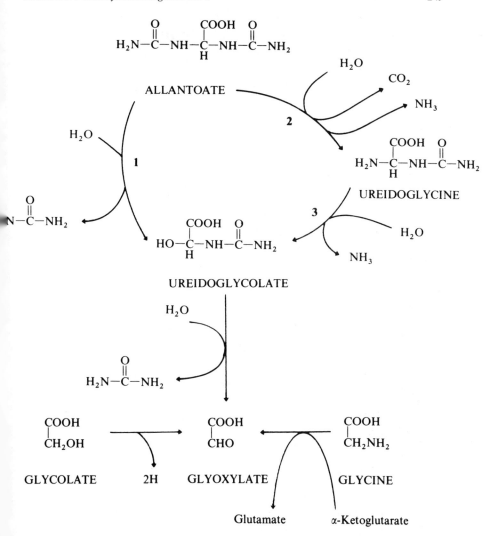

Figure 7.21. Convergent reactions giving rise to glyoxylate in *Pseudomonas*

condensation. The thiamine pyrophosphate-dependent enzyme, glyoxylate carboligase, catalyses the synthesis of the 3-carbon compound tartronic semi-aldehyde with the release of carbon dioxide (Figure 7.22). The aldehyde is reduced to D-glycerate by a NADPH-dependent dehydrogenase which has been crystallized from *P. putida* (*P. ovalis* Chester) by Gotto and Kornberg (1961). D-Glycerate is then phosphorylated to the common intermediary

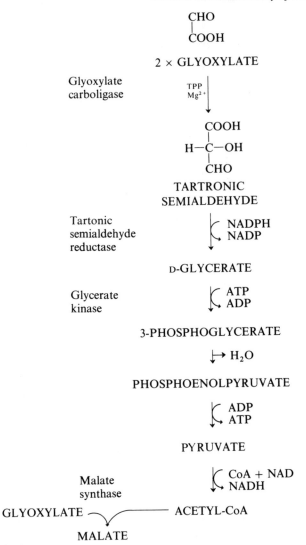

Figure 7.22. The tartronic semialdehyde pathway for the assimilation of glyoxylate in *Pseudomonas* (Dagley *et al.*, 1961; Kornberg & Gotto, 1961)

metabolite 3-phosphoglycerate. There is little doubt that the tartronic semialdehyde pathway plays an essential role in the utilization of glyoxylate by most *Pseudomonas* species. Mutant strains of *P. putida* that lack glyoxylate carboligase do not grow at the expense of glycolate or glycine and do not oxidize glyoxylate (Wu & Ornston, unpublished observations).

Kornberg and Sadler (1961) reported that malate synthase was also induced by the growth of *P. putida* or *E. coli* with metabolic precursors of glyoxylate.

It is known that in *E. coli* there are two malate synthases, one of which is coordinately induced by growth with acetate (malate synthase A) and malate synthase G, which forms about 40 % of the activity of acetate-grown cells and 90 % of the activity of glycolate-grown cells (Vanderwinkel *et al.*, 1963; Vanderwinkel & Vlieghere, 1968). Kornberg and Sadler (1961) found that a mutant lacking citrate synthase could oxidize glycolate, and could grow with glycolate as the major carbon source when the medium was supplemented with a small amount of glutamate. This led them to the conclusion that the TCA cycle was not essential for glyoxylate oxidation but that malate synthase acted as a 'condensing enzyme' for a dicarboxylic acid respiratory cycle. However, Ornston and Ornston (1969) showed that mutants of *E. coli* K12 lacking malate synthase G were still able to grow at the expense of glyoxylate and in these strains glyoxylate was presumably metabolized exclusively *via* the tartronic semialdehyde pathway and the TCA cycle. Malate synthase may serve an assimilatory function when cells are exposed simultaneously to a source of both glyoxylate and acetate. The role of malate synthase has been less extensively investigated in *Pseudomonas*. Skinner and Clarke (1968) showed that malate synthase was essential for growth on acetate and that isocitrate-lyase negative mutants (acetate-negative) of a strain of *P. aeruginosa*, which is unable to grow on glyoxylate, were able to grow in a medium which contained both glyoxylate and acetate. Under these conditions malate synthase allowed the utilization of both acetate and glyoxylate by the mutant.

Dunstan, Anthony and Drabble (1972) suggest that *Pseudomonas* AM1 metabolizes acetate by a partial oxidation to glyoxylate which then condenses with a molecule of acetyl-CoA to form malate by the action of malate synthase. Dunstan and Anthony (1973) suggest that this pathway could allow the metabolism of ethanol, malonate and β-hydroxybutyrate by this organism. Growth of *Pseudomonas* AM1 on pyruvate may also involve these reactions. Salem and his colleagues (1973a) were unable to obtain evidence for pyruvate utilization by conversion to oxaloacetate by any of the reactions known for other bacteria. An active pyruvate dehydrogenase was present in pyruvate-grown cultures which would convert pyruvate to acetyl-CoA while phosphoenolpyruvate carboxylase activity was low in cultures grown on 3-carbon compounds and no pyruvate carboxylase could be detected. This metabolic pathway provides another route for the net synthesis of 4-carbon compounds from 2-carbon and 3-carbon compounds and an additional metabolic role for malate synthase.

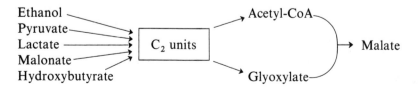

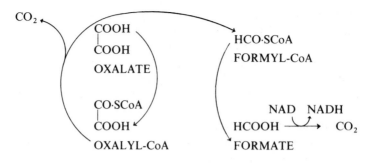

Figure 7.23. The oxidation of oxalate to formate and carbon dioxide in *Pseudomonas*

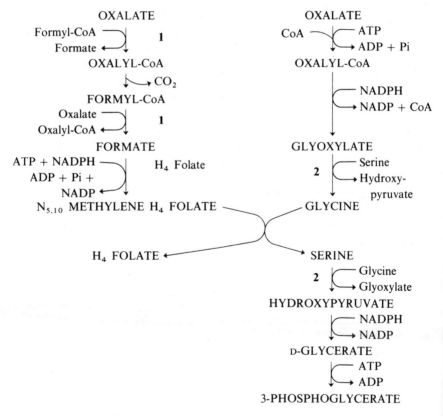

Note the economy of enzyme reactions. Both reactions **1** and **2** recur in the sequence.

Figure 7.24. The serine pathway for the reductive assimilation of oxalate in *Pseudomonas*

5.d. Growth with Oxalate

Only a few *Pseudomonas* strains are able to grow with oxalate as the carbon source and these reduce oxalate to glyoxalate by the following two reactions:

$$\text{oxalate} + \text{succinyl-CoA} \longrightarrow \text{oxalyl-CoA} + \text{succinate}$$

$$\text{oxalyl-CoA} + \text{NADPH} \longrightarrow \text{glyoxalate} + \text{NADP}$$

Different metabolic pathways may be used for the assimilation of the glyoxalate. In *P. oxaliticus* the glyoxylate produced from oxalate is converted to tartronic semialdehyde and assimilated by the reactions of the tartronic semialdehyde pathway shown in Figure 7.22. Other *Pseudomonas* strains employ the serine pathway (Figure 7.24) for the reductive assimilation of the glyoxalate produced from oxalate (Blackmore & Quayle, 1970). In the latter sequence glyoxylate is converted to glycine and a 1-carbon fragment derived from oxalate is added to form serine. The serine is converted to glycerate *via* hydroxypyruvate so that the two pathways for oxalate assimilation converge at D-glycerate.

The evolutionary basis for the different reductive pathways is not clear but for those organisms which employ the serine pathway there is the possibility of economy of genetic information since a similar metabolic sequence is used for the assimilation of 1-carbon compounds.

The oxalate-utilizing pseudomonads are able to derive metabolic energy from the oxidation of oxalate to formate and carbon dioxide, by the reactions shown in Figure 7.23 (Quayle, 1963).

5.e. Growth with 1-Carbon Compounds

Relatively few *Pseudomonas* strains are able to grow at the expense of 1-carbon compounds. The properties of these strains and the general metabolic problems inherent in 1-carbon metabolism have been discussed in recent reviews (Ribbons, *et al.*, 1970; Quayle, 1972) and will not be considered in detail here. There is at the present time widespread interest in this field since microorganisms capable of utilizing 1-carbon compounds for growth are being employed on the industrial scale for the synthesis of protein from methane and methanol.

Metabolic energy may be obtained from 1-carbon compounds by the following series of reactions.

$$\text{CH}_3\text{OH} \longrightarrow \text{HCHO} \longrightarrow \text{HCOO}^- \longrightarrow \text{CO}_2$$
$$\quad\quad 2\text{H} \quad\quad\quad 2\text{H} \quad\quad\quad 2\text{H}$$

At some stage it would be expected that the reductant produced by these reactions would be used to provide ATP by oxidative phosphophorylation.

In some instances mixed-function oxidases are known to initiate the attack on the 1-carbon compound which introduces it into the metabolic sequence. Methane is hydroxylated to methanol by a monooxygenase in *P. methanica* (Higgins & Quayle, 1970) and a mono-oxygenase produces formaldehyde from dimethylamine in *P. aminovorans* (Eady & Large, 1971).

Several different metabolic routes are employed for the assimilation of one-carbon compounds into common intermediary metabolites. Formate-grown cultures of *P. oxalaticus* oxidize formate to CO_2 and incorporate the CO_2 *via* the ribulose diphosphate cycle reactions characteristic of auto-trophes (Quayle & Keech, 1959). *P. oxalaticus* grows more rapidly with formate than with oxalate. Blackmore and Quayle (1968) demonstrated that when bacteria were exposed to mixtures of formate and oxalate, both compounds were completely oxidized and new cell carbon was assimilated autotrophically *via* ribulose disphosphate carboxydismutase. The auto-trophic mode of carbon assimilation predominates when cells are exposed to formate together with other carbon compounds, e.g., glycerol or malonate which by themselves support only slow growth. Growth substrates that permit rapid growth of *P. oxalaticus* repress the autotrophic enzymes leading to an exclusively heterotrophic pattern of metabolism.

Cultures of *P. methanica* do not form ribulose diphosphate carboxy-dismutase. These organisms have limited metabolic activities and can utilize only methane or methanol as growth substrates; the bacteria oxidize these compounds to formaldehyde, which is condensed directly with ribose-5-phosphate to form the hexose derivative allulose-6-phosphate (Figure 7.25) (Kemp & Quayle, 1967); epimerization of the allulose-6-phosphate gives rise to fructose-6-phosphate which is converted to cell carbon and also is used to regenerate ribose-5-phosphate by reactions analogous to those of the ribulose diphosphate cycle. Quayle (1972) has noted that the thiamine pyrophosphate-dependent acyloin condensation of formaldehyde with

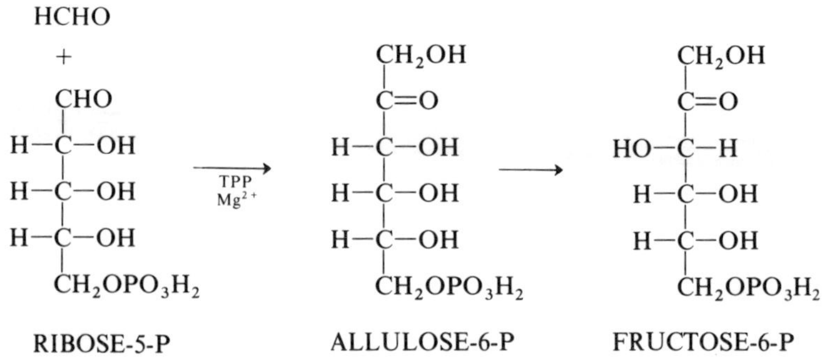

Figure 7.25. Formation of allulose-6-P during the assimilation of formaldehyde by *P. methanica*

ribose-5-phosphate is similar to the reaction catalysed by glyoxylate carboligase. These reactions bear no resemblance to the ribulose-5-phosphate carboxydismutase reaction used by *P. oxalaticus* for the assimilation of CO_2.

A number of bacteria, including some *Pseudomonas* strains, assimilate one-carbon compounds by the serine pathway (Figure 7.24). The C1 unit is condensed with glycine to give serine which is subsequently converted into glycolytic intermediates. The glycine is regenerated by a cyclic route in which a product of the serine pathway is cleaved to form two 2-carbon compounds both of which can be converted into glyoxylate. Salem and his colleagues (1973b) showed that extracts of *Pseudomonas* AM1 contain an enzyme which cleaves malyl-CoA to equimolar amounts of acetyl-CoA and glyoxylate and the high activity of this enzyme in extracts from bacteria grown with 1-carbon compounds suggests that this reaction provides the system for the regeneration of glyoxylate. The serine pathway is also employed for the assimilation of glyoxylate produced from oxalate by *Pseudomonas* AM1 (Blackmore & Quayle, 1970).

The presence of hydroxypyruvate reductase is indicative of the serine pathway and mutations which prevent the synthesis of this enzyme prevent the growth of *Pseudomonas* AM1 with methanol, methylamine or formate. Evidence of an additional pathway for the assimilation of one carbon compounds was provided by Wagner and Quayle (1972) who demonstrated that hydroxypyruvate reductase is not formed by cultures of *Pseudomonas* MS during growth with methylamine. In these bacteria, but not in *Pseudomonas* AM1, the methylamine carbon is initially fixed as the methyl group of *N*-methylglutarate and γ-methylaminoglutamate (Shaw *et al.*, 1966; Kung and Wagner, 1969). Thus the two pathways of methylamine assimilation appear to differ from the beginning. Little is known of the mechanism by which *Pseudomonas* MS converts methyl groups to common intermediary metabolites.

The diverse metabolic pathways used for the assimilation of one-carbon compounds suggests that this metabolic capacity might have arisen independently in different evolutionary groups. Serological evidence indicates that this is not necessarily the case for pathways of methanol oxidation. Primary alcohol dehydrogenase, an enzyme employed by all methanol utilizing bacteria, has been purified to homogeneity from *Methylococcus capsulatus* and *Pseudomonas* M27 (Patel *et al.*, 1972). Antisera prepared against the enzyme from the latter organism cross-reacted with the dehydrogenase from *P. methanica* as well as with the enzyme from other bacterial genera (Patel *et al.*, 1973).

6. BIBLIOGRAPHY

Amarasingham, C. R. & Davis, B. D. (1965) *Journal of Biological Chemistry*, **240**, 3664.
Ambler, R. P. (1963) *Biochemical Journal*, **89**, 349.

Ambler, R. P. (1974) *Biochemical Journal*, **137**, 3.
Ambler, R. P. & Brown, L. H. (1967) *Biochemical Journal*, **104**, 784.
Ambler, R. P. & Murray, S. (1973) *Biochemical Society Transactions*, **1**, 162.
Ambler, R. P. & Taylor, E. (1973) *Biochemical Society Transactions*, **1**, 166.
Ambler, R. P. & Wynn, M. (1973) *Biochemical Journal*, **131**, 485.
Atkinson, D. E. (1969) *Annual Review of Microbiology*, **23**, 47.
Azoulay, E. & Couchoud-Beaumont, P. (1965) *Biochimica et Biophysica Acta*, **110**, 301.
Barrett, J. T. & Kallio, R. E. (1953) *Journal of Bacteriology*, **66**, 517.
Bartsch, R. G. (1968) *Annual Review of Microbiology*, **22**, 181.
Bayly, R. C. & Dagley, S. (1969) *Biochemical Journal*, **111**, 303.
Bayly, R. C. & Wigmore, G. J. (1973) *Journal of Bacteriology*, **113**, 1112.
Bird, J. A. & Cain, R. B. (1968) *Biochemical Journal*, **109**, 479.
Blackmore, M. A. & Quayle, J. R. (1968) *Biochemical Journal*, **107**, 705.
Blackmore, M. A. & Quayle, J. R. (1970) *Biochemical Journal*, **118**, 53.
Brammar, W. J. & Clarke, P. H. (1964) *Journal of General Microbiology*, **37**, 307.
Cain, R. B. & Farr, D. R. (1968) *Biochemical Journal*, **106**, 859.
Calhoun, D. H. & Jensen, R. A. (1972) *Journal of Bacteriology*, **109**, 365.
Calhoun, D. H., Pierson, D. L. & Jensen, R. A. (1973a) *Molecular and General Genetics*, **121**, 117.
Calhoun, D. H., Pierson, D. L. & Jensen, R. A. (1973b) *Journal of Bacteriology*, **113**, 241.
Campbell, J. J. R., Smith, R. A. & Eagles, B. A. (1953) *Biochimica et Biophysica Acta*, **11**, 594.
Campbell, J. J. R. & Stokes, F. N. (1951) *Journal of Biological Chemistry*, **190**, 853.
Cánovas, J. L., Ornston, L. N. & Stanier, R. Y. (1967) *Science*, **156**, 1695.
Cánovas, J. L. & Stanier, R. Y. (1967) *European Journal of Biochemistry*, **1**, 289.
Cartwright, N. J., Holdom, K. S. & Broadbent, B. A. (1971) *Microbios*, **3**, 113.
Cartwright, N. J. & Smith, A. R. W. (1967) *Biochemical Journal*, **102**, 826.
Chakrabarty, A. M. (1972) *Journal of Bacteriology*, **112**, 815.
Chakrabarty, A. M. & Gunsalus, I. C. (1969) *Proceedings of the National Academy of Sciences, U.S.A.*, **64**, 1217.
Chamberlain, E. M. & Dagley, S. (1968) *Biochemical Journal*, **110**, 755.
Chapman, P. J. & Duggleby, R. G. (1967) *Biochemical Journal*, **103**, 76.
Clarke, P. H. (1970) *Advances in Microbial Physiology*, **4**, 179.
Clarke, P. H. (1972) *Journal of General Microbiology*, **71**, 241.
Clarke, P. H., Houldsworth, M. A. & Lilly, M. D. (1968) *Journal of General Microbiology*, **51**, 225.
Clarke, P. H. & Meadow, P. M. (1959) *Journal of General Microbiology*, **20**, 144.
Cohen, G. N., Stanier, R. Y. & Le Bras, G. (1969) *Journal of Bacteriology*, **99**, 791.
Cohen-Bazire, G. & Jolit, M. (1953) *Annales de l'Institut Pasteur*, **84**, 937.
Crawford, I. P. & Gunsalus, I. C. (1966) *Proceedings of the National Academy of Sciences, U.S.A.*, **56**, 717.
Crawford, I. P. & Yanofsky, C. (1971) *Journal of Bacteriology*, **108**, 248.
Dagley, S. (1971) *Advances in Microbial Physiology*, **6**, 1.
Dagley, S., Evans, W. C. & Ribbons, D. W. (1960) *Nature*, **188**, 560.
Dagley, S. & Gibson, D. T. (1965) *Biochemical Journal*, **95**, 466.
Dagley, S. & Trudgill, P. W. (1965) *Biochemical Journal*, **95**, 48.
Dagley, S., Trudgill, P. W. & Callely, A. G. (1961) *Biochemical Journal*, **81**, 623.
Datta, P. (1969) *Science*, **165**, 556.
Datta, P. & Gest, H. (1964) *Nature*, **203**, 1259.
Davies, J. I. & Evans, W. C. (1964) *Biochemical Journal*, **91**, 251.

Dunn, N. W. & Gunsalus, I. C. (1973) *Journal of Bacteriology*, **114**, 974.
Dunstan, P. M. & Anthony, C. (1973) *Biochemical Journal*, **132**, 797.
Dunstan, P. M., Anthony, C. & Drabble, W. T. (1972) *Biochemical Journal*, **128**, 99.
Dus, K., Sletten, K. & Kamen, M. D. (1968) *Journal of Biological Chemistry*, **243**, 5507.
Eady, R. R. & Large, P. J. (1971) *Biochemical Journal*, **123**, 757.
Ellfolk, N. & Soininen, R. (1970) *Acta Chemica Scandinavica*, **24**, 2126.
Emmer, M., de Crombrugghe, B., Pastan, I. & Perlman, R. (1970) *Proceedings of the National Academy of Sciences, U.S.A.*, **66**, 480.
Evans, W. C., Smith, B. S. W., Fernley, H. N. & Davies, J. (1972) *Biochemical Journal*, **122**, 543.
Fargie, B. & Holloway, B. W. (1965) *Genetical Research*, **6**, 284.
Feist, C. F. & Hegeman, G. D. (1969a) *Journal of Bacteriology*, **100**, 869.
Feist, C. F. & Hegeman, G. D. (1969b) *Journal of Bacteriology*, **100**, 1121.
Fewson, C. A. & Nicholas, D. J. D. (1961) *Biochimica et Biophysica Acta*, **49**, 335.
Fujisawa, H., Hiromi, K., Uyeda, M., Okuno, S., Nozaki, M. & Hayaishi, O. (1972) *Journal of Biological Chemistry*, **247**, 4422.
Gauthier, J. J. & Rittenberg, S. C. (1971) *Journal of Biological Chemistry*, **246**, 3737.
Gerhart, J. C. & Pardee, A. B. (1962) *Journal of Biological Chemistry*, **237**, 891.
Gibson, D. T., Cardini, G. E., Maseles, F. C. & Kallio, R. E. (1970) *Biochemistry*, **9**, 1631.
Gibson, D. T., Keek, J. R. & Kallio, R. E. (1968) *Biochemistry*, **7**, 3795.
Gorini, L., Gundersen, W. & Burger, M. (1961) *Cold Spring Harbor Symposium Quantitative Biology*, **26**, 173.
Gotto, A. M. & Kornberg, H. L. (1961) *Biochemical Journal*, **81**, 273.
Gray, C. T., Wimpenny, J. W. T. & Mossman, M. R. (1966) *Biochimica et Biophysica Acta*, **117**, 33.
Gryder, R. M. & Adams, E. (1969) *Journal of Bacteriology*, **97**, 292.
Gunsalus, I. C. (1972) in *Degradation of Synthetic Organic Molecules in the Biosphere*, National Academy of Sciences, Washington, U.S.A.
Gunsalus, I. C., Bertland, A. U. & Jacobson, L. A. (1967) *Archiv. für Mikrobiologie*, **59**, 113.
Gunsalus, I. C., Conrad, H. E., Trudgill, P. W. & Jacobson, L. A. (1965) *Israel Journal of Medical Science*, **1**, 1099.
Gunsalus, I. C., Gunsalus, C. F., Chakrabarty, A. M., Sikes, S. & Crawford, I. P. (1968) *Genetics*, **60**, 419.
Gunsalus, I. C., Tyson, C. A., Tsai, R. L. & Lipscomb, J. D. (1971) *Chemico-Biological Interactions*, **4**, 75.
Hamilton, W. A. & Dawes, E. A. (1959) *Biochemical Journal*, **71**, 25P.
Hamlin, B. T., Ng, F. M. W. & Dawes, E. A. (1967) *Microbial Physiology Continuous Culture Proceedings 3rd International Symposium*, 211.
Hansen, R. W. & Hayaishi, J. A. (1962) *Journal of Bacteriology*, **83**, 679.
van Hartingsveldt, J., Marinus, M. & Stouthamer, A. H. (1971) *Genetics*, **67**, 469.
van Hartingsveldt, J. & Stouthamer, A. H. (1973) *Journal of General Microbiology*, **74**, 97.
Hartline, R. A. & Gunsalus, I. C. (1971) *Journal of Bacteriology*, **106**, 468.
Hayaishi, O. (1962) in *Oxygenases*, (Ed. O. Hayaishi), Academic Press, London and New York, p. 1.
Hayaishi, O. (1966) *Bacteriological Reviews*, **30**, 720.
Hegeman, G. D. (1966a) *Journal of Bacteriology*, **91**, 1140.
Hegeman, G. D. (1966b) *Journal of Bacteriology*, **91**, 1155.

Hegeman, G. D. (1966c) *Journal of Bacteriology*, **91**, 1161.
Higgins, S. J. & Mandelstam, J. (1972a) *Biochemical Journal*, **126**, 901.
Higgins, S. J. & Mandelstam, J. (1972b) *Biochemical Journal*, **126**, 917.
Higgins, I. J. & Quayle, J. R. (1970) *Biochemical Journal*, **118**, 201.
Holloway, B. W. (1969) *Bacteriological Reviews*, **33**, 419.
Hopper, D. J., Chapman, P. J. & Dagley, S. (1971) *Biochemical Journal*, **122**, 29.
Horio, T. (1958) *Journal of Biochemistry*, **45**, 195.
Horio, T., Higashi, T., Sasagawa, M. Kusai, K., Nakai, M. & Okunuki, K. (1960) *Biochemical Journal*, **77**, 194.
Horio, T., Higashi, T., Yamanaki, T., Matsubara, H. & Okunuki, K. (·1961) *Journal of Biological Chemistry*, **236**, 944.
Horio, T. & Kamen, M. D. (1970) *Annual Review of Microbiology*, **24**, 399.
Hosakawa, K. (1970) *Journal of Biological Chemistry*, **245**, 5304.
Hosakawa, K. & Stanier, R. Y. (1966) *Journal of Biological Chemistry*, **241**, 2453.
Howell, L. G., Spector, T. & Massey, V. (1972) *Journal of Biological Chemistry*, **247**, 4340.
Isaac, J. H. & Holloway, B. W. (1968) *Journal of Bacteriology*, **96**, 1732.
Jacobson, L. A., Bartholomaus, R. C. & Gunsalus, I. C. (1966) *Biochemical and Biophysical Research Communications*, **24**, 955.
Jensen, R. A., Calhoun, D. H. & Stenmark, S. L. (1973) *Biochimica et Biophysica Acta*, **293**, 256.
Jensen, R. A., Nasser, D. S. & Nester, E. W. (1967) *Journal of Bacteriology*, **94**, 1582.
Jensen, R. A. & Nester, E. W. (1966) *Journal of Biological Chemistry*, **241**, 3365.
Jerina, D. M., Daly, J. W., Jeffrey, A. M. & Gibson, D. T. (1971) *Archives of Biochemistry and Biophysics*, **142**, 394.
Johnson, B. F. & Stanier, R. Y. (1971) *Journal of Bacteriology*, **107**, 476.
Jones, M. V. & Hughes, D. E. (1972) *Biochemical Journal*, **129**, 755.
Kamen, M. D. & Horio, T. (1970) *Annual Review of Biochemistry*, **39**, 673.
Karlsson, J. L. & Barker, H. A. (1948) *Journal of Biological Chemistry*, **175**, 913.
Katagiri, M., Maeno, H., Yamamoto, S., Hayaishi, O., Kitae, T. & Oae, S. (1965) *Journal of Biological Chemistry*, **240**, 3414.
Kay, W. W. & Gronlund, A. F. (1969a) *Journal of Bacteriology*, **97**, 273.
Kay, W. W. & Gronlund, A. F. (1969b) *Biochimica et Biophysica Acta*, **193**, 444.
Kay, W. W. & Gronlund, A. F. (1969c) *Journal of Bacteriology*, **98**, 116.
Kay, W. W. & Gronlund, A. F. (1969d) *Journal of Bacteriology*, **100**, 276.
Kay, W. W. & Gronlund, A. F. (1971) *Journal of Bacteriology*, **105**, 1039.
Kefauver, M. & Allison, F. E. (1957) *Journal of Bacteriology*, **73**, 8.
Kemp, M. B. & Hegeman, G. D. (1968) *Journal of Bacteriology*, **96**, 1488.
Kemp, M. B. & Quayle, J. R. (1967) *Biochemical Journal*, **102**, 94.
Kobayashi, S., Kuno, S., Itada, N. & Hayaishi, O. (1964) *Biochemical and Biophysical Research Communications*, **16**, 556.
Kodama, T. & Shidara, S. (1969) *Journal of Biochemistry*, **65**, 351.
Kogut, M. & Podoski, E. P. (1953) *Biochemical Journal*, **55**, 800.
Kojima, Y., Fujisawa, H., Nakazawa, A., Nakazawa, T., Kanetsuna, F., Taniuchi, H., Nozaki, M. & Hayaishi, O. (1967) *Journal of Biological Chemistry*, **242**, 3270.
Kornberg, H. L. (1966a) in *Essays in Biochemistry*, (Eds. P. N. Campbell & G. D. Greville), Vol. 2, Academic Press, London, p. 1.
Kornberg, H. L. (1966b) *Biochemical Journal*, **99**, 1.
Kornberg, H. L. & Grotto, A. M. (1961) *Biochemical Journal*, **78**, 69.
Kornberg, H. L. & Madsen, N. B. (1958) *Biochemical Journal*, **68**, 549.
Kornberg, H. L. & Sadler, J. R. (1961) *Biochemical Journal*, **81**, 503.

Kundig, W., Ghosh, S. & Roseman, S. (1964) *Proceedings of the National Academy of Sciences, U.S.A.*, **52**, 1067.

Kung, H. F. & Wagner, C. (1969) *Journal of Biological Chemistry*, **244**, 4136.

Lawrence, D. A., Smith, D. A. & Rowbury, R. J. (1968) *Genetics*, **58**, 473.

Leidigh, B. J. & Wheelis, M. L. (1973) *Journal of Molecular Evolution*, **2**, 235.

Lessie, T. G. & Neidhardt, F. C. (1967a) *Journal of Bacteriology*, **93**, 1337.

Lessie, T. G. & Neidhardt, F. C. (1967b) *Journal of Bacteriology*, **93**, 1800.

Lessie, T. C. & Vander Wyk, J. C. (1972) *Journal of Bacteriology*, **110**, 1107.

Lewis, J. A. & Ames, B. N. (1972) *Journal of Molecular Biology*, **66**, 131.

Lode, E. T. & Coon, M. J. (1971) *Journal of Biological Chemistry*, **246**, 791.

McKenna, E. J. & Coon, M. J. (1970) *Journal of Biological Chemistry*, **245**, 3882.

Magasanik, B. (1961) *Cold Spring Harbor Symposium in Quantitative Biology*, **26**, 249.

Mandelstam, J. & Jacoby, G. A. (1965) *Biochemical Journal*, **94**, 569.

Marinus, M. G. & Loutit, J. S. (1969a) *Genetics*, **63**, 547.

Marinus, M. G. & Loutit, J. S. (1969b) *Genetics*, **63**, 557.

Marshall, V. P. & Sokatch, J. R. (1972) *Journal of Bacteriology*, **110**, 1073.

Maurer, R. & Crawford, I. P. (1971) *Journal of Bacteriology*, **106**, 331.

May, S. W. & Abbott, B. J. (1973) *Journal of Biological Chemistry*, **248**, 1725.

Meagher, R. B., McCorkle, G. M., Ornston, M. K. & Ornston, L. N. (1972) *Journal of Bacteriology*, **111**, 465.

Mee, B. J. & Lee, B. T. O. (1969) *Genetics*, **62**, 687.

Midgley, M. & Dawes, E. A. (1973) *Biochemical Journal*, **132**, 141.

Mukkada, A. J., Long, G. L. & Romano, A. H. (1973) *Biochemical Journal*, **132**, 155.

Murray, K., Duggleby, C. J., Sala-Trepat, J. M. & Williams, P. A. (1972) *European Journal of Biochemistry*, **28**, 301.

Nason, A. & Takahashi, H. (1958) *Annual Review of Microbiology*, **12**, 203.

Ng, F.M-W. & Dawes, E. A. (1973) *Biochemical Journal*, **132**, 129.

Nishizuka, Y., Ichiyama, S., Nakamura, S. & Hayaishi, O. (1962) *Journal of Biological Chemistry*, **237**, PC268.

Novick, A. & Weiner, M. (1957) *Proceedings of the National Academy of Sciences, U.S.A.*, **43**, 553.

Nozaki, M., Kagamiyama, H. & Hayaishi, O. (1963) *Biochemische Zeitschrift*, **338**, 582.

Ohta, Y. & Ribbons, D. W. (1970) *Federation of European Biochemical Societies Letters*, **11**, 189.

Ornston, L. N. (1966) *Journal of Biological Chemistry*, **241**, 3800.

Ornston, L. N. & Ornston, M. K. (1969) *Journal of Bacteriology*, **98**, 1098.

Ornston, M. K. & Ornston, L. N. (1972) *Journal of General Microbiology*, **73**, 455.

Ornston, L. N. & Stanier, R. Y. (1964) *Nature*, **204**, 1279.

Palleroni, N., Doudoroff, M., Stanier, R. Y., Solanes, R. E. & Mandel, M. (1970) *Journal of General Microbiology*, **60**, 215.

Palleroni, N. J. & Stanier, R. Y. (1964) *Journal of General Microbiology*, **35**, 319.

Patel, R. N., Bose, H. R., Mandy, W. J. & Hoare, D. S. (1972) *Journal of Bacteriology*, **110**, 570.

Patel, R. N., Mandy, W. J. & Hoare, D. S. (1973) *Journal of Bacteriology*, **113**, 937.

Peterson, J. A. (1970) *Journal of Bacteriology*, **103**, 714.

Peterson, J. A., Basu, D. & Coon, M. J. (1966) *Journal of Biological Chemistry*, **241**, 5162.

Phibbs, P. V. & Eagon, R. G. (1970) *Archives of Biochemistry and Biophysics*, **138**, 470.

Pichinoty, F., Azoulay, E., Couchoud-Beaumont, P., Le Minor, L., Rigano, G., Bigliadi-Rouvier, J. & Piechaud, M. (1969) *Annales de l'Institut Pasteur*, **116**, 27.

Poillon, W. N., Maeno, H., Koike, K. & Fiegelson, P. (1969) *Journal of Biological Chemistry*, **244**, 3447.

Quayle, J. R. (1963) *Biochemical Journal*, **89**, 492.

Quayle, J. R. (1972) in *Advances in Microbial Physiology*, (Eds. A. H. Rose & D. W. Tempest), Vol. 7, Academic Press, London and New York, p. 119.

Quayle, J. R. & Keech, D. B. (1959) *Biochemical Journal*, **72**, 631.

Rann, D. L. & Cain, R. B. (1969) *Biochemical Journal*, **114**, 77P.

Reiner, A. M. (1971) *Journal of Bacteriology*, **108**, 89.

Reiner, A. M. (1972) *Journal of Biological Chemistry*, **247**, 4960.

Reiner, A. M. & Hegeman, G. D. (1971) *Biochemistry*, **10**, 2530.

Rheinwald, J. G., Chakrabarty, A. M. & Gunsalus, I. C. (1973) *Proceedings of the National Academy of Sciences, U.S.A.*, **70**, 885.

Ribbons, D. W. (1970) *Archiv für Mikrobiologie*, **74**, 103.

Ribbons, D. W. (1971) *Federation of European Biochemical Societies Letters*, **12**, 161.

Ribbons, D. W., Harrison, J. E. & Wadzinski, A. M. (1970) *Annual Review of Microbiology*, **24**, 135.

Ribbons, D. W. & Senior, P. J. (1970) *Biochemical Journal*, **117**, 28.

Robert-Gero, M., Poiret, M. & Cohen, G. N. (1970) *Biochimica et Biophysica Acta*, **206**, 17.

Robert-Gero, M., Poiret, M. & Stanier, R. Y. (1969) *Journal of General Microbiology*, **57**, 207.

Romano, A. H., Eberhard, S. J., Dingle, S. J. & McDowell, T. D. (1970) *Journal of Bacteriology*, **104**, 808.

Rosenberg, S. L. (1971) *Journal of Bacteriology*, **108**, 1257.

Rosenberg, S. L. & Hegeman, G. D. (1969) *Journal of Bacteriology*, **99**, 353.

Sala-Trepat, J. M. & Evans, W. C. (1971) *European Journal of Biochemistry*, **20**, 400.

Sala-Trepat, J. M., Murray, K. & Williams, P. A. (1972) *European Journal of Biochemistry*, **28**, 347.

Salem, A. R., Wagner, C., Hacking, A. J. & Quayle, J. R. (1973a) *Journal of General Microbiology*, **76**, 375.

Salem, A. R., Hacking, A. J. & Quayle, J. R. (1973b) *Journal of General Microbiology*, **76**, xii.

Sawula, R. V. & Crawford, I. P. (1972) *Journal of Bacteriology*, **112**, 797.

Seubert, W. & Remberger, U. (1961) *Biochemische Zeitschrift*, **334**, 401.

Sharrock, M., Munck, E., Debrunner, P. G., Lipscomb, J. D., Marshall, V. & Gunsalus, I. C. (1973) *Biochemistry*, **12**, 258.

Shaw, W. V., Tsai, L. & Stadtman, E. R. (1966) *Journal of Biological Chemistry*, **241**, 935.

Skinner, A. J. & Clarke, P. H. (1968) *Journal of General Microbiology*, **50**, 183.

Smyth, P. F. & Clarke, P. H. (1972) *Journal of General Microbiology*, **73**, ix.

Stadtman, E. R. (1963) *Bacteriological Reviews*, **27**, 170.

Stanier, R. Y. (1947) *Journal of Bacteriology*, **54**, 339.

Stanier, R. Y. (1951) *Annual Review of Microbiology*, **5**, 35.

Stanier, R. Y., Palleroni, N. J. & Doudoroff, M. (1966) *Journal of General Microbiology*, **43**, 159.

Stevenson, I. L. & Mandelstam, J. (1965) *Biochemical Journal*, **96**, 354.

Suda, M., Hayaishi, O. & Oda, Y. (1950) *Medical Journal of Osaka University*, **2**, 21.

Teng, N., Kotowycz, G., Calvin, M. & Hosakawa, K. (1971) *Journal of Biological Chemistry*, **246**, 5448.

von Tigerstrom, M. & Campbell, J. J. R. (1966) *Canadian Journal of Microbiology*, **12**, 1025.

Tiwari, N. P. & Campbell, J. J. R. (1969) *Canadian Journal of Microbiology*, **15**, 1095.
Trijbels, F. & Vogels, G. D. (1967) *Biochimica et Biophysica Acta*, **132**, 115.
Tsai, S.-S., Brown, K. K. & Gaudy, E. T. (1971) *Journal of Bacteriology*, **108**, 82.
Ueda, T., Lode, E. T. & Coon, M. J. (1972) *Journal of Biological Chemistry*, **247**, 2109.
Vanderwinkel, E., Liard, P., Ramos, F. & Wiame, J. M. (1963) *Biochemical and Biophysical Research Communications*, **12**, 157.
Vanderwinkel, E. Vlieghere, M. (1968) *European Journal of Biochemistry*, **5**, 81.
Wagner, C. &. Quayle, J. R. (1972) *Journal of General Microbiology*, **72**, 185.
Waltho, J. A. & Holloway, B. W. (1966) *Journal of Bacteriology*, **92**, 35.
Weitzman, P. D. J. & Dunmore, P. (1969) *Federation of European Biochemical Societies Letters*, **3**, 265.
Weitzman, P. D. J. & Jones, D. (1968) *Nature*, **219**, 270.
Wheelis, M. L. & Ornston, L. N. (1972) *Journal of Bacteriology*, **109**, 790.
Wheelis, M. L. & Stanier, R. Y. (1970) *Genetics*, **66**, 245.
Wong, D. T. O. & Ajl, S. J. (1956) *Journal of the American Chemical Society*, **78**, 3230.
Wu, C.-H., Eisenbraun, E. J. & Gaudy, E. T. (1970) *Biochemical and Biophysical Research Communications*, **39**, 976.
Wu, C.-H., Ornston, M. K. & Ornston, L. N. (1972) *Journal of Bacteriology*, **109**, 796.
Yamamoto, S., Nakazawa, T. & Hayaishi, O. (1972) *Journal of Biological Chemistry*, **247**, 3434.
Yamanaka, T. (1964) *Nature*, **204**, 253.
Yamanaka, T. (1967) *Nature*, **213**, 1183.
Yamanaka, T. & Okunuki, K. (1963) *Biochimica et Biophysica Acta*, **67**, 379.
Yano, K., Higashi, N. & Arima, K. (1969) *Biochemical and Biophysical Research Communications*, **34**, 1.
Yu, C. A. & Gunsalus, I. C. (1969) *Journal of Biological Chemistry*, **244**, 6149.
Zubay, G., Schwartz, D. & Beckwith, J. (1970) *Proceedings of the National Academy of Sciences, U.S.A.*, **66**, 104.

CHAPTER 8

Metabolic Pathways and Regulation: II

PATRICIA H. CLARKE and NICHOLAS ORNSTON

1. EXOENZYMES AND THE UTILIZATION OF MACROMOLECULES

Most biochemical studies with *Pseudomonas* have been carried out with growth substrates of low molecular weight that are metabolized by intracellular (or membrane-bound) enzymes and in some cases specific transport systems may also be involved. Some strains are known to produce extracellular enzymes that cleave macromolecules (proteins, polysaccharides, etc.) to oligometric or monomeric units that may be transported across the bacterial membrane. In many cases the extracellular location of the enzymes has been inferred from the consequences of their activity around regions of growth on plates. For example, extracellular lipases have been identified routinely on the basis of the production of calcium oleate from the detergent 'Tween' 80 in zones surrounding patches of bacterial growth (Sierra, 1957). As emphasized by Pollock (1962) such evidence does not prove that the enzymes are extracellular because the detected activity may be attributed to intracellular enzymes liberated by the lysis of some of the cells. Nevertheless, the experimental documentation required for the unambiguous proof of the extracellular location of enzymes is too cumbersome to be established during the screening of large numbers of bacterial strains. In these cases, the application of techniques giving merely presumptive evidence for extracellular enzymes has proved to be of great taxonomic value. The ability to produce enzymes that hydrolyse gelatin on plates, for example, distinguishes strains of *P. aeruginosa* and *P. fluorescens* from those of *P. putida* which do not carry out this reaction (Stanier, Palleroni & Doudoroff, 1966).

Some *Pseudomonas* enzymes have been demonstrated unequivocally to be extracellular and in a few cases detailed studies of their biochemistry and physiology have been made. Morihara and his associates have crystallized an extracellular protease from *P. aeruginosa* (Morihara, 1963) and presented evidence that it is an ellipsoid protein of 48,400 daltons molecular weight (Inouye *et al.*, 1963). Subsequent studies by Morihara (1964) revealed that this enzyme is but one of three physically separable proteases that may be produced by strains of *P. aeruginosa*. The three enzymes differ in their kinetic properties as well as in the manner in which their synthesis is controlled.

Pectate lyases (polygalacturonic acid 4,5 eliminases, EC 4.2.99.3) which produce 4,5-galacturonides, play an important part in the biological dissimilation of pectin. As might be expected, the enzymes are inducible in a number of pathogenic and non-pathogenic fluorescent pseudomonads (Fuchs, 1965; Zucker & Hankin, 1970). The pectate lyases appear to be relatively amenable to purification. Direct comparison of their primary structures might yield some insight into the taxonomic relationships among pathogenic and non-pathogenic pseudomonads.

An extracellular α-amylase from *P. saccharophila* has been crystallized and its physicochemical properties characterized (Markovitz *et al.*, 1956). More recently a different extracellular amylase from *P. stutzeri* was purified by Robyt and Ackerman (1971). Unlike α-amylase the latter enzyme produces amylotetrose oligomeric units from starch. Selective enrichment with poly-β-hydroxybutyrate, an insoluble bacterial carbon reserve material, yielded a new *Pseudomonas* species, *P. lemoignei* (Delafield *et al.*, 1965b). This organism produces two physicochemically and serologically distinct extracellular poly-β-hydroxybutyrate depolymerases (Lusty & Doudoroff, 1966). The extracellular enzymes cleave the polymer to β-hydroxybutyrate dimers which are further metabolized by intracellular enzymes (Delafield *et al.*, 1965a). The depolymerases are excreted during the growth of *P. lemoignei*, but the rate of production of the enzymes increases dramatically as cultures enter the stationary phase. The production of the enzymes is repressed by β-hydroxybutyrate, and in the presence of β-hydroxybutyrate the cultures of *P. lemoignei* remove the depolymerase activity from the medium (Delafield *et al.*, 1965b). The mechanism of this remarkable regulatory phenomenon has not been explored further.

2. CATABOLIC PATHWAYS FOR COMPOUNDS UTILIZED AS CARBON SOURCES

2.a. Monosaccharides and Hydroxy Acids

Although the pseudomonads are biochemically very versatile, few species are able to metabolize many of the sugars and related compounds. *P. acidovorans* and *P. testosteroni* do not even utilize glucose which is a growth substrate for most of the other *Pseudomonas* species. However, various strains have been described which can grow with glucose, gluconate, galactose, mannose, fructose, arabinose, ribose and xylose. The species that has contributed most to these studies is *P. saccharophila*.

The Embden–Meyerhof pathway appears to be completely absent from *Pseudomonas* and most of the hexoses and their derivatives are metabolized by the reactions of the Entner–Doudoroff pathway which was first established in *P. saccharophila* (Entner & Doudoroff, 1952). Figure 8.1 shows the inter-conversions which have been shown to occur in the pathways for the oxidation of D-mannose, D-fructose, D-glucose and D-gluconate. The three hexoses are phosphorylated and glucose-6-phosphate is oxidized *via* the intermediate formation of 6-phosphogluconate lactone to 6-phosphogluconate. This compound can either undergo the reactions of the Entner–Doudoroff pathway, or be converted by oxidation and decarboxylation to ribulose-5-phosphate and enter the pentose phosphate pathway. The reactions of the pentose phosphate cycle can furnish pentoses for biosynthesis and provide reduced nucleotide coenzymes, particularly NADPH, but complete assimilation of the hexoses involves the Entner–Doudoroff pathway reactions which produce ultimately the intermediates which enter the tricarboxylic acid cycle.

The key enzymes of the Entner–Doudoroff pathway are those which convert 6-phosphogluconate to 2-keto-3-deoxy-6-phosphogluconate (6-PG dehydratase) and the following enzyme which catalyses an aldolase-type cleavage of 2-keto-3-deoxy-6-phosphogluconate (KDPG) to give pyruvate and glyceraldehyde-3-phosphate (KDPG aldolase) shown in Figure 8.1.

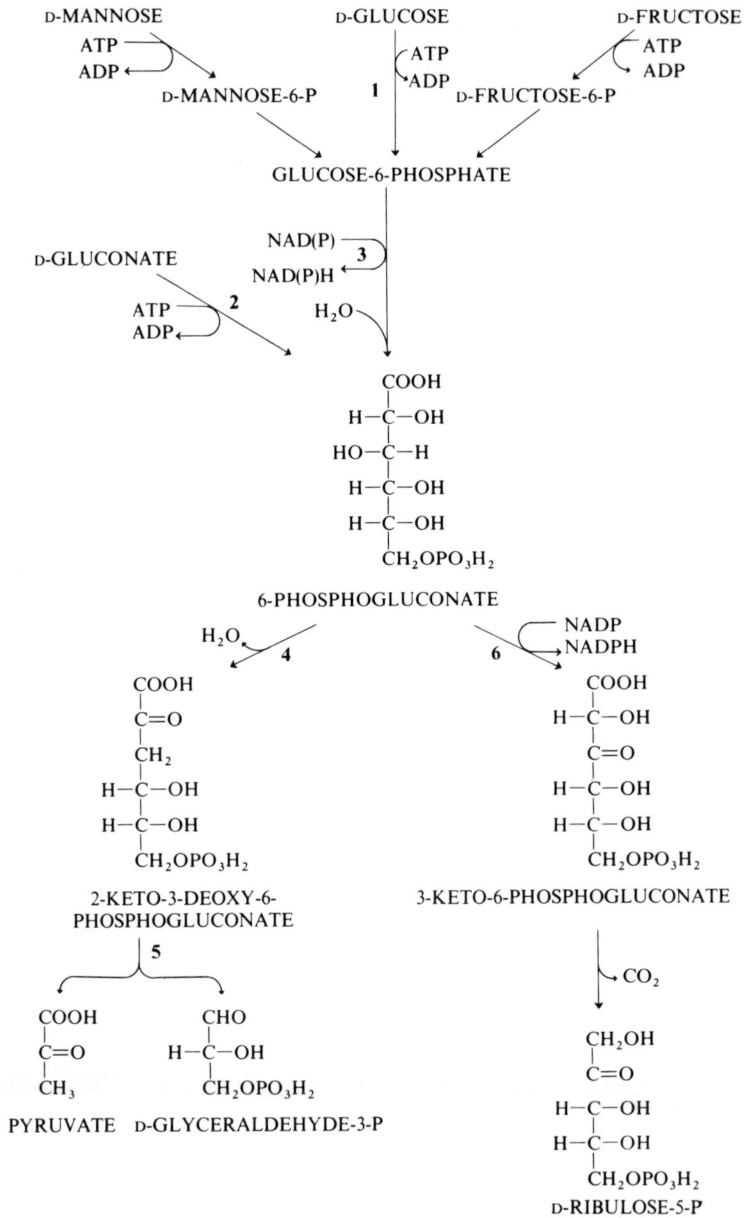

Enzymes: 1. glucokinase; 2. gluconokinase; 3. glucose-6-phosphate dehydrogenase; 4. 6-phosphogluconate dehydratase; 5. 2-keto-3-deoxy-6-phosphogluconate aldolase; 6. 6-phosphogluconate dehydrogenase. (6-Phosphogluconate lactone is an intermediate in the oxidation of glucose-6-phosphate to 6-phosphogluconate.)

Figure 8.1. The Entner–Doudoroff pathway reactions for the utilization of D-glucose, D-gluconate, D-fructose and D-mannose by *Pseudomonas* species

A critical metabolic control point occurs at the level of 6-phosphogluconate. Lessie and Neidhart (1967a) showed that the glucose-6-phosphate dehydrogenase of *P. aeruginosa* was inhibited by ATP and Lessie and Vander Wyk (1972) showed that *P. cepacia* has multiple forms of both glucose-6-phosphate dehydrogenase and 6-phosphogluconate dehydrogenase. One of the glucose-6-phosphate dehydrogenases, and one of the 6-phosphogluconate dehydrogenases, is active with both NAD and NADP and is inhibited by ATP. This form of control could allow the regulation of the flow of 6-phosphogluconate into either the pentose pathway or the Entner–Doudoroff pathway and thus regulate the supply of such key metabolites as pyruvate, tricarboxylic acid cycle intermediates and pentose phosphates. The second glucose-6-phosphate dehydrogenase and the second 6-phosphogluconate dehydrogenase are both NADP dependent and are not inhibited by ATP.

The enzymes of the Entner–Doudoroff pathway are induced by growth with glucose, gluconate or glycerol, but not with citrate, succinate or pyruvate (Hamlin *et al.*, 1967; Lessie & Niedhardt, 1967a). In *P. aeruginosa* the enzymes induced include glucokinase, glucose-6-phosphate dehydrogenase, 6-phosphogluconate (6-PG) dehydratase and 2-keto-3-deoxy-6-phosphogluconate (KDPG) aldolase. However, phosphorylation need not precede the enzymatic transformations of the Entner–Doudoroff pathway. With *P. fluorescens*, Quay, Friedman and Eisenberg (1972) were unable to detect any glucokinase activity in either glucose or gluconate-grown cells. This had been reported earlier and had been taken to indicate that glucose is not necessarily phosphorylated before undergoing further metabolic transformation. A membrane-associated glucose oxidase was found to be present in glucose-grown cells but the activity was very low in gluconate-grown cells. Both glucose and gluconate-grown cells contained the two enzymes, gluconokinase and 6-PG dehydratase, but in a mutant lacking glucose oxidase these two enzymes could be induced by gluconate but not by glucose. These findings led them to the conclusion that gluconate, rather than glucose, is the inducer of the Entner–Doudoroff pathway enzymes in this strain. Other enzymes induced included a particulate gluconate oxidase, 6-phosphogluconate dehydrogenase and glucose-6-phosphate dehydrogenase. Quay, Friedman and Eisenberg (1972) suggest the scheme shown in Figure 8.2 for the catabolism of glucose by *P. fluorescens*.

Galactose can be metabolized by *P. saccharophila* and as shown in Figure 8.3 it is converted into 2-keto-3-deoxygalactonate before any phosphorylation step occurs (DeLey & Doudoroff, 1957). The aldol cleavages of 2-keto-3-deoxy-6-phosphogluconate (Figure 8.1) and of 2-keto-3-deoxy-6-phosphogalactonate (Figure 8.3) are quite similar, the substrates differ only in the orientation of the hydroxyl group about the carbon-4 atom. The aldolases that catalyse these reactions share many physical and kinetic properties (Shuster & Doudoroff, 1967), but possess absolute specificities and do not

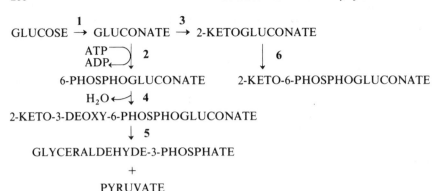

Enzymes: 1. glucose oxidase; 2. gluconokinase; 3. gluconate oxidase; 4. 6-phosphogluconate dehydratase; 5. 2-keto-3-deoxy-6-phosphogluconate aldolase; 6. 2-ketogluconate kinase.

Figure 8.2. Catabolism of glucose in *P. fluorescens*

give any serological cross-reactions. The enzymes also differ in their regulation, the 2-keto-3-deoxy-6-phosphogalactonate aldolase is inducible while the 2-keto-3-deoxy-6-phosphogluconate aldolase is constitutive in *P. saccharophila*. Nevertheless the possibility remains that the enzymes are the products of homologous structural genes which have a common origin.

In many respects the catabolism of D-arabinose is similar to that of D-galactose in *P. saccharophila* (Palleroni & Doudoroff, 1957). Both sugars are converted by a similar sequence of reactions to the corresponding 2-keto-3-deoxy derivatives. The cleavage of 2-keto-3-deoxy-D-arabonate takes place however without prior phosphorylation and is a relatively complex reaction. Rather than producing pyruvate and glycolaldehyde, the predicted products of a simple aldolase reaction, the extracts of D-arabinose-grown cells require NAD to carry out an oxidative cleavage of 2-keto-3-deoxy-D-arabonate to produce pyruvate and glycolate. Thus a single enzyme appears to perform both an aldol cleavage and an oxidation. Subsequent investigations by Jeffcoat, Hassall and Dagley (1969b) have shown how a single enzyme may carry a covalently bound ketose through two sequential transformations. It is not known if the oxidative cleavage of 2-keto-3-deoxy-D-arabonate is catalysed by a similar mechanism.

There are no phosphorylation reactions in the first stages of the catabolic pathways for D-fucose or L-arabinose utilization by an unidentified pseudomonad MSU-1 studied by Dahms and Anderson (1969; 1972a,b,c,d). As shown in Figure 8.5 the reactions in these parallel pathways (b and c in Figure 8.5) are similar to those discussed previously for glucose, galactose and D-arabinose shown in Figures 8.1, 8.3 and 8.4. The monosaccharide substrate is oxidized to a lactone; this is hydrolysed to a polyol carboxylic

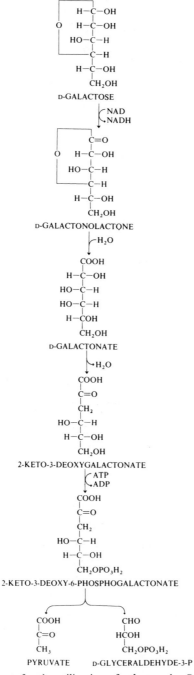

Figure 8.3. The pathway for the utilization of galactose by *P. saccharophila*

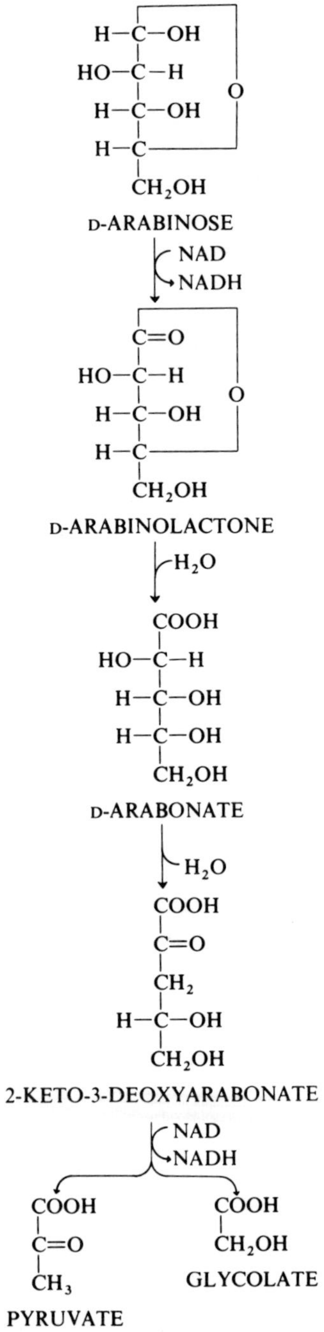

Figure 8.4. The oxidative pathway for the utilization of D-arabinose by *P. saccharophila*

acid and dehydrated to produce a 2-keto-3-deoxy derivative. Aldol cleavage of the 2-keto-3-deoxy acid gives rise to pyruvate and D-lactaldehyde (from D-fucose) or glycolaldehyde (from L-arabinose). The parallel reactions in Figure 8.5 appear to be catalysed by a single set of enzymes with a specificity sufficiently broad to permit them to operate in either the five-carbon or the six-carbon substrates. An addition enzyme induced by growth with D-fucose, catalyses the NAD-dependent oxidation of D-fucanofuranose to D-fucono-γ-lactone which hydrolyses spontaneously to D-fuconate (reaction series b in Figure 8.5 (Dahms & Anderson, 1972a).

Weimberg and Doudoroff (1955) found that L-arabinose was metabolized to α-ketoglutarate in *P. saccharophila.* Subsequent investigations by Weimberg (1961) established that D-xylose, D-arabinose and L-arabinose are converted to the corresponding pentonic acids and subsequently to α-ketoglutarate by extracts of appropriately induced cultures of *P. fragi.* Dagley and Trudgill (1965) confirmed these results and identified α-ketoglutarate semialdehyde as a common intermediate in the pathways thus establishing the convergent pathways shown in Figure 8.6. These authors also demonstrated that the hydroxy-hexanoic acids, D-glucarate and D-galactarate, are catabolized to α-ketoglutarate semialdehyde (Figure 8.6) by cultures of *P. acidovorans.* This reaction sequence is initiated by a reaction typical of polyol catabolism in *Pseudomonas,* dehydration to give a keto-deoxy product. An extraordinary conversion that is both a dehydration and a decarboxylation, converts 4-deoxy-5-ketoglucarate to α-ketoglutarate semialdehyde (Figure 8.6). Jeffcoat *et al.* (1969b) purified the enzyme that catalyses this reaction and provided evidence indicating that a lysine residue in the protein formed a Schiff's base with the substrate. The catalytic activities of the enzyme could be readily interpreted as the consequence of the Schiff's base formation.

In Enterobacteriaceae the hydroxy-hexanoic acids also undergo dehydration reactions but the catabolic pathway diverges sharply from the *Pseudomonas* sequence at this point. Blumenthal and Fish (1963) demonstrated that induced cultures of *E. coli* subject 4-deoxy-5-ketoglucarate to an aldolase reaction giving rise to D-glycerate and tartronic semialdehyde. The same pathway was found in *Klebsiella (Aerobacter) aerogenes,* indicating that it is probably widely employed by the Enterobacteriaceae (Jeffcoat *et al.,* 1969a).

A number of different catabolic pathways converge upon α-ketoglutarate semialdehyde in *Pseudomonas.* In addition to those shown in Figure 8.6, hydroxyproline is converted to α-ketoglutarate semialdehyde by a sequence of reactions described by Gryder and Adams (1969) (Figure 8.23).

The utilization of tartrates by pseudomonads provides an interesting study in stereospecificity of enzymes and pathways. Shilo (1957) isolated strains which were specific for the different tartrate isomers. These strains attack the tartrate by specific dehydratases which convert L-, D- or *meso*-tartrate

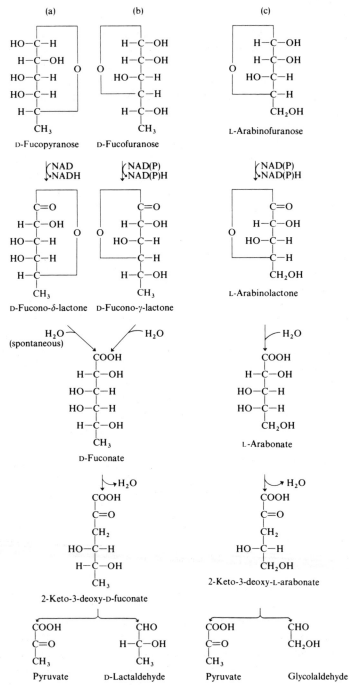

Figure 8.5. Pathways for the utilization of D-fucose and L-arabinose in *Pseudomonas* MSU-1

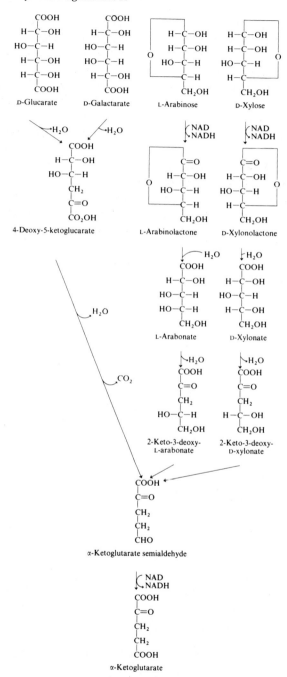

Figure 8.6. Oxidative pathways converging upon α-ketoglutarate semialdehyde in *Pseudomonas*

to oxaloacetate in each case. The stereospecific dehydratases are induced by their substrates and in addition, each tartrate isomer induces a stereospecific permease (Shilo & Stanier, 1957). Hurlbert and Jakoby (1965) found that *P. putida* attacks L-tartrate in the same way to produce oxaloacetate but converts *meso*-tartrate to D-glycerate (Dagley & Trudgill, 1965; Kohn & Jakoby, 1968a). On the other hand *P. acidovorans* converts both L- and *meso*-tartrate to D-glycerate. The enzymes of this pathway were isolated and studied by Kohn and Jakoby (1968a,b,c,d) and Kohn *et al.* (1968). The first enzyme is a tartrate dehydrogenase which converts tartrate to oxaloglycolate. From *P. putida* an NAD-dependent enzyme was crystallized which had both malate dehydrogenase and *meso*-tartrate dehydrogenase activity. A second enzyme was crystallized which had L- and *meso*-tartrate dehydrogenase activity and was also NAD dependent. The next step in the pathway is a reductive decarboxylation of oxaloglycolate to D-glycerate (Figure 8.7). This enzyme was also crystallized and found to catalyse the NADH- or NADPH-dependent decarboxylation of oxaloglycolate and in addition D-glycerate formation from hydroxypyruvate. Kohn and Jakoby

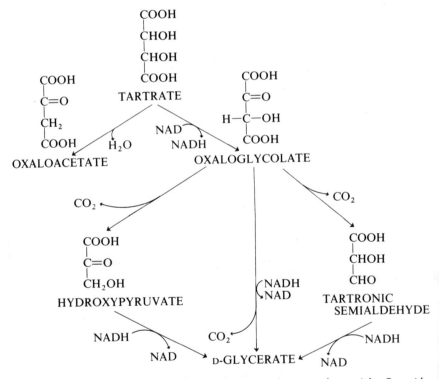

Figure 8.7. Pathways for the assimilation of tartrates (L-, D- and *meso*-) by *P. putida*, *P. acidovorans* and other *Pseudomonas* species

(1968d) isolated also a crystalline hydroxypyruvate reductase from *P. acidovorans*. Since oxaloglycolate can undergo non-enzymic decarboxylation to produce hydroxypyruvate and tartronic semialdehyde, all the interconversions outlined in Figure 8.7 can occur by known reactions.

2.b. Aliphatic Acids and Hydrocarbons

Acetate is readily utilized by most *Pseudomonas* species. In Chapter 7 we discussed the function of the anaplerotic reactions of the glyoxylate cycle, in the synthesis of the C_4 dicarboxylic acids from acetate, which allows the replenishment of the pool of tricarboxylic acid cycle intermediates withdrawn for biosynthesis. The same considerations apply to compounds which give rise directly to acetate or acetyl-CoA in metabolism. The presence of the glyoxylate cycle enzymes is presumptive evidence for a metabolic pathway involving acetyl-CoA as the main intermediate, but is not conclusive since isocitrate lyase may be derepressed gratuitously. Many pseudomonads can grow with higher homologues of monocarboxylic fatty acids and dicarboxylic acids. The accumulated evidence suggests that these acids are metabolized by β-oxidation following activation to acyl-thioesters.

Chapman and Duggleby (1967) showed that *P. aeruginosa* grows on all the fatty acids from C_2 to C_{10}. Mutants lacking isocitrate lyase were unable to grow with the even-numbered fatty acids which give rise exclusively to the 2-carbon fragments by β-oxidation. The isocitrate lyase mutants were however able to grow with the odd-numbered fatty acids which would be expected to yield one 3-carbon fragment for each molecule of the fatty acid metabolized. It is known (Skinner & Clarke, 1968), that isocitrate lyase is not required for growth with propionate and other 3-carbon acids.

The metabolism of the dicarboxylic acids from C_3 to C_{10} was also examined by Chapman and Duggleby (1967) who showed that while the wild-type strain of *P. aeruginosa* was able to utilize the whole range, the isocitrate lyase mutants grew on the even-numbered dicarboxylic acids but not the odd-numbered. This is compatible with β-oxidation producing 2-carbon fragments and malonyl-CoA from the odd-numbered dicarboxylic acids, and 2-carbon fragments together with a 4-carbon fragment from the even-numbered dicarboxylic acids. Hoet and Stanier (1970a) showed that in *P. fluorescens* a single acyl thiokinase was responsible for the activation of pimelate (C_7), suberate (C_8), azaleate (C_9) and sebacate (C_{10}) and that this enzyme was induced by growth with any of these acids. In addition, dehydrogenases that oxidize the α,β-unsaturated acyl-CoA esters and the β-hydroxy acyl-CoA esters (Figure 8.8) were induced. They found that adipate (C_6) was activated by a different mechanism involving an adipate: succinyl-CoA transferase. The product of β-oxidation of adipate is β-ketoadipate which is a well-known intermediate of the '*ortho*' cleavage pathway for the catabolism

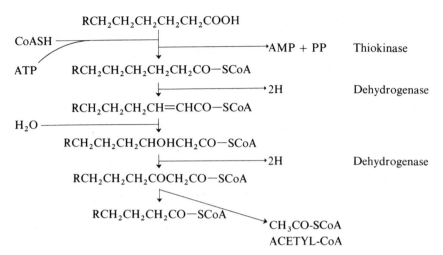

Figure 8.8. Pathway for β-oxidation of mono- and dicarboxylic acids

of aromatic compounds. Hoet and Stanier (1970b) showed that *P. fluorescens* produces two enzymes with β-ketoadipate:succinyl-CoA transferase activity. Transferase I is induced by growth with β-ketoadipate and appears to function in the aromatic pathway, while transferase II is induced by growth with adipate and appears to belong to the pathway for adipate utilization.

In a strain of *P. aeruginosa*, Hoet and Stanier (1970b) found that only one β-ketoadipate:succinyl-CoA transferase was produced and that was induced by growth with suberate and azaleate as well as with adipate. Further, two of the enzymes of the protocatechuate branch of the β-keto-adipate pathway which are considered to be induced by β-ketoadipate (see Chapter 7, section 3.b) were found to be present in cultures grown with suberate and azaleate. This led to the suggestion that the two latter dicar-boxylic acids are also metabolized via β-ketoadipate which then gives rise to gratuitous induction of the two aromatic pathway enzymes. The meta-bolism of the dicarboxylic acids in *P. aeruginosa* clearly merits further investigation.

A *Pseudomonas* strain studied by Cooper and Kornberg (1964) grows on the branched-chain dicarboxylic acid, itaconate. Washed suspensions of bacteria grown with itaconate are also able to oxidize mesaconate and citramalate by the reactions shown in Figure 8.9. Itaconate is converted to itaconyl-CoA by itaconyl-CoA synthetase and also by the itaconate: succinyl-CoA transferase. The products of the pathway are pyruvate and acetyl-CoA which enter the reactions of the tricarboxylic acid cycle. Itaconate:succinyl-CoA transferase, itaconyl-CoA hydratase and citra-malyl-CoA lyase are found in bacteria grown with itaconate, citramalate and

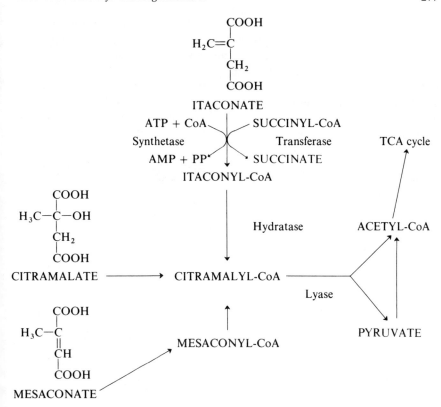

Figure 8.9. Metabolic pathway for branched-chain dicarboxylic acids in *P. acidovorans*

methyl succinate but not in bacteria grown with malate and pyruvate. Itaconyl-CoA synthetase was present in bacteria grown with any of these compounds so that this enzyme appears to be constitutive.

The straight-chain aliphatic hydrocarbons are metabolized in the natural environment by several bacterial species, including pseudomonads (Foster, 1962; McKenna & Kallio, 1965; van der Linden & Thijsse, 1965). *P. aeruginosa* grows with hexadecane and dodecane as carbon sources and this distinguishes it from the other fluorescent *Pseudomonas* species (Stanier *et al.*, 1966). Strains have been isolated from enrichment culture with single hydrocarbons providing the sole carbon source for growth and also from media containing kerosene or other mixtures of several different hydrocarbons. Trust and Millis (1970) found that *Pseudomonas* strains isolated from kerosene enrichment grew on the alkanes from C_6 to C_{20} but grew better with mixtures than with single alkanes.

The first step in alkane assimilation is the hydroxylation of the terminal methyl group (Gholson, Baptist & Coon, 1963). In Chapter 7 we discussed the role of oxygenases in introducing molecular oxygen directly into organic molecules as the first step for several catabolic pathways (see Chapter 7, section 2.c). Peterson, Basu & Coon (1966) showed that the hydroxylation of octane required a three-component system consisting of (1) rubredoxin, (2) a flavoprotein reductase (NAD-rubredoxin reductase) and (3) the ω-hydroxylase. The three proteins have been isolated from *P. oleovorans* and the first two purified to homogeneity. The system is also able to carry out the ω-hydroxylation of fatty acids by the mechanism shown in Figure 8.10.

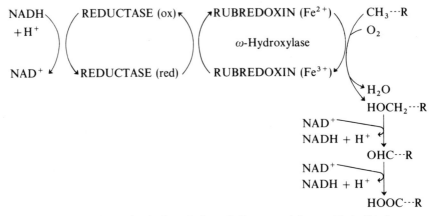

Figure 8.10. Reactions of ω-hydroxylation of alkanes and fatty acids in *P. oleovorans* (Peterson *et al.*, 1967)

The incorporation of molecular oxygen into the substrate molecule produces a primary alcohol which is then oxidized sequentially by two NAD-dependent dehydrogenases to give a carboxylic acid. If a fatty acid is the initial substrate the product is a dicarboxylic acid. Kusunose, Kusunose and Coon (1964a,b) identified suberate as a product of octane oxidation by a pseudomonad grown with hexane. It would therefore be possible for an alkane to undergo a double hydroxylation to produce a dicarboxylic acid.

The later stages in the metabolism of alkanes are the same as those for the corresponding fatty acids involving activation to acyl-thioesters and β-oxidation. Trust and Millis (1971) found that extracts of pseudomonads grown with hexadecanoate contained several acyl thiokinases with different but overlapping specificities. There were: C_{11} to C_{19} acyl thiokinase, C_4 to C_{14} thiokinase, C_2 to C_6 acyl thiokinase, acetyl thiokinase and acetokinase. These organisms were able to grow with most of the monocarboxylic fatty acids up to C_{20} and also on some of the dicarboxylic acids from C_3 to C_{10}. Growth with hexadecane and dodecane induced levels of isocitrate lyase that were comparable to those found in acetate-grown cultures, which would be expected for a β-oxidation pathway producing 2-carbon fragments.

The branched-chain hydrocarbons, the alkylalkanes, are less amenable to microbial degradation. McKenna (1972) reported that strains of *P. aeruginosa* that grow well with hexadecane did not grow with the alkylalkanes, dipropyl-dodecane or 2,6,10,14-tetramethylpentadecane (pristane) which is found widely in natural environments. Van der Linden and Thijsse (1965) reported that *P. aeruginosa* grew sufficiently well with 2-methylhexane as sole carbon source to study the possible metabolic routes. Although they were able to detect both 2-methyl- and 5-methylhexanoates, the pathway leading to 5-methylhexanoate was the preferred one so that this branched chain alkane was oxidized mainly from the 'long-chain end'.

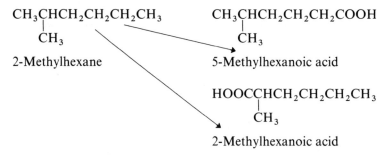

$$CH_3CHCH_2CH_2CH_2CH_3$$
$$\quad CH_3$$

2-Methylhexane

$$CH_3CHCH_2CH_2CH_2COOH$$
$$\quad CH_3$$

5-Methylhexanoic acid

$$HOOCCHCH_2CH_2CH_2CH_3$$
$$\qquad CH_3$$

2-Methylhexanoic acid

2.c. Aromatic Compounds

A very large number of aromatic compounds can serve as growth substrates for various *Pseudomonas* species. We have described many of the pathways and their enzymes in Chapter 7, since the metabolism of aromatic compounds was used extensively to illustrate the discussion of the mechanisms known for the regulation of catabolic pathways. In this chapter we shall be concerned with comparisons between some of the enzyme reactions themselves.

The most detailed genetic and biochemical studies are those which have been carried out on the β-ketoadipate pathways of *P. putida* and *P. aeruginosa* (see Chapter 7 for details). Considerable economy of genetic information is achieved by the dissimilation of metabolites through convergent pathways which enable the later enzymes of the pathway to participate in the utilization of so many different growth substrates. Figure 8.11 shows a few of the organic compounds which are utilized by means of the β-ketoadipate pathway, involving 'ortho' cleavage of a dihydric phenol as the first step of the common pathways reached by the activities of the peripheral enzymes on the growth substrates. For other pathways, corresponding metabolic sequences are catalysed by enzymes of broad specificity and this permits a limited amount of genetic information to be used for the dissimilation of a wide range of related compounds. The enzyme reactions, and the successive steps in the metabolism of aromatic compounds by pseudomonads, illustrate

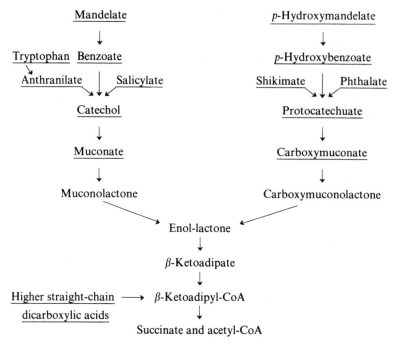

Figure 8.11. Metabolic convergence in the β-ketoadipate pathway. Underlined compounds may be used as growth substrates in the natural enviroment

very well the elegant solutions which have evolved in nature to carry out chemical transformations of this interesting group of organic compounds. An excellent review by Dagley (1971) gives a detailed analysis of these pathways.

Frequently similar metabolic mechanisms are used for the dissimilation of different metabolites. Examples of these are shown in Figures 8.12 and 8.13. In Figure 8.12 we show a number of different derivatives of benzoate and catechol with chloro, fluoro or carboxy substituents in the 4-position of the aromatic ring. For each of these compounds '*ortho*' oxygenative cleavage is the first step in the dissimilation of the catechols. In the case of 4-chlorocatechol (Evans *et al.*, 1971), the product is β-chloromuconate; a remarkable substitution reaction replaced the halide atom with the oxygen of a carboxyl group thus forming a lactone (Figure 8.12a). Hydrolysis of the lactone gives rise to β-hydroxymuconate; its tautomer, maleyl acetate, is converted to fumarate and acetate by isomerization and hydrolysis. The utilization of fluoromuconate appears to go through a somewhat different route Harper & Blakeley, 1971) as shown in Figure 8.12b. As with chlorocatechol, oxygenation and dehalogenation give rise to a lactone, γ-carboxymethylene

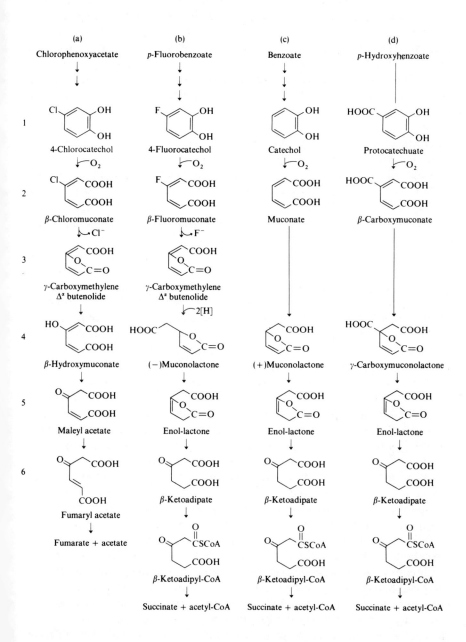

Figure 8.12. Similar catabolic pathways for dissimilation of aromatic compounds in *Pseudomonas*. 'ortho' ring cleavage

Δ^{α} butenolide. Reduction of this compound produces (−)muconolactone which is metabolized to succinate and acetyl-CoA *via* β-ketoadipate.

Oxygenative ring fission of catechol and protocatechuate (Figures 8.12c and d) gives rise to muconate and β-carboxymuconate respectively (Evans & Smith, 1951; MacDonald, Stanier & Ingraham, 1954). Lactones are formed from these compounds by the addition of a carboxyl group oxygen across a double bond (Sistrom & Stanier, 1954; Ornston & Stanier, 1966) rather than the substitution reactions shown in columns (a) and (b) of Figure 8.12. The muconolactone formed from catechol is dextro-rotatory, unlike the laevo-rotatory muconolactone derived from fluorocatechol. Endocyclic rearrangement of the double bond in muconolactone (or its γ-carboxy derivative) gives rise to β-ketoadipate enol-lactone (Ornston & Stanier, 1966), which is hydrolysed to β-ketoadipate and converted to succinate and acetyl-CoA *via* β-ketoadipyl-CoA.

Gentisate, like most substrates of aromatic ring cleaving oxygenases, possesses two hydroxyl groups but differs from most of the diphenolic substrates in that the hydroxyl groups are *para* rather than *ortho* to each other (Figure 8.13). Ring fission of gentisate in *P. acidovorans* produces maleylpyruvate; isomerization of this metabolite produces fumarylpyruvate which is hydrolysed to fumarate and pyruvate (Lack, 1959; 1961; Wheelis, Palleroni & Stanier, 1967). The catabolism of 3-methylgentisate in another pseudomonad is initiated by similar oxygenation but the subsequent metabolic steps differ (Hopper, Chapman & Dagley, 1968).

The reactions shown in Figures 8.12 and 8.13 appear to be catalysed by enzymes with fairly narrow substrate specificities (Ornston, 1966a,b), so the recurrence of similar metabolites and types of reaction cannot be attributed to the action of a single set of enzymes operating on a wide range of chemically similar substrates as is the case with the cresols (see Chapter 7, Figure 7.15). Rather, the similarities observed in the pathways outlined in Figures 8.12 and 8.13 point to two major considerations about the evolutionary origins of the structural genes for the enzymes of the catabolic pathways. The first of these is that the evolutionary processes are confined within the laws of chemistry and that there is a limited number of chemical solutions to metabolic problems. It is not surprising that independent evolutionary processes led to quite similar biochemical solutions to similar metabolic problems. Columns (b) and (c) in Figure 8.12 give an example of such convergent metabolic evolution in which muconolactone was selected as a catabolite that could be converted by relatively few enzymatic steps to common intermediary metabolites. When produced from fluorocatechol the muconolactone is laevo-rotatory, whereas it is dextro-rotatory when formed as an intermediate in the dissimilation of catechol. The different optical isomers of muconolactone indicate that a similar catabolic pathway evolved quite independently rather than developed as a modification of a pre-existing metabolic sequence for the utilization of muconolactone.

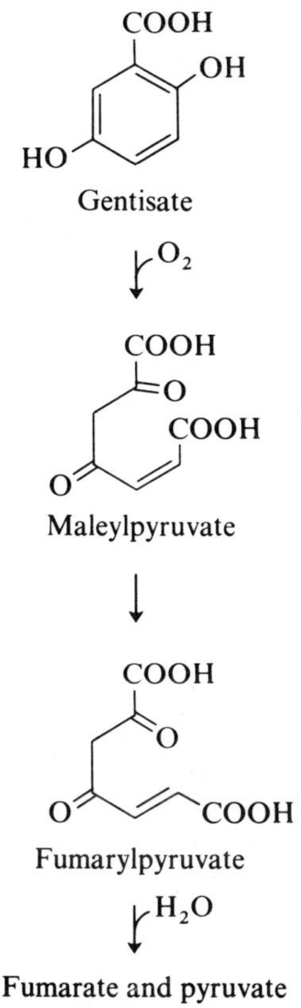

COOH
OH

HO

Gentisate

O_2

COOH
O
COOH

O

Maleylpyruvate

COOH
O

O COOH

Fumarylpyruvate

H_2O

Fumarate and pyruvate

Figure 8.13. Catabolism of gentisate by *Pseudomonas acidovorans*

The second consideration raised by catabolic similarities is that in some cases genes for enzymes with similar functions may be derived from common ancestral genes. Evidence for such evolutionary homology has come from comparisons of the properties of enzymes catalysing similar transformations in the catabolism of muconate and carboxymuconate to enol-lactone in *P. putida* (Figure 8.14). The two lactonizing enzymes (MLE and CMLE in Figure 8.14) possess similar molecular sizes (about 200,000 daltons as judged by gel filtration), sub-unit sizes (about 40,000 daltons) and crystalline structure (square plates). The same dipeptide, Val Met, is found in the first amino acid residues of the two enzymes (Meagher & Ornston, 1973; Patel, Meagher & Ornston, 1973). The two enzymes that give rise to enol-lactone (MI and CMD in Figure 8.14), also appear to be the products of homologous genes. They form single needle-like crystals and their molecular sizes (about 93,000 daltons as estimated by gel filtration) and sub-unit sizes (11,000 to 13,000 daltons) are quite similar (Meagher & Ornston, 1973; Parke, Meagher & Ornston, 1973).

Enzyme   Enzyme

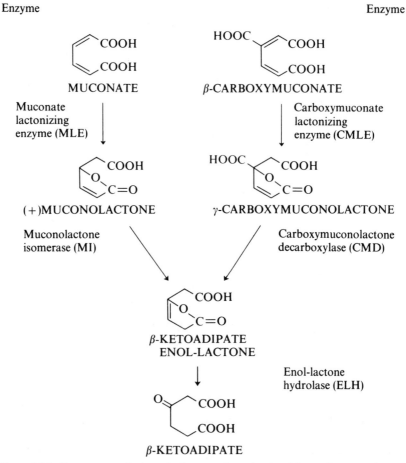

Figure 8.14. Enzymes catalysing similar reactions in the β-ketoadipate pathway of bacteria

The unusually small sub-unit size of MI and CMD is shared by another enzyme that catalyzes the migration of a double bond, the ketosteroid isomerase of *P. testosteroni*. The complete amino acid sequence of the latter protein has been reported (Benson, Jarabak & Talalay, 1971).

Horowitz (1945; 1965) has proposed that genes for enzymes catalysing sequential metabolic reactions may have arisen by the modification of homologous genes that were formed by genetic duplication. Certainly such evolutionary events must have been infrequent. Several authors (Wu, Lin & Tanaka, 1968; Hegeman & Rosenberg, 1970) have stressed that a far more likely process is the 'borrowing' of genes for enzymes with analogous function from other pathways. Nevertheless, some lines of evidence indicate that in a few instances, new gene functions may have been acquired by a process similar to that envisaged by Horowitz. The enol-lactone hydrolase (ELH in Figure 8.14) of *P. putida* has a sub-unit size of 11,000 daltons. The amino terminal sequence of the enzyme is Met Leu Leu, whereas the amino terminal sequence of the muconolactone isomerase (MI) from the same organism is Met Leu Phe (Patel, Meagher, McCorkle & Ornston, unpublished observations).

The existence of so many enzymes for the metabolism of aromatic compounds whose reactions have already been studied in some detail, make them ideal subjects for the investigation of the evolution of catabolic pathways.

3. UTILIZATION OF ALIPHATIC AMIDES

3.a. Metabolism of Amides: Occurrence of Aliphatic Amidases

Several species of *Pseudomonas* are able to utilize amides for growth. The first step is the hydrolysis of the amide and the release of ammonia. It is therefore likely that more amides will be available as nitrogen than as carbon sources since the fate of the acyl moiety will depend on the presence or absence of the relevant catabolic pathway enzymes. Acetamide has been examined in more detail than other amides and poses few metabolic problems since acetate is readily metabolized as a carbon source for growth (Figure 8.15). Growth in minimal medium with acetamide providing either the carbon or nitrogen source, or both, is presumptive evidence for the presence of an acetamidase. Following reports that *P. aeruginosa* strains produced an alkaline reaction in glucose + acetamide medium (Buhlman, Vischer & Bruhin, 1961), various methods have been devised to use acetamide media to discriminate among the fluorescent species. Stanier and his colleagues (1966) found that all the strains which they had identified as *P. aeruginosa* grew on

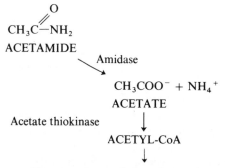

$$CH_3C\overset{O}{\overset{\|}{-}}NH_2$$
ACETAMIDE

Amidase

$$CH_3COO^- + NH_4^+$$
ACETATE

Acetate thiokinase

ACETYL-CoA

Tricarboxylate and glyoxylate cycle reactions

Figure 8.15. The metabolic pathway for acetamide in *P. aeruginosa*

minimal agar plates with acetamide as the carbon source. A similar growth response was observed with some of the strains assigned to *P. putida* biotype A and strains of *P. acidovorans* and *P. cepacia* (*multivorans*). None of their strains of *P. putida* biotype B or *P. testosteroni* grew on acetamide. Various complex media used for the identification of pseudomonads include acetamide as one of the components and the results suggest that acetamidase activity is a usual characteristic of *P. .aeruginosa* strains. Mossel and Zadelhoff (1971), for example, reported that only 1·2 % of 500 recent isolates of *P. aeruginosa* appeared to lack an acetamidase.

P. aeruginosa grows on acetamide and propionamide and the aliphatic amidase is induced in media containing these two amides. Butyramide does not support growth and does not induce amidase synthesis. On the other hand, some strains of *P. putida* biotype A and *P. acidovorans* can grow on butyramide as well as on acetamide and propionamide. *P. cepacia* resembles *P. aeruginosa* and does not grow on butyramide. The aliphatic amidases produced by these four species are similar in substrate profile. They hydrolyse propionamide at 2 to 3 times the acetamide rate, and butyramide at only 2 to 3 % of the acetamide rate. However, this rate of butyramide hydrolysis might be sufficient to allow growth if sufficient enzyme were to be synthesized by the bacteria. Clarke (1972) found that the different amide growth phenotypes of the four *Pseudomonas* species were mainly due to differences in the ways in which the amides affected the regulation of amidase synthesis. Butyramide induces amidase synthesis in some strains of *P. putida* biotype A and *P. acidovorans*, although the rate of amidase synthesis is lower than when acetamide is used as the inducer. With *P. aeruginosa* and *P. cepacia* strains however, butyramide not only fails to induce but can compete with the inducer amides and repress amidase synthesis.

Similarities between the amidase proteins produced by these *Pseudomonas* species can be observed by immunodiffusion cross-reactions. Cell-free extracts prepared from 8 different strains of *P. aeruginosa* cross-reacted completely in immunodiffusion tests with antiserum prepared against a purified preparation of amidase from the wild-type strain *P. aeruginosa* PAC1, suggesting very close similarities in the amidases produced; *P. putida* biotype A extracts also gave strong cross-reaction, but with marked spurring; *P. acidovorans* extracts gave strong cross-reactions with very marked spurring with both *P. aeruginosa* and *P. putida* extracts. Preparations from *P. cepacia* gave only weak cross-reactions with the *P. aeruginosa* antiserum suggesting that the aliphatic amidase from this species differs in composition and sequence from the *P. aeruginosa* amidase rather more extensively than do the enzymes from *P. putida* and *P. acidovorans*.

There are relatively few accounts of the production of aliphatic amidases by other microorganisms. Halpern and Grossowicz (1957) found that *Mycobacterium phlei* hydrolysed formamide and acetamide and Kimura

(1959) observed that *Mycobacterium smegmatis* had acetamidase activity. The latter observation was extended by Draper (1967) who obtained evidence to suggest that *Mycobacterium smegmatis* produces two inducible aliphatic amidases, one specific for formamide and one with a broad substrate range, optimal for butyramide. Hynes and Pateman (1970) found that *Aspergillus nidulans* produces a specific formamidase, and a second aliphatic amidase which attacks acetamide and some other amides.

A second amidase has been identified in some *P. putida*, *P. acidovorans* and *P. cepacia* strains which enables growth on phenylacetamide. These species unlike *P. aeruginosa*, are able to metabolize phenyl acetate, so that they can use phenylacetamide as both carbon and nitrogen source for growth. *P. putida* A90 possesses only the phenylacetamidase and mutants of both *P. putida* A87 and *P. cepacia* 716 have been isolated which lack acetamidase and retain phenylacetamidase activity. It is possible that these two amidases arose by a process of gene duplication of a single amidase gene and then evolved separately. These naturally occurring phenylacetamidases have substrate specificities which are broadly similar to some of the mutant phenylacetamidases, evolved from the *P. aeruginosa* acetamidase by a series of mutations aimed at selecting enzymes with altered substrate specificities (Betz & Clarke, 1972; 1973).

3.b. *P. aeruginosa* Aliphatic Amidase: Enzyme Properties and Regulation

The enzyme produced by *P. aeruginosa* is the only one so far investigated in detail, it belongs to the group of acyl transferases and is able to catalyse the following series of reactions:

Amide hydrolysis	$RCONH_2 + H_2O$	$\rightarrow RCOO^- + NH_4^+$
Ester hydrolysis	$RCOOR' + H_2O$	$\rightarrow RCOO^- + H^+$ $+ R'OH$
Amide transfer	$RCONH_2 + NH_2OH + H^+$	$\rightarrow RCONHOH + NH_4^+$
Acid transfer	$RCOO^- + NH_2OH + H^+$	$\rightarrow RCONHOH + H_2O$
Ester transfer	$RCOOR' + NH_2OH$	$\rightarrow RCONHOH + R'OH$

Maximum enzyme activity is obtained with $R = C_2H_5$ for amide hydrolysis; $R = CH_3$ for amide, acid and ester transferase; R and $R' = CH_3$ for ester hydrolysis and transferase.

The only reaction likely to be of physiological importance is amide hydrolysis. The transferase reaction with hydroxylamine as the acceptor for amides, acids or esters as substrates, is however useful for assay of enzyme activity and is the method which has been used for most of the studies on the kinetics of amidase synthesis. Methods used to measure amide hydrolase activity by assaying the ammonia released include the Conway micro-diffusion method and a method using ninhydrin (Brown, Brown & Clarke, 1969). Esterase activity is very low and although a number of low molecular weight esters can be substrates for the hydrolase or transferase reactions the apparent K_m values are in the range of 500 mM so that the esterase activity of the enzyme is probably irrelevant to its biological function (McFarlane, Brammar & Clarke, 1965). Studies with the purified enzyme have shown that it can also hydrolyse acylhydroxamates and acylhydrazides. Enzyme activity is inhibited by urea and by acetaldehyde ammonia (Findlater & Orsi, unpublished).

The molecular weight of the purified enzyme was determined as 200,000 by sedimentation equilibrium studies. It can be dissociated into six identical sub-units for which molecular weights of 33,000–35,000 were obtained by electrophoresis in SDS/acrylamide gels and by filtration on 'Sephadex' G200 in SDS/PMB solution. The hexameric sub-unit structure was confirmed by obtaining six bands on SDS/acrylamide gel electrophoresis after treatment with the cross-linking reagent dimethyl suberimidate. The number of tryptic peptides obtained is also compatible with a hexameric structure. The N-terminal amino acid is methionine and the C-terminal amino acid is alanine (Brown *et al.*, 1973).

The inducer specificity is distinct from the substrate specificity. The best inducers are the substrates, acetamide, propionamide and glycolamide and some N-substituted amides such as N-methyl- and N-acetylacetamide which are not substrates. In addition, lactamide is a very poor substrate but an excellent inducer and can be used for gratuitous enzyme induction. If acetamide is regarded as the parent molecule, it is easy to see how many relatively simple acetamide analogues can be tested for substrate and inducer specificities (Table 8.1 and Figure 8.16). In addition to urea, iodoacetamide and acrylamide are enzyme inhibitors and glycineamide inhibits hydrolase activity of whole cells by preventing the uptake of propionamide. Butyramide, cyanoacetamide and thioacetamide are classed as amide analogue repressors. They compete with inducer amides for an amide binding site (presumably on the regulator protein), and prevent enzyme synthesis in the way in which 2-nitrophenyl-β-D-fucoside prevents the induction of β-galactosidase by thiogalactoside inducers (Jayaraman, Müller-Hill & Rickenberg, 1966). Table 8.1 gives a qualitative comparison of aliphatic amides and various substituted derivatives as amidase substrates, enzyme inhibitors, inducers and amide analogue repressors.

Amidase, like other catabolic enzymes, is subject to catabolite repression and succinate is one of the strongest catabolite repressors and pyruvate one of the least. The rate of synthesis of amidase in pyruvate medium, by inducible or constitutive strains, is enhanced by the addition of cyclic AMP, which also

Table 8.1. Comparison of amides as substrates, inducers, enzyme inhibitors and amide analogue repressors

Amide		Hydro-lase substrate	Inducer	Enzyme in-hibitor	Ana-logue repressor
Formamide	$HCONH_2$	+	+	·	+
Acetamide	CH_3CONH_2	+ +	+ + +	·	·
Propionamide	$CH_3CH_2CONH_2$	+ + +	+ +	·	·
Butyramide	$CH_3CH_2CH_2CONH_2$	tr.	−	·	+
iso-Butyramide	CH_3 $\diagdown$ $CHCONH_2$ $\diagup$ CH_3	−	−	·	+
Valeramide	$CH_3(CH_2)_3CONH_2$	−	·	·	+
Hexanoamide	$CH_3(CH_2)_4CONH_2$	−	·	·	+
N-Methylformamide	$HCONHCH_3$	−	+ + +	−	·
N-Ethylformamide	$HCONHC_2H_5$	−	+	·	·
N-Methylacetamide	$CH_3CONHCH_3$	−	+ + +	−	·
N-Ethylacetamide	$CH_3CONHC_2H_5$	−	+ +	·	·
N-Acetylacetamide	$CH_3CONHCOCH_3$	−	+ + +	·	·
N-Phenylacetamide	$CH_3CONHC_6H_5$	−	−	−	+
N-Dimethylacetamide	$CH_3CON(CH_3)_2$	−	−	−	−
N-Methylpropionamide	$CH_3CH_2CONHCH_3$	−	+ +	·	·
N-Ethylpropionamide	$CH_3CH_2CONHC_2H_5$	−	+ +	·	·
Cyanoacetamide	$CH_2CNCONH_2$	−	−	·	+
Fluoroacetamide	CH_2FCONH_2	+	ϕ	·	·
Iodoacetamide	CH_2ICONH_2	−	ϕ	·	·
Glycine amide	$CH_2NH_2CONH_2$	−	−	−	+
Sarcosine amide	$CH_2NH(CH_3)CONH_2$	−	−	−	+
Glycolamide	$CH_2OHCONH_2$	+ +	+	−	+
Acrylamide	$CH_2 = CHCONH_2$	+ +	ϕ	+	·
β-OH-propionamide	$CH_2OHCH_2CONH_2$	−	−	−	+
Lactamide	$CH_3CHOHCONH_2$	+	+ + +	−	·
Pyruvamide	$CH_3COCONH_2$	−	·	·	·
Fumaramide	$CHCONH_2$ $\|\|$ $CHCONH_2$	−	−	−	·
Malonamide	$\diagup CONH_2$ CH_2 $\diagdown CONH_2$	−	−	−	·
Methyl carbamate	CH_3OCONH_2	−	+ +	−	·
Urea	NH_2CONH_2	−	−	+	·
Benzamide	$C_6H_5CONH_2$	−	·	−	+
Thioacetamide	CH_3CSNH_2	−	−	−	+

+ + +, + +, +, tr., = relative activities; −, = no activity detected; ·, not tested; ϕ, growth inhibited. Data from Kelly & Clarke (1962) revised later.

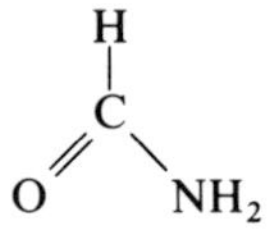

Formamide
(1)

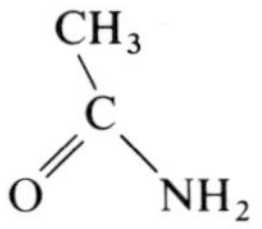

Acetamide
(2)

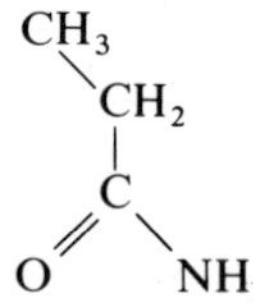

Propionamide
(3)

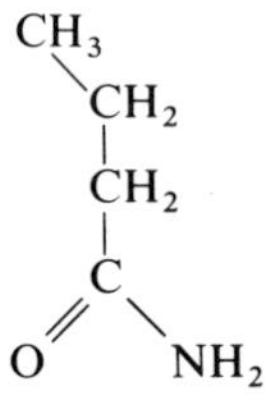

Butyramide
(4)

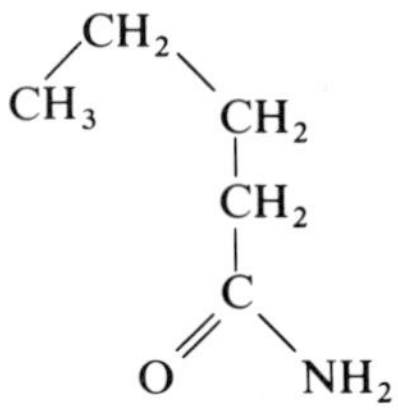

Valeramide
(5)

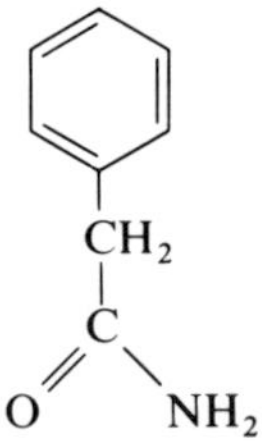

Phenylacetamide
(6)

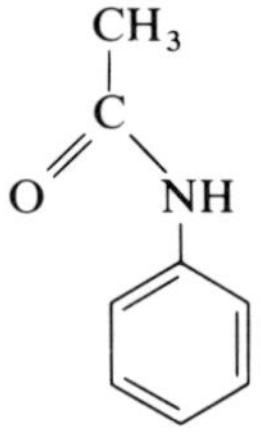

Acetanilide
(7)

Figure 8.16

relieves the partial catabolite repression produced by the weaker repressing compounds such as lactate (Smyth & Clarke, 1972). This suggests that for this enzyme system there may be a stimulation of gene transcription involving c-AMP and an activator protein as for the enzymes of sugar catabolism in *Escherichia coli* (Emmer *et al.*, 1970).

3.c. Mutant Isolation

Positive selection methods have been based on the exploitation of amide substrate and inducer specificities and the catabolite repression control. Regulator mutants are readily isolated spontaneously, or after mutagen treatment, on succinate/formamide plates containing a low concentration of formamide as the nitrogen source, and succinate as the carbon source. Formamide is a poor substrate and very poor inducer, and succinate produces

Table 8.2. Amidase regulator mutants

Mutant class	Selection medium	Reference
Formamide-inducible	Succinate/formamide	Brammar, Clarke & Skinner (1967)
Semi-constitutive	Succinate/formamide	Brammar, Clarke & Skinner (1967)
Magnoconstitutive Butyramide-sensitive	Succinate/formamide	Brammar, Clarke & Skinner (1967)
Magnoconstitutive Butyramide-resistant	Succinate/formamide	Brammar, Clarke & Skinner (1967)
Magnoconstitutive Butyramide-resistant	Butyramide	Brown & Clarke (1970)
Catabolite repression-resistant	Succinate/lactamide	Brown & Clarke (1972)

catabolite repression. This selection procedure gives some mutants with altered inducer specificities which are more readily induced by formamide than is the wild type. Most of the mutants which appear are constitutive and may be either fully constitutive or semi-constitutive (Table 8.2). The fully constitutive group can be divided into two classes on the basis of their response to butyramide. Some of them are unable to grow on butyramide plates and amidase synthesis in various growth media is severely repressed by butyramide. Others are resistant to butyramide repression and produce sufficient enzyme to be able to grow on butyramide plates. Such butyramide-resistant constitutive mutants can also be isolated directly from butyramide plates either from the wild-type inducible strain, or from constitutive butyramide-sensitive strains (Brown & Clarke, 1970). It is reasonable to suppose that most, if not all, of these mutations occur in the amidase regulator gene *ami*R. Some mutants have been isolated which are acetamide-positive and constitutive at temperatures below about 30 °c and acetamide-negative at higher temperatures. The phenotypic properties of these regulator mutants are in accord with a positive control of the amidase structural gene but this remains to be confirmed by dominance tests with partial diploids (Farin & Clarke, 1974). All these regulatory characters are cotransduced at high frequency, >90 % with the amidase-positive character, which indicates very close linkage of the *ami*R gene and *ami*E, the amidase structural gene. It is possible that some of the semi-constitutive mutants with low basal enzyme levels are operator mutants, but the present state of genetic analysis does not permit any distinction to be made and meanwhile they have all been assigned to *ami*R.

Catabolite repression-resistant mutants can be isolated from succinate/lactamide plates with lactamide, a poor substrate but a very good inducer, as nitrogen source and succinate, a strong catabolite repressor, as carbon source.

The catabolite repression resistance character of mutants isolated by this method has always been found to be unlinked to the two amidase genes and the exact nature of these mutations is not known. For practical purposes catabolite repression-resistant, constitutive mutants are useful for obtaining large amounts of enzyme since they can produce amidase amounting to 10 % of the total cell protein.

Amidase-negative mutants can be isolated by a positive selection method on pyruvate plates containing fluoroacetamide. *P. aeruginosa* is not very sensitive to growth inhibition by fluoroacetate produced by the hydrolysis of fluoroacetamide, so that it is necessary to start either with a constitutive strain maximally derepressed, or with an inducible strain which has been fully induced (Clarke & Tata, 1973). The mutants selected by this method include *ami*E structural gene mutants, temperature-sensitive regulator mutants and mutants in the promotor region with a very reduced rate of transcription. The promotor region of the *lac* operon of *E. coli* is known to be the site of the catabolite repression effect (Silverstone *et al.*, 1969) and promotor mutants for the amidase gene were identified by selecting for catabolite repression-resistant revertants from a number of mutants with a lactamide negative phenotype. From 116 acetamide-negative mutants, derived from a butyramide-resistant constitutive strain, 5 gave revertants which grew on plates with butyramide as the nitrogen source and succinate as the carbon source. Growth on this medium requires amidase synthesis to be resistant to catabolite repression. Transduction analysis showed that the catabolite repression-resistance character was cotransduced at a frequency of 100 % with the amidase-positive character so that these mutants were more likely to have catabolite repression-resistance mutations at the promotor site than in a gene determining a c-AMP dependent activator protein. The 5 mutants differed in their rates of enzyme synthesis but one had a very high rate of synthesis in all media and this could possibly be described as a 'super-promotor' mutant (Smyth & Clark, 1972).

The substrate specificity of the wild-type enzyme is fairly narrow with regard to aliphatic amides and the hydrolase rates for formamide, acetamide, propionamide and butyramide are in the ratio $0.2:1:3:0.02$ (Figure 8.16, **1, 2, 3, 4**). Brown, Brown and Clarke (1969) isolated a mutant producing an altered enzyme (B amidase) with a greater affinity for butyramide as a substrate.

This was the first step in the selective evolution of this enzyme. Mutants were subsequently isolated producing V-type amidases and able to grow on valeramide (Figure 8.16, **5**) and then a group of mutants able to grow on phenylacetamide (Figure 8.16, **6**) and producing Ph-type amidases (Betz & Clarke, 1972) (Table 8.3). These mutants form a series in which the maximum size of the side-chain of the amide which can be used as a growth substrate has been successively increased. The mutant enzymes have the same molecular weights and sub-unit structures as the wild-type A amidase (M. J.

Table 8.3. Mutant amidases

Selection medium	Strain number	Amidase type	Reference
None (wild-type)	PAC1	A	Kelly & Clarke (1962)
Butyramide	PAC351	B	Brown, Brown & Clarke (1969)
Valeramide	PAC360	V9	Brown, Brown & Clarke (1969)
Phenylacetamide	PAC377	PhB3	Betz & Clarke (1972)
	PAC388	PhV1	Betz & Clarke (1972)
	PAC389	PhV2	Betz & Clarke (1972)
	PAC391	PhA	Betz & Clarke (1972)
	PAC392	PhF	Betz & Clarke (1972)
Acetanilide	PAC366	AI3	Brown & Clarke (1972)

Smyth, unpublished). The B amidase gives a complete cross-reaction in immunodiffusion tests against antiserum to the wild-type enzyme and differs slightly in electrophoretic mobility on starch gels. The V amidases are a heterogenous group and less stable than the A and B amidases. The Ph amidases differ from one another in substrate profiles and thermal stabilities and were derived from different parental strains. The V and Ph amidases give diffuse bands in immunodiffusion tests against antiserum to A amidase and this may be due to dissociation into sub-units. The relative hydrolase activities of some of the V and Ph amidases are compared with those of B amidase in Table 8.4.

A different class of mutants isolated by Brown & Clarke (1972) are able to utilize the *N*-substituted amide, acetanilide (*N*-phenylacetamide, Figure 8.16, **7**). These mutants hydrolyse acetanilide to produce acetate and aniline and grow well on plates in which acetanilide provides the carbon source. They are also able to grow with acetanilide as both carbon and nitrogen source, but growth is slower and the colonies become brown in colour with the accumulation of aniline products.

$$CH_3CONH\!-\!\langle\ \rangle + H_2O \rightarrow CH_3COOH + NH_2\!-\!\langle\ \rangle$$

The AI enzymes produced by these mutants are less thermostable than the A and B enzymes, but more stable than any of the V and Ph enzymes. Amidase AI3 gives a complete cross-reaction, with a single precipitin band, with the antiserum to A or B amidase. The structural difference between AI3 amidase and the wild-type enzyme resides in a single amino acid substitution in which a threonine residue is replaced by an isoleucine residue. The hexapeptides isolated from a tryptic-chymotryptic digest of the wild-type A amidase and the mutant AI3 amidase, contained the following sequences of amino acids.

A amidase   Ser-Leu-*Thr*-Gly-Glu-Arg
AI3 amidase   Ser-Leu-*Ile* -Gly-Glu-Arg

Table 8.4. Relative hydrolase activities of *Pseudomonas aeruginosa* mutants: washed suspensions of bacteria (from Betz & Clarke, 1972)

Strain no.	Amidase type	Amide substrates[a]						Butyramide hydrolysed (μmoles/ min/mg bacteria)
		F	A	P	B	V	Ph	
PAC351	B6	75	370	1200	100	tr.	0	3·0
PAC360	V9	ND	45	200	100	8	0	8·0
PAC377	PhB3	0	0	100	100	190	97	0·34
PAC388	PhV1	tr.	tr.	150	100	220	100	0·50
PAC389	PhV2	ND	23	153	100	136	55	2·09
PAC391	PhA	ND	21	111	100	140	56	1·06
PAC392	PhF	0	25	206	100	130	52	3·0

ND, = not determined; tr., = trace activity.
[a] F = formamide; A = acetamide; P = propionamide; B = butyramide; V = valeramide; Ph = phenylacetamide.

3.d. Genetic Data

The amidase genes *ami*R, *ami*E comprise a regulation unit that does not include any genes for a related function such as acetate or propionate metabolism. A constitutive acetamide permease was demonstrated by Brammar, McFarlane and Clarke (1966) but no mutants lacking permease activity have been isolated. Transduction with the pseudomonad phage F116, isolated by Holloway, Egan & Monk (1960), was used to determine linkage of the structural and regulator genes (Brammar, Clarke & Skinner, 1967 (see also Chapter 6). Mutants with an acetamide-negative phenotype include not only amidase-negative strains but also some of the strains producing altered enzymes in which the specificity has changed to such an extent that there is no appreciable hydrolysis of acetamide. Transductional crosses can be carried out between any two acetamide-negative mutants with the recombinant transductants selected on acetamide plates. However, an acetamide-positive phenotype may be due to the production of wild-type A amidase or one of the mutant enzymes, such as B amidase, which has good activity for both acetamide and butyramide. The *ami*R gene is so closely linked to the *ami*E gene that it can be used as the outside marker in 3-factor crosses. Fine structure mapping of the amidase structural gene has been carried out by selection of acetamide-positive transductants which are subsequently analysed for both amide substrate phenotype and regulator phenotype (Betz, Brown, Clarke & Day, 1974). The B amidase phenotype was recovered from crosses in which one of the parents was an acetamide-negative mutant producing an altered enzyme, which had been derived by a second mutational step from strain PAC351 which produces the mutant B amidase. It could therefore be concluded that these acetamide-negative mutants, strain PAC356, producing a V-type amidase and PAC377, producing a phenyl-acetamide amidase (see Table 8.3), had retained the original mutation from wild-type A amidase to mutant B amidase.

The biochemical versatility of the pseudomonads suggests that they are still capable of adapting to new growth substrates. In Chapter 9 the results of the studies on the experimental evolution of *P. aeruginosa* amidase will be discussed in relation to the biochemical evolution of the *Pseudomonas* species.

4. BIOSYNTHESIS AND CATABOLISM OF AMINO ACIDS

4.a. Amino Acid Metabolism

Pseudomonas species are non-exacting with respect to amino acids with the exception of *P. maltophilia* which requires methionine (Stanier *et al.*, 1966). The biosynthetic pathways have not all been examined in detail but results obtained mostly with *P. aeruginosa* and *P. putida*, indicate that they differ very little from those established for other groups of bacteria. However, there are marked differences in regulation both of enzyme activity and of enzyme synthesis (see Chapter 7, section 4 and the following sections on individual pathways).

A wide range of amino acids can be utilized as the sole carbon and nitrogen source for growth. Jacoby (1964) showed that *P. putida* could grow on any of 18 amino acids, and that the oxidative enzymes were induced by growth on their substrates and repressed by the presence of glucose in the medium. Kay and Gronlund (1969b) found that *P. aeruginosa* grew on a similar range of amino acids. Among the amino acids which could be utilized as a nitrogen but not as a carbon source, were methionine, threonine and cysteine. Some strains of *P. aeruginosa* grow poorly, or not at all, on particular amino acids but appear to possess most of the enzymes of the required pathway. This will be discussed with reference to the lysine and histidine pathways later in this chapter. Unlike the biosynthetic pathways, the degradative pathways for an amino acid may be different in two different *Pseudomonas* strains and more than one pathway may be present in one strain. Some of the D-amino acids can be utilized, and in the case of lysine, although racemases may be present, there are separate pathways which degrade the L- and D-amino acids respectively (Miller & Rodwell, 1971). Enzyme specificity for the catabolic pathways is greater than inducer specificity and in some cases one isomer induces the enzymes of the pathway for the degradation of the other isomer of the amino acid (Marshall & Sokatch, 1972). It has occasionally been suggested that the catabolic enzymes might play some part in controlling the biosynthetic pathways by removing excess products from the amino acid pool. In the few cases where this hypothesis has been examined it has been found to be untrue, and no mutants with defects in caabolic pathways have been found to over-produce the end-products of the biosynthetic pathway.

In discussing the genetics of both biosynthetic and catabolic pathways for amino acids it is essential to discriminate clearly between the genes con-concerned. Since the auxotrophic mutants were isolated first the biosynthetic genes have become recognized as *his*, *lys*, *pro*, etc., while it is convenient to denote the genes required for the utilization of these amino acids as *hut*, *lut*, *put*, etc. The amino acids that have been examined in most detail with respect to the biochemistry and genetics of the pathways will be discussed individually, but the amino acid permeases will be considered together in the following section.

4.b. Amino Acid Permeases

Kay and Gronlund (1969a) found that *P. aeruginosa* grown on glucose possessed active transport systems for 18 of the common amino acids. These were inhibited by azide, were saturated at high amino acid concentrations and allowed equilibration with the amino acid pools. It thus appeared that *P. aeruginosa* had a wide range of constitutive amino acid permeases which had high affinities for the amino acid transported. (Apparent K_m values were in the range 10^{-7} to 10^{-6} M). It was somewhat surprising that the amino acid permeases were constitutive since the catabolic pathway enzymes were known to be inducible. The permeases also appeared to be stable since washed cells lost very little activity during carbon or nitrogen starvation (Kay & Gronlund, 1969b). They suggested that the stability of the constitutive permeases would be advantageous since in an environment which was very depleted in nutrients the permeases would be available for scavenging amino acids present at low concentrations. Kay and Gronlund (1969c) also found that the level of one of the permeases was increased if the amino acid it transported was present in the medium and the advantages of this type of control became clear. Cells grown on proline had proline-transport activities 10-fold greater than glucose-grown cells so that when proline became available, the high affinity transport system already present at the basal level was rapidly increased. Proline could then be utilized since the proline degradative enzymes were induced coordinately with the proline permease. Wild-type *P. aeruginosa* grown on proline also had increased glutamate permease and this was presumed to be induced by glutamate produced by proline degradation, since it was not present in *put⁻* cells grown on proline. Repression of proline permease occurred when glucose was present in the medium so that regulation of proline permease synthesis was like that of a catabolic enzyme, induced by its substrate and repressed by glucose. The proline analogue 3,4-dehydroproline inhibits growth of *P. aeruginosa* and is also a competitive inhibitor of proline transport. Resistant mutants isolated by Kay and Gronlund (1969d) still possessed the wild-type level of proline permease so that the mutation conferring resistance had not affected permease regulation of specificity. However, bacteria grown in the presence of low concentration of 3,4-dehydroproline had repressed levels of proline permease. If, as the authors suggest, 3,4-dehydroproline inhibits an early enzyme of the proline biosynthetic pathway, then the amount of proline in the metabolic pool could be very low. If the constitutive permease is in fact induced by endogenous proline then this could account for the repression of permease synthesis. Alternatively, the 3,4-dehydroproline could act as 'analogue repressor' in the way that butyramide represses synthesis of acetamidase (see this Chapter, section 3.b). Since proline permease and the proline catabolic enzymes were coordinately induced (Kay & Gronlund, 1969c) then it would be expected that they would also be coordinately repressed. Kay and Gronlund (1969d) report that the rate of proline oxidation was reduced in 3,4-dehydroproline-grown cells but this may have been a consequence of the reduced permease activity.

Transport of proline in *P. fluorescens* is by a specific transport system with an apparent K_m of 5×10^{-6} M probably similar to the proline permease of *P. aeruginosa*, but proline can also be transported by a permease common to L-α-alanine and β-alanine. Hechtman and Scriver (1970) found that L-α-alanine itself is also transported by a specific alanine permease and by another permease which transports both L-alanine and L-phenylalanine.

Two aromatic permeases with overlapping specificities for the aromatic amino acids were found in *P. aeruginosa* by Kay and Gronlund (1971). The apparent K_m values were similar for both systems and the values found were 4.4×10^{-7} M for phenylalanine, 5.4×10^{-7} M for tyrosine and 1.4×10^{-6} M for tryptophan. Each aromatic amino acid inhibited the uptake of the other two. Results obtained for the kinetics of uptake, competition by the other substrates and with mutants which had altered aromatic amino acid transport systems, suggested that the two permeases had different functions. One mutant was resistant to 5-fluorotryptophan and was defective in system II which transported tryptophan > phenylalanine > tyrosine. Another mutant was isolated as a slow grower on tyrosine and this was thought to have a defect in system I which transported phenylalanine > tyrosine > tryptophan. In *P. acidovorans* the tryptophan transport system is highly specific and cannot transport the other aromatic amino acids although like *P. aeruginosa* system II it is inhibited by 5-fluorotryptophan (Rosenfeld & Fiegelson, 1969a).

A complex transport system for the basic amino acids was found in *P. putida* by Fan, Miller and Rodwell (1972). Cells grown on arginine possessed an inducible arginine-specific permease with a low K_m of 5.2×10^{-8} M. Cells grown on lysine were induced for two permeases one of which transported the diamino acids, lysine and ornithine with apparent K_m values of about 10^{-7} M and was inhibited by the D-amino acids. Another lysine-induced permease transported lysine and arginine and was inhibited by the D-amino acids and by the arginine homologues, L-homoarginine and L-α-amino-γ-guanidinobutyrate. This third permease had a lower affinity for its substrates and the K_m values were about 5×10^{-6} M.

Table 8.5 lists amino acid permeases reported for various *Pseudomonas* species.

Table 8.5. Amino acid permeases of *Pseudomonas* species

Species	References	Amino acids transported
P. aeruginosa	1, 2, 3	Constitutive permeases for 18 amino acids, stable, high affinities for amino acids transported. K_m range 10^{-7}–10^{-6} M
	4	Permease specific for L-proline. Constitutive basal level induced 10-fold by proline
	5	2 permeases for aromatic amino acids differing in specificities
P. putida	6	3 permeases for basic amino acids. 2 lysine-inducable permeases with differing specificities; arginine-inducible permease transports only arginine (K_m 5.2×10^{-8} M)
P. fluorescens	7	4 neutral amino acid permeases with overlapping specificities
P. acidovorans	8	Tryptophan permease transports tryptophan only, inhibited by 5-fluorotryptophan

1, 2, 3. Kay & Gronlund (1969a,b,d).   4. Kay & Gronlund (1969c).
5. Kay & Gronlund (1971).   6. Fan *et al.* (1972).
7. Hechtman & Scriver (1970).   8. Rosenfeld & Fiegelson (1969b).
Evidence for other amino acid transport systems which have not been investigated in detail, may be found in papers concerned with catabolic pathways.

4.c. Amino Acids Synthesized from Aspartate

The pathways for the biosynthesis of the aspartate family of amino acids, lysine, threonine and methionine, are regulated quite differently in *Pseudomonas* species from *Escherichia coli*, although the enzyme reactions appear to be the same (Figure 8.17). The key points at which the rates of synthesis of these amino acids are controlled are (1), the reaction in which aspartyl phosphate is synthesized from aspartate and (2), the reaction in which homoserine is synthesized from aspartic semialdehyde. The first reaction is required for the synthesis of all these amino acids and in *E. coli* there are three

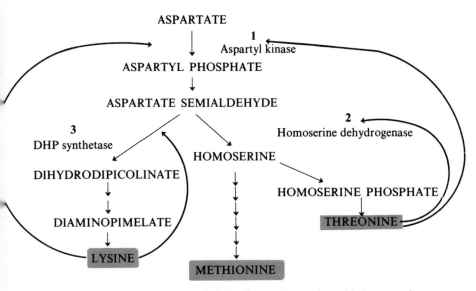

Enzyme 1 also repressed by lysine, methionine and threonine added separately.

Figure 8.17. Feedback inhibition of the enzymes for the branched pathway for the biosynthesis of lysine, methionine and threonine from aspartate

distinct aspartokinases regulated independently by one or other of the three amino acids which are the end-products of this branched pathway. Asparto-kinase I is inhibited by threonine and repressed by threonine (plus isoleucine). Threonine is the precursor for the synthesis of isoleucine but this part of the pathway will be considered later. Aspartokinase I is a multifunctional protein which also carries homoserine dehydrogenase activity (Cohen, Stanier & Le Bras, 1969). Homoserine dehydrogenase I is therefore regulated in the same way and is subject to both feedback inhibition and repression by threonine. Aspartokinase III is inhibited by lysine and is also repressed by lysine, but homoserine dehydrogenase is not required for lysine bio-synthesis and lysine has no effect on activity. In the presence of both lysine and threonine two of the aspartokinase isoenzymes will be inhibited but the remaining aspartokinase will be functional and allow the synthesis of methionine. Aspartokinase II is not under feedback inhibition control but is repressed by methionine. In *E. coli* strain K12, but not strain B, the methionine-repressible aspartokinase II also has homoserine dehydrogenase activity. This pattern of control occurs throughout the Enterobacteriaceae and similar classes of isoenzymes were found by Cohen and his colleagues (1969) in *Salmonella*, *Serratia*, *Klebsiella* and *Proteus*. It was already known that other types of regulation existed for the control of branched pathways.

Datta and Gest (1964) had found that a single aspartokinase of *Rhodopseudo-monas capsulatus* was subject to concerted feedback inhibition by its end-products and a similar system was described for *Bacillus polymyxa* by Paulus and Gray (1964).

The questions to be asked about *Pseudomonas* species were whether (1) single enzymes or several isoenzymes occurred at the branch points, (2) the specificity of inhibition and repression controls and (3) whether any of the enzymes were multifunctional. The survey by Cohen and his associates (1969) examined these points for representative strains of several species. They found that three strains of *Aeromonas* species possessed isoenzymes for aspartokinase and thus had regulatory systems resembling those of entero-bacteria. All the *Pseudomonas* species had a single aspartokinase regulated by concerted feedback inhibition by lysine and threonine. In cell extracts the aspartokinases of *P. aeruginosa*, *P. putida* and *P. fluorescens* were markedly inhibited when both lysine and threonine were present at a concentration of 3×10^{-4} M, but at higher concentrations each amino acid alone could produce up to 90 % inhibition in some strains. These findings show how important it is to test potential feedback inhibitors over a range of concentra-tions. The aspartokinase activities of *P. acidovorans* and *P. testosteroni* were rather less sensitive to inhibition by threonine and lysine when added singly, but together they produced concerted feedback inhibition at 10^{-3} M. Strains belonging to *P. cepacia*, *P. stutzeri* and *P. methanica* behaved like the fluores-cent group with respect to lysine and theonine inhibition. One of the consequences of having a single aspartokinase regulated by concerted feed-back inhibition by lysine and threonine is that there may not be enough residual activity to allow methionine synthesis. Datta and Gest (1964) had shown that growth of *Rhodopseudomonas capsulatus* is severely inhibited by threonine plus lysine and that this growth inhibition could be relieved by methionine. Cohen and his coworkers (1969) found that growth of *P. acidovorans* was inhibited by lysine plus threonine and that this effect, thought to be due to concerted inhibition of aspartokinase, was reversed by methionine.

The homoserine dehydrogenase activities of all the *Pseudomonas* species were inhibited by threonine although the *P. testosteroni* strains were less sensitive to inhibition than the other species. The acidovorans group specifically require NAD as the coenzyme for serine dehydrogenase although the enzymes from the other species are more active with NADP and can also use NAD. The growth inhibition of *P. acidovorans* by threonine alone, would seem to be due mainly to inhibition of homoserine dehydrogenase activity. Partial purification of the enzymes from *P. acidovorans* and *P. testosteroni* indicated that these are separate and distinct enzymes and that multi-functional proteins carrying both aspartokinase and homoserine dehydro-genase activities do not occur in these species. This was confirmed with a purified preparation of aspartokinase from *P. putida* by Robert-Gero, Poiret and Cohen (1970). The enzyme was partially inhibited by either threonine or lysine, added alone, and complete inhibition was obtained by a combination of 30 μm threonine and 400 μm lysine.

There is little experimental evidence for repression of the enzymes of this pathway but Robert-Gero, Poiret and Cohen (1970) found that when *P. putida* was grown in a minimal medium with glutamate as the carbon source the synthesis of aspartokinase was repressed by the addition to the medium of lysine, threonine or methionine. The most severe repression was produced by methionine and the residual aspartokinase synthesized under these conditions was insensitive to inhibition by threonine and lysine. The pure enzyme is known to lose its sensitivity to these inhibitors during storage, so that the aspartokinase produced under repressing conditions may be in a conformational state which masks the binding sites for the inhibitor amino acids.

4.d. Methionine

The growth reponse of auxotrophic mutants to compounds known to be involved in methionine biosynthesis in other organisms indicated that the biosynthetic pathway is probably the same as that in *Escherichia coli* (Calhoun & Feary, 1969). Cross-feeding among the methionine auxotrophs was detectable but weak, suggesting that only small quantities of inter-mediates are accumulated and excreted. The methionine-requiring mutants comprised four unlinked transduction groups and the cysteine-requiring mutants fell into at least three groups. The pathway is shown in Figure 8.18. The gene designations are those established for *E. coli* but the data of Calhoun and Feary (1969) would suggest that the same gene–enzyme relationships occur in *P. aeruginosa*. Methionine is reported to be utilized by *P. aeruginosa* as a nitrogen source but not a carbon source (Kay & Gronlund, 1969b).

4.e. Threonine

Stanier and his colleagues (1966) found that all strains of *P. cepacia* (*multi-vorans*) could utilize threonine as a carbon source for growth although none of the fluorescent species could do so. The threonine catabolic pathway may be the same as that reported for *Arthrobacter* (McGilvray & Morris, 1969). In this organism threonine is oxidized to α-amino-β-ketobutyrate which then undergoes CoA-dependent cleavage to give acetyl-CoA and glycine. Lessie and Whiteley (1969) showed that growth of *P. cepacia* on threonine induced the synthesis of threonine dehydrogenase and the addition of citrate resulted in repression. These authors demonstrated convincingly that threonine deaminase has no catabolic function since mutants lacking threonine

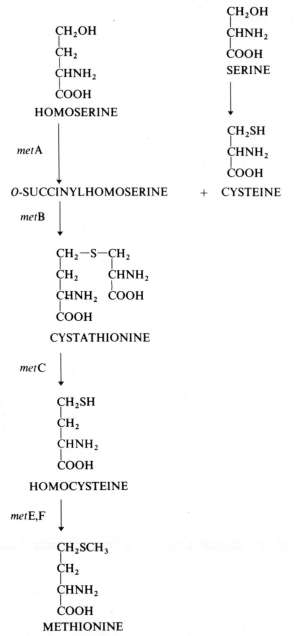

The growth response of methionine auxotrophs of *P. aeruginosa* corresponded to defects in *met* genes as above. Mutants could be assigned to 4 transduction groups.

Figure 8.18. Pathway for the biosynthesis of methionine. Data from Calhoun and Feary (1969)

deaminase grew normally on threonine, provided that the cultures were supplemented with isoleucine, while strains lacking threonine dehydrogenase were completely unable to grow on threonine.

4.f. Lysine

Dihydropicolinate synthetase, the first enzyme of the lysine branch of the pathway, is inhibited by lysine in strains of *P. putida* and *P. acidovorans* (Hermann *et al.*, 1972). Mutants of *P. acidovorans* with a dihydropicolinate synthetase insensitive to lysine inhibition, and double mutants which also had a desensitized aspartokinase, showed lack of control of lysine synthesis and excreted lysine into the culture medium. Mutants of *P. aeruginosa* requiring lysine for growth, and lacking either diaminopimelate decarboxylase or epimerase, accumulated diaminopimelate in the growth medium (Clarkson & Meadow, 1971) which would suggest that similar feedback controls occur in *P. aeruginosa*.

The pathway for L-lysine degradation in *P. putida* is shown in Figure 8.19. The inducer of at least the first two enzymes, lysine oxygenase and 5-aminovaleramidase, is L-lysine and not as for some other amino acid catabolic pathways one of the early intermediates. Lysine oxygenase was not repressed by various carbon compounds tested, but was activated by its substrate L-lysine and inhibited by various ions and several carbon compounds particularly citrate (Vandecasteele & Hermann, 1972). As a consequence, the lysine catabolic enzymes are only synthesized when exogenous lysine is present as inducer, or endogenous lysine is formed as a result of the breakdown of the feedback control on biosynthesis. Further, lysine oxygenase is only fully active when the L-lysine concentration reaches a high enough level. This would suggest that the characteristics of the two pathways are such that the rate of biosynthesis of lysine is not regulated by its catabolic pathway. On the other hand, if the regulation of biosynthesis breaks down, *P. putida* can cannibalize the excess lysine while *P. acidovorans*, which does not have a lysine degradative pathway, loses lysine to the environment. From the point of view of the experimenter the existence of active catabolic pathways adds to the difficulties of elucidating controls of biosynthetic pathways.

A second catabolic pathway functions in the degradation of D-lysine *via* pipecolate (Figure 8.19). A strain of *P. putida*, which was isolated on pipecolate, was induced for lysine degradation by growth on pipecolate. Inhibition of lysine racemase showed that the immediate substrate for this pathway was D-lysine and not L-lysine (Miller & Rodwell, 1971). Both pathways were present in anotner strain of *P. putida*, studied by Chang and Adams (1971) who showed that a mutant lacking glutamate:5-aminovalerate transaminase was totally unable to grow on L-lysine. There was some induction of the D-pathway enzymes, but they suggest that the activity of the

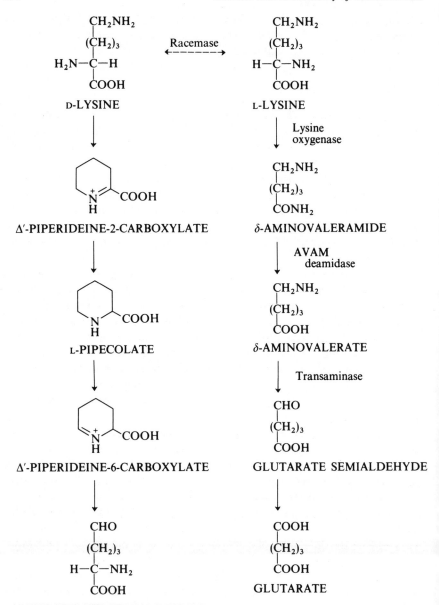

Δ'-PIPERIDEINE-2-CARBOXYLATE   δ-AMINOVALERAMIDE

L-PIPECOLATE   δ-AMINOVALERATE

Δ'-PIPERIDEINE-6-CARBOXYLATE   GLUTARATE SEMIALDEHYDE

                                     GLUTARATE

α-AMINOADIPATE SEMIALDEHYDE

Note that D-lysine is metabolized by the 'cyclic' pathway and L-lysine by the lysine oxygenase pathway. Lysine racemase allows some interconversion of the lysine isomers.

Figure 8.19. Catabolic pathways for lysine degradation in *P. putida*. Data from Miller and Rodwell (1971) and Chang and Adams (1971)

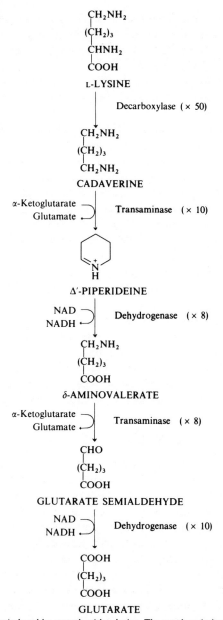

The enzymes are induced by growth with L-lysine. The numbers in brackets show the increased activities of induced cultures compared with control cultures grown with glucose.

Figure 8.20. Catabolic pathway for lysine degradation in *P. aeruginosa*. (Guest & Fothergill, unpublished)

lysine racemase was insufficient for the L-lysine to be channelled into the D-lysine pathway. The existence of catabolic pathways for both L-lysine and D-lysine, and the link between them provided by lysine racemase, gives a metabolic potential for switching to the alternative pathway under strong selection pressure. However, in the strains so far described L-lysine is catabolized *via* 5-aminovaleramide and D-lysine *via* the other pathway involving pipecolate.

Guest & Fothergill (unpublished) have obtained evidence for another pathway of L-lysine degradation in *P. aeruginosa* strain PAC1. Figure 8.20 shows the enzyme reactions which decarboxylate L-lysine to produce cadaverine and eventually convert the intermediates to glutarate. The wild-type strain PAC1 grows very poorly on lysine and this is thought to be due to a defect in the lysine decarboxylase activity, the first enzyme of the pathway. Mutants which grew well on lysine were readily obtained and thus the lysine catabolic pathway, like the histidine catabolic pathway, appears to be latent in this strain. The enzymes are all inducible and the activities were 8-fold to 50-fold greater for cultures grown on lysine as carbon source compared with those grown on succinate. Guest and Fothergill could not detect any L-lysine oxygenase, L-lysine oxidase or the enzymes of the pipe-colate pathway in this strain of *P. aeruginosa*. However, both the decarboxy-lase and the oxygenase pathway enzymes were present in a *P. fluorescens* strain, while in strains of *P. putida* the oxygenase pathway was present but not the decarboxylase pathway. Investigation of other strains of these species may confirm these findings of species specificity in the catabolic routes employed by *Pseudomonas* strains for the catabolism of lysine.

4.g. Isoleucine, Leucine and Valine

The first enzyme of the isoleucine biosynthetic pathway is threonine deaminase. This enzyme is required specifically for isoleucine biosynthesis and in *P. aeruginosa*, as in *E. coli*, it is sensitive to feedback inhibition by isoleucine. The later enzymes of the isoleucine biosynthetic pathway are required for the biosynthesis of valine and leucine (Figure 8.21) and also pantothenate. Marinus and Loutit (1969b) showed that threonine deaminase was inhibited by isoleucine and that this could be reversed by valine. Acetohydroxy acid synthetase, the first enzyme of the common pathway was inhibited by valine and to a lesser extent by isoleucine and leucine. This pattern of feedback inhibition also occurs in *Acinetobacter* (Twarog, 1972).

The synthesis of the enzymes leading from threonine to isoleucine is regulated in *E. coli* by multivalent repression by the four end-products isoleucine, leucine, valine and pantothenate. There is some evidence that amino acyl-*t*-RNA may be involved and Hatfield and Burns (1970) have proposed a model in which leucyl-*t*-RNA, bound to an immature form of

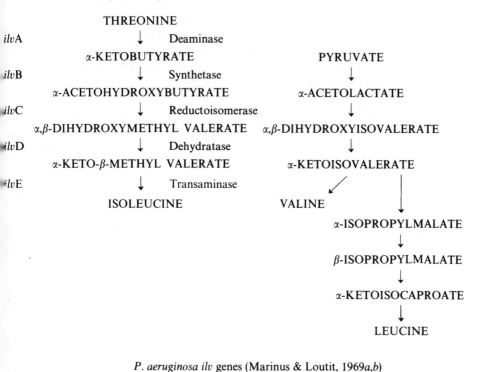

P. *aeruginosa ilv* genes (Marinus & Loutit, 1969a,b)

Figure 8.21. Pathway for the biosynthesis of the branched-chain amino acids, isoleucine and valine

threonine deaminase, may act as a repressor of these enzymes. In *E. coli* the genes for threonine deaminase *ilv*A, α-hydroxy acid synthetase *ilv*B, aceto-hydroxy acid isomeroreductase *ilv*C, dihydroxy acid dehydrase *ilv*D, and transaminase B *ilv*E, form a cluster comprising two or three operons with common regulation. Marinus and Loutit (1969a) found that the *ilv*B and *ilv*C loci in *P. aeruginosa* mapped closely together but that *ilv*D mapped at a site about 25 minutes away. No mutants were obtained for *ilv*A or *ilv*E. Acetohydroxy acid synthetase and the reductoisomerase responded to multivalent repression by isoleucine, leucine, valine and pantothenate but the other enzymes in the pathway were unaffected. Acetohydroxy acid synthetase and reductoisomerase were synthesized coordinately and certain regulator mutations affected both enzymes. These findings suggest that *ilv*B-*ilv*C may form an operon under multivalent repression. However a report from Horvath and colleagues (Horvath, Varga & Szentirmai, 1964) stated that cultures grown in the presence of isoleucine had higher levels of aceto-hydroxy acid synthetase and dihydroxy acid synthetase and suggested that isoleucine acted as an inducer for these enzymes.

The regulation of the interlocking system of enzymes for the biosynthesis of the branched-chain amino acids is complex and the same might be said of the catabolic enzymes. As shown in Figure 8.22 from Marshall and Sokatch (1972) *P. aeruginosa* and *P. putida* convert D-valine to 2-ketoisovalerate by the activity of D-amino acid dehydrogenase which is also induced by growth on L-valine, L-leucine, or L-isoleucine and has a fairly broad specificity. In *P. putida*, but not *P. aeruginosa*, much higher levels of this enzyme were obtained after growth on D-valine, suggesting a lower specificity of induction in *P. putida*. On the other hand, L-valine is a substrate for the constitutive branched-chain transaminase which probably also functions in biosynthesis. The enzyme of step 3, (Figure 8.22) isobutyryl-CoA dehydrogenase, was also constitutive. The other enzymes measured were inducible but were regulated

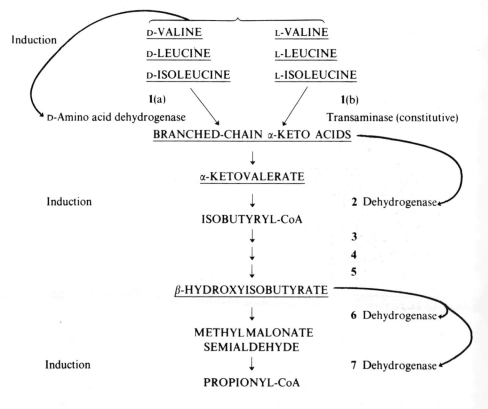

Compounds acting as inducers are underlined and arrows indicate the enzymes induced.
Figure 8.22. Pathway for the catabolism of the branched-chain amino acids (Marshall & Sokatch, 1972)

in different ways. The enzyme for step 2 (Figure 8.22), the branched-chain α-keto acid dehydrogenase also attacks 2-keto-3-methyl valerate and 2-ketoisocaproate, and appears to be induced by any of these keto acids. The results suggested a common pathway for the first two steps of the degradation of all three branched-chain amino acids.

The enzymes for steps 6 and 7 (Figure 8.22) in valine degradation, 3-hydroxyisobutyrate dehydrogenase and methyl malonate semialdehyde dehydrogenase, were synthesized coordinately and could be induced by growth on valine, isobutyrate or 3-hydroxyisobutyrate, but not by growth on isoleucine or leucine. These two enzymes therefore appear to form a separate regulation group and to be produced in valine medium as a result of sequential induction by an intermediate of the pathway. Figure 8.22 summarizes the regulatory patterns suggested by Marshall and Sokatch (1972).

4.h. Proline and Hydroxyproline

Proline-requiring auxotrophs have been isolated from *P. aeruginosa* and proline markers have been located on the chromosome at 3 min (*pro*-71), 28 min (*pro*-70) and 40 min (*pro*-73) (Pemberton and Holloway, 1972) but the enzyme defects have not been identified (see Chapter 5). Mutants resistant to growth inhibition by 3,4-dehydroproline have been isolated by Potts (unpublished). In *E. coli* this analogue (and azetidine-2-carboxylic acid which is less effective on *P. aeruginosa*) mimics proline in effecting feedback inhibition of one of the early enzymes (Tristram & Thurston, 1966). The proline biosynthetic pathway as established in *E. coli* is shown in Figure 8.23.

Proline is utilized as a carbon source for growth by *P. aeruginosa, P. putida, P. acidovorans, P. testosteroni* and *P. cepacia* (Stanier *et al.*, 1966). The catabolic pathway is presumed to be the same as that established for other bacteria (Figure 8.24). Δ'-Pyrroline-5-carboxylic acid is an intermediate of both biosynthetic and degradative pathways and very few steps separate proline from glutamate which is both the biosynthetic precursor and the catabolic product. Kay and Gronlund (1969b) found that proline induced proline permease and the enzymes of the catabolic pathway. Glutamate permease was also induced in cells grown on proline but it was not induced in a *put⁻* mutant so that it is likely that glutamate is the real inducer of its own permease. Some of the mutants resistant to 3,4-dehydroproline isolated by Potts were unable to grow on proline and may have been permease mutants.

P. putida is also able to utilize hydroxyproline by the pathway shown in Figure 8.25. The four enzymes of the pathway are induced by hydroxy-L-proline and by the first intermediate, allohydroxy-D-proline. These two compounds also induce the synthesis of a transport system for hydroxy-proline. The final product of the pathway is α-ketoglutarate which enters the TCA cycle (Gryder & Adams, 1969).

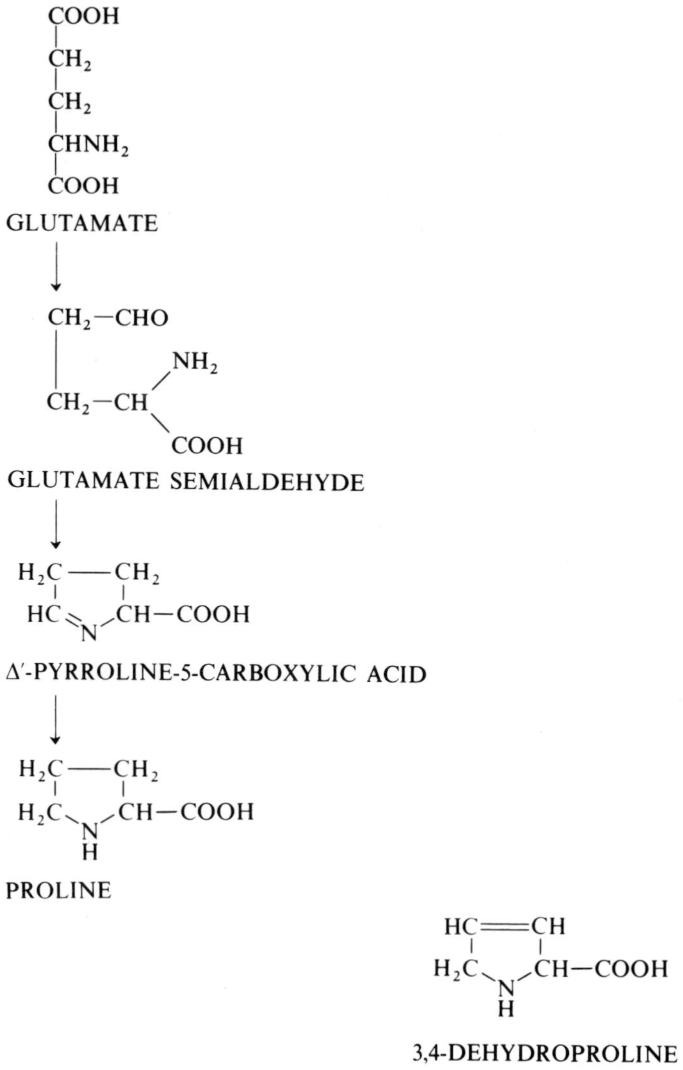

COOH
|
CH$_2$
|
CH$_2$
|
CHNH$_2$
|
COOH

GLUTAMATE

↓

CH$_2$—CHO
|
CH$_2$—CH ⟨ NH$_2$ / COOH

GLUTAMATE SEMIALDEHYDE

↓

H$_2$C——CH$_2$
| |
HC≈N CH—COOH

Δ′-PYRROLINE-5-CARBOXYLIC ACID

↓

H$_2$C——CH$_2$
| |
H$_2$C—N(H) CH—COOH

PROLINE

HC══CH
| |
H$_2$C—N(H) CH—COOH

3,4-DEHYDROPROLINE

Figure 8.23. The biosynthetic pathway for proline

PROLINE
↓ Proline oxidase
Δ′-PYRROLINE-5-CARBOXYLIC ACID (5-PC)
↓ 5-PC Dehydrogenase
GLUTAMATE

Figure 8.24. The catabolic pathway for proline in *Pseudomonas* is probably the same as that for *E. coli* shown above

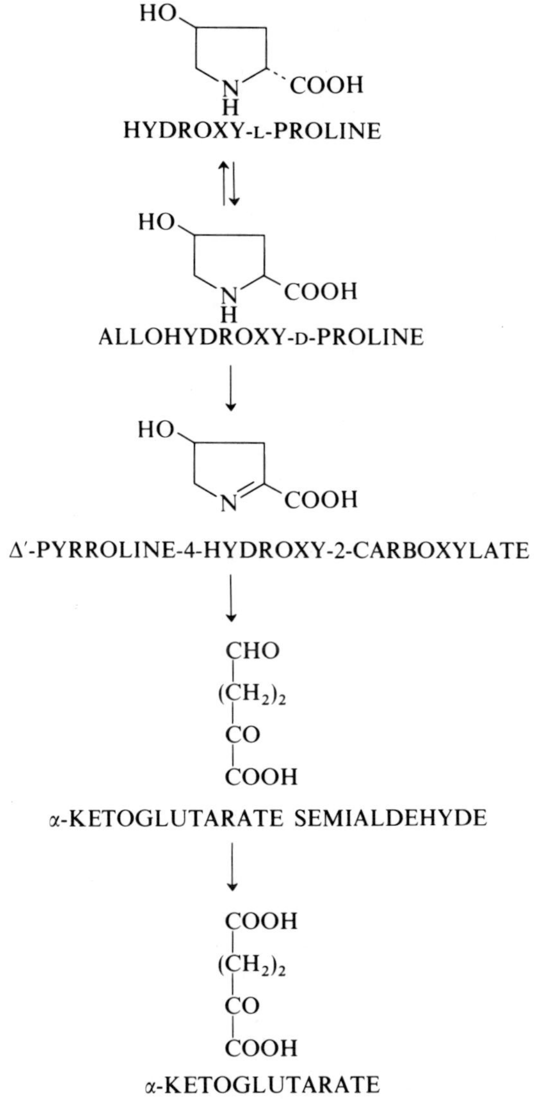

HYDROXY-L-PROLINE

ALLOHYDROXY-D-PROLINE

Δ'-PYRROLINE-4-HYDROXY-2-CARBOXYLATE

$$
\begin{array}{c}
CHO \\
| \\
(CH_2)_2 \\
| \\
CO \\
| \\
COOH
\end{array}
$$

α-KETOGLUTARATE SEMIALDEHYDE

$$
\begin{array}{c}
COOH \\
| \\
(CH_2)_2 \\
| \\
CO \\
| \\
COOH
\end{array}
$$

α-KETOGLUTARATE

Figure 8.25. The catabolic pathway for the degradation of hydroxyproline in *P. putida* (Gryder & Adams, 1969)

4.i. Arginine

Arginine is synthesized from glutamate by the sequence of reactions shown in Figure 8.26. *Pseudomonas* species carry out step 5, from *N*-acetylornithine to ornithine, by a transacetylation reaction with glutamate which results in the production of ornithine and *N*-acetylglutamate (Udaka, 1966). This

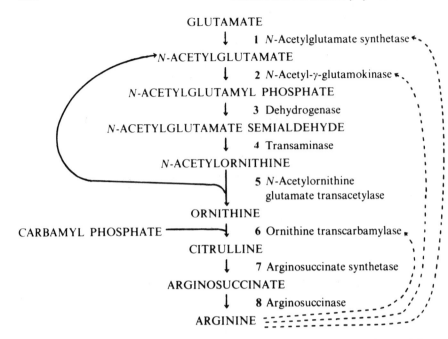

Enzyme 1 and enzyme 2 are inhibited by arginine. Enzyme 6 is repressed by arginine.

Figure 8.26. The biosynthetic pathway for arginine in *P. aeruginosa* (Isaac & Holloway, 1972; Leisinger *et al.*, 1972)

reaction also occurs in *Micrococcus glutamicus* as well as in *Saccharomyces, Neurospora* and *Chlamydomonas*. In the Enterobacteriaceae and *Bacillus* species, ornithine is formed by the action of a specific *N*-acetylornithine deacylase. In *Escherichia coli* the first enzyme of the pathway, *N*-acetylglutamate synthetase, is subject to feedback inhibition by arginine and this determines the rate of synthesis of arginine. The reaction in which *N*-acetylglutamate is formed, at the expense of the deacetylation of *N*-acetylornithine, effectively bypasses enzyme 1. Udaka (1966) found that organisms which possess ornithine transacetylase have another feedback inhibition site. Enzyme 2, *N*-acetylglutamate 5-phosphotransferase, is subject to feedback inhibition by arginine only in organisms which possess ornithine transacetylase and not in those which possess *N*-acetylornithine deacylase, Isaac and Holloway (1972) showed that both arginine and citrulline inhibited the activity of enzyme 2 in cell-free extracts of *P. aeruginosa* and that ornithine also inhibited to some extent. Inhibition by arginine of *N*-acetylglutamate 5-phosphotransferase was confirmed by Leisinger, Haas and Hegarty (1972) who used a highly purified preparation of this enzyme. Udaka (1966) had pointed out that the bacteria which possess ornithine transacetylase might

Table 8.6. L-Arginine and L-lysine analogues and effects on *Pseudomonas* species

Compound	References	Effect
Canavanine	1, 2	Weak inhibition of growth of *P. aeruginosa*
Indospicine (1,2-amino-6-amidinohexanoate)	2, 5	Inhibits growth and both *N*-acetylglutamate synthetase and *N*-acetylglutamate 5-phosphotransferase in *P. aeruginosa*
Amino ethyl cysteine	3	Inhibits growth of *P. putida*. Resistant mutants with aspartokinase insensitive to inhibition by theonine and lysine
	3	Inhibits growth of *P. acidovorans*. Resistant mutants with dihydropicolinate synthetase resistant to lysine inhibition
D-Arginine, D-lysine, D-ornithine, L-homoarginine, L-α-amino-γ-guanidinobutyrate	4	Inhibit arginine and lysine common transport system in lysine-induced cells

1. Kay & Gronlund (1969d). 2. Leisinger *et al.* (1972).
3. Hermann *et al.* (1972). 4. Fan *et al.* (1971).
5. Haas, Kurer & Leisinger (1972).

be a more highly evolved group, since this step would enable them to synthesize arginine with less expenditure of energy. They would require enzyme 1 only to start the sequence of reactions since thereafter *N*-acetylglutamate would be produced by the transacetylation cycle. Haas, Kurer and Leisinger (1972) have purified enzyme 1 from *P. aeruginosa* and they found that this enzyme was also inhibited by arginine so that a very tight control by feedback inhibition operates on the rate of operation of this pathway. However, the significance of enzyme 1 is not completely clear since Chou and Gunsalus (1971) state that it is not required for the growth of *P. putida*.

The arginine analogue, canavanine, inhibits growth of *E. coli* in a complete minimal medium but is much less effective against *P. aeruginosa* and can be used as a carbon source for growth. On the other hand, indospicine (1,2-amino-6-amidinohexanoic acid) is more effective as a growth inhibitor for *P. aeruginosa* and the inhibition is more marked when the culture is growing slowly with ornithine as the carbon source than when it is grown in a glucose medium. Indospicine interferes with incorporation of arginine into proteins and also inhibits the feedback-sensitive enzyme 2 (Leisinger *et al.*, 1972) (Table 8.6).

Arginine auxotrophic mutants of *P. aeruginosa* could be separated by transductional analysis into seven separate linkage groups (Feary *et al.*, 1969) indicating that the genes were fairly widely scattered around the chromosome. In *Escherichia coli* the nine genes for this pathway do not form a tight cluster but comprise six groups one of which includes the structural genes for four of the enzymes (Gorini, Gundersen & Burger, 1961). The genes for arginine biosynthesis in *E. coli* are under the control of a single regulator gene *arg*R and in strain K12 all the enzymes are repressed by arginine. It was a very interesting observation that arginine had a slight inductive effect on some of the biosynthetic enzymes of strain B. The *arg*R genes of these two strains are allelic and a mutation in the regulator gene of strain B gave mutants which were repressible by arginine (Jacoby & Gorini, 1967) Isaac and Holloway (1972) found little variation in the levels of enzymes 2, 5 and 8 (*N*-acetylglutamate 5-phosphotransferase, *N*-acetylornithine glutamate transacetylase and arginosuccinase) when *P. aeruginosa* was grown under conditions of deprivation or excess of arginine. Ornithine transcarbamylase was the only enzyme which changed significantly in amount and was de-repressed during exponential growth in minimal medium and severely repressed when arginine was added to the medium. The derepressed level of ornithine transcarbamylase was no higher in arginine bradytrophs than in the wild type. Voellmy and Leisinger (1972) obtained 50-fold repression of ornithine transcarbamylase and 2-fold repression of *N*-acetylglutamate semialdehyde dehydrogenase by arginine in batch culture, and could observe repression or depression of these two enzymes when a double auxotroph was grown either with limiting leucine or with limiting arginine. *N*-Acetylornithine transaminase was induced 15-fold by arginine, while there was little or no change in the specific activities of the other enzymes of the pathway.

Many *Pseudomonas* species can utilize arginine as the sole carbon and nitrogen source for growth. The initial reactions of arginine degradation in *P. aeruginosa* and *P. fluorescens* are given in Figure 8.27 and involve three enzymes, arginine deiminase, ornithine transcarbamylase and carbamate kinase (Stalon *et al.*, 1972). These reactions also occur in *Streptococcus faecalis* and *Clostridium welchii* and provide for ATP synthesis by an

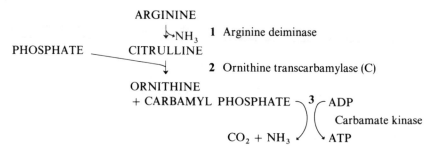

Figure 8.27. Arginine catabolic pathway in *P. aeruginosa* and *P. fluorescens* (Stalon *et al.*, 1967: 1972)

anaerobic process. Other microorganisms degrade arginine by reactions involving arginase and urease in a manner analogous to the urea cycle of animals. Wiame and his colleagues have shown that the catabolic ornithine transcarbamylases (C) of *P. aeruginosa* and *P. fluorescens* differ from the enzymes operating in the biosynthesis of arginine. OTCase (C) has a neutral pH optimum while the biosynthetic enzyme OTCase (B) has a broad optimum around pH 8·5. The catabolic OTCase of *P. aeruginosa* is induced by arginine and the biosynthetic enzyme is repressed, but a strain of *P. fluorescens* isolated from a citrulline enrichment culture was constitutive for all three catabolic enzymes (Stalon *et al.*, 1967). Ornithine trans-carbamylase (C) is subject to catabolite repression coordinately with the other enzymes and maximum yields were obtained at the end of exponential growth on a limiting amount of citrate.

The catabolic enzyme cannot function *in vivo* for citrulline synthesis; a mutant which lacked the biosynthetic enzyme, although it still produced the catabolic enzyme, became auxotrophic for arginine. Equally, the bio-synthetic enzyme is unable to function *in vivo* for the phosphorolysis of citrulline. It is clear that these two enzymes have become highly specialized in function and at one time it was thought that this might be explained on the basis of compartmentalization. Stalon and his colleagues (1972) have now established that these two enzymes work in unidirectional and opposite ways and that this is determined solely by the regulatory properties of the enzyme proteins. The reaction in the direction of citrulline synthesis can be measured *in vitro* with both enzymes but this has physiological significance only for the biosynthetic enzyme (Figure 8.28). A detailed study of the enzyme kinetics showed that the catabolic enzyme exhibits very high cooperativity for carbamoyl phosphate and this rules out an anabolic function for this enzyme with carbamoyl phosphate as a substrate. The biosynthetic enzyme is saturated with carbamoyl phosphate at concentrations at which the catabolic enzyme has negligible activity. Further, phosphate which is one of the substrates of the catabolic enzyme in its physiological direction is an activator and ATP, a product of the catabolic reaction, reduces enzyme activity. The importance of these regulatory controls in the *in vivo* control of arginine metabolism was confirmed by the isolation of a mutant which was able to use its catabolic OTCase to synthesize citrulline *in vivo*. This mutant was derived from a mutant lacking OTCase (B) and had an arginine require-ment for growth. Mutation to arginine independence in this case involved an alteration in ornithine transcarbamylase (C). The altered OTCase (C) exhibited reduced cooperatively towards carbamoyl phosphate compared with the wild-type enzyme and was able to function at a low rate as an ana-bolic enzyme. The mutant OTCase (C) had the same molecular weight as the wild-type OTCase (C), 400,000 daltons, whereas the anabolic OTCase (B) had a molecular weight of 100,000 daltons. *Halobacterium salinarium* also

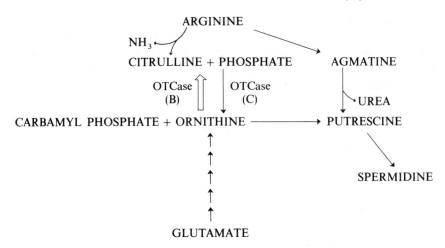

The catabolic ornithine transcarbamylase is induced by arginine (OTCase (C)) and the biosynthetic enzyme is repressed (OTCase (B)). Agmatine and putrescine inhibit OTCase (C).

Figure 8.28. Interactions of the arginine biosynthetic and catabolic pathways in *P. aeruginosa* and *P. fluorescens* (Stalon *et al.*, 1972)

degrades arginine by this route but possesses only one ornithine trans-carbamylase. *Bacillus licheniformis* on the other hand appears to have two OTCases but also has arginase, urease and pyrroline dehydrogenase so would not be expected to need a catabolic OTCase (Laishley & Bernlohr, 1968).

Miller and Rodwell (1971) found that another degradative pathway for arginine occurred in *P. putida*. After an initial reaction in which arginine is deaminated to produce α-ketoarginine, the pathway proceeds *via* decarboxy-lation to produce γ-guanidinobutyrate which then loses urea to give γ-aminobutyrate (Figure 8.29). These enzymes were induced by growth on arginine and the synthesis of the first enzymes appeared to be coordinate.

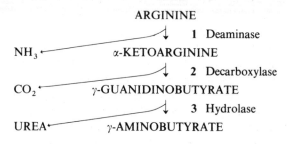

Figure 8.29. Arginine catabolic pathway in *P. putida* (Rodwell & Miller, 1971)

Induction of a specific arginine transport system preceded the induction of the arginine catabolic enzymes. The enzyme for step 3, γ-guanidinobutyrate amidinohydrolase, was induced by growth on arginine or γ-guanidino-butyrate, but not by γ-aminobutyrate or by urea plus malate. It is thus induced by its substrate and not by either of the products and is regulated separately from the two earlier enzymes. The amidinohydrolase was purified by Chou and Rodwell (1972) and shown to have a high specificity; the only other compounds which were hydrolysed were δ-guanidinovalerate and ε-guanidinocaproate.

Pseudomonads can attack the naturally occurring polyamines and diamines (Stanier *et al.*, 1966) Razin and his associates (1959) found that *P. aeruginosa* oxidized spermine, spermidine, putrescine and cadaverine by means of inducible enzymes and Jakoby and Fredericks (1959) isolated a fluorescent pseudomonad which grew on putrescine or pyrrolidine as sole carbon and nitrogen source. Putrescine can be formed by the decarboxylation of ornithine, or from arginine *via* agmatine (Figure 8.28). Putrescine can therefore be regarded as intermediate of the arginine catabolic pathway and it is interesting to note that Stalon and coworkers (1972) observed that both putrescine and agmatine inhibited the catabolic ornithine trans-carbamylase. However, the polyamines appear to have an essential role in bacterial growth (Rubenstein *et al.*, 1972) and these reactions may have a biosynthetic as well as a catabolic importance. Mass (1972) has mapped genes for spermidine synthesis in *Escherichia coli.* Spermidine is synthesized from putrescine formed either from ornithine, or from arginine *via* agmatine but in the presence of exogenous arginine the synthesis of ornithine is prevented by feedback inhibition and repression. Mutants were isolated. which required putrescine for growth in the presence of arginine. Some of the mutants had enzyme defects for arginine decarboxylase and others for agmatine ureohydrolase.

4.j. Histidine

The biosynthesis of histidine is presumed to follow the same pathway as in other microorganisms. This biosynthetic pathway has been studied in most detail in *Salmonella typhimurium*; a single operon comprising nine structural genes determines the ten enzyme activities of the pathway (Hartman *et al.*, 1971, Martin *et al.*, 1971) (Figure 8.30). The histidine operon is controlled by feedback inhibition by histidine of the first enzyme of the pathway, and by repression of synthesis of all the enzymes. The regulatory control exercised by several genes and mutations in any one of these may give derepressed mutants. Charged histidinyl-*t*-RNA, rather than histidine itself, appears to be responsible for repression and the first enzyme may be involved in repression (Kovach *et al.*, 1969; Rothman-Dienes & Martin, 1971; Singer & Smith, 1972).

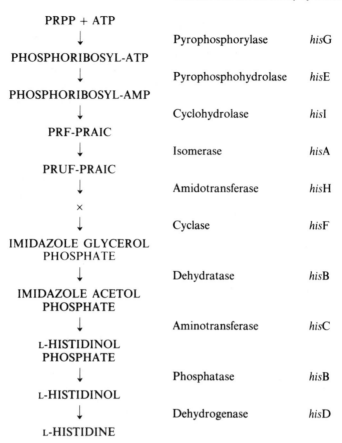

PRPP + ATP		
↓	Pyrophosphorylase	*his*G
PHOSPHORIBOSYL-ATP		
↓	Pyrophosphohydrolase	*his*E
PHOSPHORIBOSYL-AMP		
↓	Cyclohydrolase	*his*I
PRF-PRAIC		
↓	Isomerase	*his*A
PRUF-PRAIC		
↓	Amidotransferase	*his*H
×		
↓	Cyclase	*his*F
IMIDAZOLE GLYCEROL PHOSPHATE		
↓	Dehydratase	*his*B
IMIDAZOLE ACETOL PHOSPHATE		
↓	Aminotransferase	*his*C
L-HISTIDINOL PHOSPHATE		
↓	Phosphatase	*his*B
L-HISTIDINOL		
↓	Dehydrogenase	*his*D
L-HISTIDINE		

P. aeruginosa his genes are found in 5 separate transduction linkage groups (Mee & Lee, 1967).

PRPP = 5-phosphoribosylpyrophosphate, PRF-PRAIC = *N*-(5'-phosphoribosylform-imino)-5-amino-1-(5''-phosphoribosyl)imidazole-4-carboxamide, PRUF-PRAIC = *N*-(5'-phosphoribulosylformimino)-5-amino-1-(5''-phosphoribosyl)imidazole-4-carbox-amide.

Figure 8.30. Pathway for the biosynthesis of histidine established in *Salmonella typhimurium* and presumed to occur in *Pseudomonas*

This tight clustering of histidine genes does not occur in *P. aeruginosa*. Mee and Lee (1967) found that 107 histidine auxotrophs could be placed in five separate transduction groups. Separation of the histidine biosynthetic genes into several clusters has also been recorded for *Streptomyces coelicolor*, *Bacillus subtilis*, *Saccharomyces* and *Neurospora* (Chapman & Nester, 1969). The genes and enzymes of the pathway have not been completely assigned to the five transduction groups of *P. aeruginosa* but Mee (1969) was able to make the following tentative allocations from assays for accumulation of intermediates, enzyme activities and response to histidine analogues.

Gene	Enzyme step	Enzyme
*his*I	(i) 10	histidinol dehydrogenase
	(ii) before 4	
*his*IIa	4, 5 or 6	
*his*IIb	7	IGP dehydrogenase
*his*III	4, 5 or 6	
*his*IV	4, 5 or 6	
*his*V	8 (or 9)	transaminase (phosphatase)

The finding that some, but not all, of the *his*I mutants responded to histidinol was interesting since it is known that in *Neurospora* the *his*-3 gene is responsible for the synthesis of a multifunctional protein catalysing three steps in the histidine biosynthetic pathway, reaction 2, 3 and 10 (Minson & Creaser, 1969) and a similar genetic situation is thought to exist in *Saccharomyces cerevisiae* and *Aspergillus nidulans*. It is possible that the *his*I region of *P. aeruginosa* also determines a multifunctional protein. Fine structure mapping (Mee & Lee, 1969) showed that mutations could be ordered into three groups; a central group associated with inability to grow on histidinol and two groups on either side of this region for which the ability to grow on histidinol was retained. The *his*I mutants which were incapable of growth on histidinol lacked histidinol dehydrogenase activity. No feedback inhibition-resistant or derepressed mutants of *P. aeruginosa* have been obtained and the regulation of the histidine biosynthetic enzymes of pseudomonads is not known. The histidine analogue 3-amino-1,2,4-triazole inhibits growth of the wild-type *P. aeruginos* and amino-triazole resistant mutants mapped in region *his*IIb with the gene for imidazole glycerol phosphate dehydratase (Mee, 1969). This is the step which is affected by the analogue in *S. typhimurium* (Table 8.7).

Mee (1969) attempted to map the 5 *his* loci on the *P. aeruginosa* chromosome by interrupted mating. Tentative assignments suggested that *his*III was an early marker, before 10 minutes; *his*I followed by *his*IV and *his*II entered between 10 and 20 minutes; *his*V entered very much later after 60 minutes. Linkage data suggested that *his*I and *his*V were well separated and *his*II,III and *his*IV might be closer together. Pemberton and Holloway (1972) located 3 separate chromosomal sites for a series of histidine auxotrophic markers, at 7, 10 and 13 minutes (see Chapter 5). One of the mutants *his*-5075, at the 13 minute site, was known to belong to transduction group *his*I. The exact location of the histidine biosynthetic genes on the chromosome remains to be elucidated.

The last enzyme of the histidine biosynthetic pathway is histidinol dehydrogenase. Organisms capable of utilizing histidinol for growth might be

Table 8.7. Histidine analogues and effects on *Pseudomonas* species

Compound	References	Effect
1,2,L-triazole-3-alanine (TRA)	1	Slight inhibition of *P. aeruginosa*. Good inhibition in presence of AMT
3-Amino-1,2,4-triazole (AMT)	1	Inhibits growth of *P. aeruginosa*. Mutants obtained from AMT + TRA plates resistant to AMT. Affects IGP dehydratase
Imidazolylpropionate (dihydrourocanate)	2, 4	Non-substrate inducer for histidase and urocanase in *P. aeruginosa*. Inhibits urocanase
	3	Growth substrate for *P. testosteroni*. Transported by same system as urocanate which induces transport system and dehydrogenase

1. Mee (1969).   2. Lessie & Niedhardt (1967).
3. Coote & Hassall (1973b).   4. George & Phillips (1970).

expected to dehydrogenate histidinol to histidine for metabolism *via* the histidine catabolic pathway. Dhawale, Creaser and Loper (1972) found that *Arthrobacter histidinolovorans* produced a catabolic histidinol dehydrogenase which was induced by growth in histidinol medium. It could be distinguished from a second histidinol dehydrogenase which was invariably present and presumed to have a biosynthetic role. The wild-type strain of *P. aeruginosa* PAO1 was unable to grow on histidinol but after mutagenic treatment a strain was isolated which could utilize histidinol as a carbon and nitrogen source. When this mutant was grown on histidinol the activity of histidinol dehydrogenase was 60-fold greater than for cultures grown in glucose + ammonium salts medium. Only a single histidinol dehydrogenase could be detected on polyacrylamide gels with extracts prepared from cultures grown either on glucose + ammonium salts or on histidinol. Both the mutant and the wild type appeared to have low basal levels of histidinol dehydrogenase which were sufficient to allow histidine biosyntheis. In minimal salt medium with histidinol as the carbon and nitrogen source, high levels of histidinol dehydrogenase were produced by the mutant corresponding to a catabolic role for the enzyme. The addition of glucose, histidine or glutamate to the histidinol medium repressed synthesis of the dehydrogenase. The nature of the mutation allowing strain PAO1 to grow on histidinol is not clear but it has resulted in an alteration in the regulation of synthesis of a biosynthetic enzyme to allow it to behave as a typical catabolic enzyme under induction and catabolite repression controls.

The degradative histidine pathways have been studied in *P. aeruginosa* (Lessie & Neidhardt, 1967b). *P. putida* (Leidigh & Wheelis, 1973) and *P. testosteroni* (Coote & Hassall, 1973a,b). The pathway for pseudomonads is shown in Figure 8.31. This is similar to that established for *Salmonella typhimurium* (Brill & Magasanik, 1969), *Bacillus subtillis* (Chasin & Magasanik, 1968) and *Klebsiella aerogenes* (Neidhardt & Magasanik, 1957) with the exception that *N*-formiminoglutamate is converted directly to glutamate and formamide by these latter species. Pseudomonads use two enzyme steps, reactions 4 and 5, to convert *N*-formiminoglutamate to formylglutamate and then to formate plus glutamate.

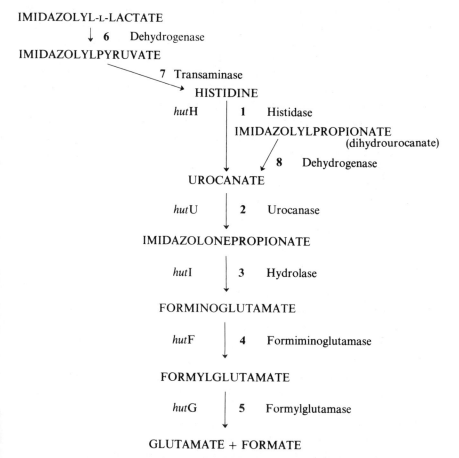

Steps 1 to 5 occur in *P. aeruginosa*, *P. putida* and *P. fluorescens*. Steps 6 to 8 have been described only in *P. testosteroni* (Lessie & Neidhardt, 1967; Coote & Hassall, 1973a,b).

P. putida, *hut* genes (Leidigh & Wheelis, 1973)

Figure 8.31. Catabolic pathway for histidine in *Pseudomonas*

Lessie and Neidhardt (1967b) showed that *P. aeruginosa* grown on histidine or urocanate had induced levels of the enzymes for steps 1, 2 and 4, histidase, urocanase, and formiminoglutamase. The induced levels were between 15-fold and 400-fold greater than those of succinate-grown cultures. Imidazolylpropionate (dihydrourocanate) was a gratuitous inducer of both histidase and urocanase. The three enzymes were all repressed by succinate, but formiminoglutamase was less severely repressed than the two earlier enzymes. Spontaneous revertants of some mutants unable to grow on histidine exhibited decreased sensitivity to catabolite repression by succinate and it is possible that some of these could have had promotor site mutations. A histidase-negative strain was induced by urocanate to synthesize urocanase, and this observation confirmed the view that urocanate rather than histidine is the inducer of the first two enzymes. Urocanate is also the inducer in *Salmonella typhimurium* and *Klebsiella aerogenes* although for *Bacillus subtilis* the inducer is histidine itself. Induction by urocanate requires the wild-type strain to be able to produce a low basal level of histidase and this appears to be so, since a urocanase-negative mutant grown on glycerol or succinate produced histidase and formiminoglutamase constitutively, although not at very high levels, presumably by the formation of small amounts of urocanate from endogenous histidine.

Hassall and his colleagues have isolated a number of non-fluorescent pseudomonads by enrichment culture on imidazolylpropionate or imidazolyllactate. The degradation of these compounds was studied in a strain identified as *P. testosteroni* (Coote & Hassall, 1973a). Apart from the initial reactions, these two compounds are metabolized by the usual histidine pathway enzymes. Imidazolyl-L-lactate is dehydrogenated to imidazolylpyruvate which is then transaminated to form histidine (reactions 6 and 7 in Figure 8.31). Imidazolylpropionate is dehydrogenated to urocanate. These imidazoyl derivatives are not utilized by *P. aeruginosa* or *P. putida* but it was not stated whether other strains of *P. testosteroni* or of *P. acidovorans* are able to use them for growth. Coote and Hassal (1973b) concluded that urocanate was the inducer for histidase, urocanase, histidine-2-ketoglutarate transaminase, N-formiminoglutamate dehydrogenase and the common transport system for urocanate and imidazolylpropionate. The enzymes for steps 1, 2 and 7 appeared to be coordinately induced but synthesis of the transaminase would be gratuitous unless imidazolyl-L-lactate or imidazolylpyruvate were present. The later enzymes for steps 4 and 5, formiminoglutamase and formylglutamase were induced both by urocanate and their substrates, and Coote and Hassall (1973b) suggested that there might be isoenzymes for these reactions. The dehydrogenase for step 8 was separately regulated and was induced by its substrate. A mutant blocked in the utilization of imdazolylpropionate produced only basal levels of histidase and urocanase and it was concluded that this compound is not able to induce the histidine pathway enzymes in *P. testosteroni*.

The genetics of the histidine degradative pathway has been examined in most detail in *P. putida*. Wheelis and Stanier (1970) found that the genes for histidine utilization were linked to genes for some of the enzymes of the aromatic catabolic pathways. Leidigh and Wheelis (1973) obtained mutants for most of the enzymes of the pathway and showed that they comprised a gene cluster which mapped between the aromatic pathway genes *pca*BDE and *pca*A and could be cotransduced with them. Constitutive mutations mapped in the same gene cluster. The genes and the enzymes affected by the mutations were:

Gene	Step	Enzyme	Inducer
*hut*H	1	histidase	urocanate
*hut*U	2	urocanase	urocanate
*hut*I	3	imidazolonepropionate hydrolase	
*hut*F	4	formiminoglutamase	
*hut*C	constitutivity		

Leidigh and Wheelis (1973) commented on the marked non-coordinancy of synthesis of the products of *hut*H, *hut*U, *hut*I and *hut*F. This contrasts with the findings of Coote and Hassall (1973b) who concluded that for *P. testosteroni* at least histidase and urocanase were coordinately regulated. The gene *hut*C in *P. putida* mapped at a site between *hut*U and *hut*H but Leidigh and Wheelis (1973) were unable to define its exact function. They suggest that it could be an operator site or specify a repressor and that other data are compatible with a negative control for these enzymes.

P. aeruginosa strain PAC1 is unlike most *P. aeruginosa* strains in that it is unable to grow with histidine as a carbon source although it can use it as a nitrogen source. Mutants can be isolated from PAC1 which are constitutive for histidase and urocanase, although still unable to utilize histidine as the sole growth substrate. Mutants which are *hut*[+] inducible can be isolated from PAC1 by spontaneous mutation and constitutive *hut*[+] mutants can be isolated from the strains that are constitutive for histidase and urocanase. The mutation to *hut*[+] in the PAC strain requires a mutation in a histidine transport system. The wild type and the various *hut*[-] constitutive mutants described above, can transport histidine at a rate sufficient for utilization as a nitrogen source but not sufficient to utilization as the sole carbon source for growth (J. R. Potts, personal communication).

Meiss, Brill and Magasanik (1969) found that although *Salmonella typhimurium* strain LT2 possessed the genes for histidase and urocanase, it was unable to utilize histidine as a nitrogen source. For this strain a mutation to allow histidine utilization was a rare event and may have involved a mutation at a promoter site.

The mechanism of catabolite repression of the histidine catabolic enzymes is still obscure. Lessie and Neidhardt (1967b) found with *P. aeruginosa* that repression of histidase was directly related to the growth rate, so that those compounds which supported a high growth rate produced most repression. Similar observations were made by Coote and Hassall (1972b) for *P. testosteroni* and they also found that when histidine was the sole nitrogen source there was no repression by succinate.

4.k. Aromatic Amino Acids

The biosynthesis of the aromatic amino acids, like the aspartate group, presents the problems of regulating the activities and rates of synthesis of enzymes of branched pathways. For the aromatic pathway it is again found that the solution reached during the evolution of *Escherichia coli*, and the other members of the Enterobacteriaceae, is for the production of iso-functional enzymes regulated independently by feedback inhibition by the final products, while *Pseudomonas* species have evolved regulatory systems for the control of single enzymes at branch points. The first and common step of the pathway is carried out by DAHP (3-deoxy-D-arabinoheptulosonate-7-phosphate) synthetase and for this step in *E. coli* there are two isoenzymes which are sensitive to feedback inhibition and repression by their respective end-products, phenylalanine and tyrosine, and a third isoenzyme which is repressed, but not inhibited, by methionine. Two isoenzymes regulated by phenylalanine and tyrosine occur at the second branch point for the con-version of chorismate to prephenate and tyrosine, phenylalanine and tryptophan also exercise feedback inhibition and repression control of the enzymes which are specifically required for their own biosynthesis.

Jensen, Nasser and Nester (1967) compared the control systems for DAHP synthetase in 32 different genera. The *Pseudomonas* species all appeared to have single enzymes and the predominant pattern was in-hibition by tyrosine with in some cases very much weaker inhibition by phenylalanine or tryptophan as well. *P. aeruginosa*, *P. putida*, *P. fluorescens*, *P. cepacia* and *P. stutzeri* strains behaved similarly but the DAHP synthetases of *P. acidovorans* and *P. testosteroni* were inhibited significantly by phenyl-alanine as well as by tyrosine. There was no concerted feedback inhibition and the inhibition produced by the two amino acids together was less than cumulative. *Aeromonas*, on the other hand, gave additive inhibition for phenylalanine and tyrosine which would be expected if there were two isoenzymes present independently regulated by these amino acids. These findings fit in with the observations of Cohen and his colleagues (1969) for the regulation of aspartokinase activity in *Aeromonas* and indicate that these organisms are not very closely related to *Pseudomonas*. The DAHP syn-thetase of the *Hydrogenomonas* species was inhibited by both phenylalanine and tyrosine in a similar way to that of *P. acidovorans*, but for *Hydrogeno-monas* the inhibition by the two amino acids together was cumulative.

Feedback inhibition of a common enzyme by any single one of the end-products of a branching biosynthetic pathway would allow total inhibition of growth by that product. However, tyrosine does not inhibit growth of the pseudomonads and this could be due to either very rapid removal of excess tyrosine by degradative systems or to a more complex form of feedback regulation than appeared at first sight. Jensen, Calhoun and Stenmark (1973) showed that DAHP synthetase from *P. aeruginosa* is sensitive to inhibition by tryptophan and phenylpyruvate in addition to tyrosine. A combination of the three inhibitors produced a cumulative, or less than cumulative, effect and a single enzyme carried the three inhibitor specificities. It was very interesting that phenylpyruvate, rather than phenylalanine, was the effector molecule, and Jensen, Calhoun and Stenmark (1973) suggest that it might be a more stable metabolite than phenylalanine itself. Inhibition by tyrosine or tryptophan appeared to be competitive with respect to phosphoenolpyruvate, and phenylpyruvate inhibition appeared to be competitive with respect to the other substrate of the enzyme, erythrose-4-phosphate. This system provides an elegant variation on the possibilities for feedback regulation of the aromatic pathway.

In *P. putida* the tryptophan group of enzymes were studied by Crawford and Gunsalus (1966) who showed that three of the enzymes were inducible, anthranilate synthetase (AS), phosphoribosyl transferase (PRT) and indoleglycerol phosphate synthetase (InGPS). The enzyme for step 3 phosphoribosylanthranilate isomerase (PRAI) appeared to be constitutive, while the final enzyme tryptophan synthetase was induced by its substrate, indoleglycerol phosphate (Figure 8.32). Maurer and Crawford (1971) found that three indole analogues, 5-methylindole (5MI), 5-fluoroindole (5FI) and 7-methylindole (7MI) inhibited growth of *P. putida* as did 5-fluorotryptophan (5FT) although concentrations of 200 μg/ml were required to achieve this result. Mutants were obtained which were resistant to these analogues and most of the 5FT-resistant mutants, and some of the 5MI- and 5FI-resistant mutants, were found to accumulate anthranilate (Table 8.8). In these mutant strains there was over-production of the repressible enzymes which were synthesized in 20-fold excess of the level found for the wild-type strain. These results would be compatible with a mutation in a regulator gene which controlled the expression of the inducible group of enzymes.

The genes for tryptophan biosynthesis in *P. aeruginosa* do not form a single cluster but were located in three separate transduction linkage groups (Fargie & Holloway, 1965). A similar arrangement of the tryptophan genes was found in *P. putida* by Gunsalus and his colleagues (1968) (Figure 8.32). The *trp*ABD linkage group was cotransduced with certain fluorophenylalanine and streptomycin resistance markers and this linkage also occurs for one of the *trp* loci of *P. aeruginosa* (Waltho & Holloway, 1966). The regulator mutants isolated in *P. putida* by Maurer and Crawford (1971)

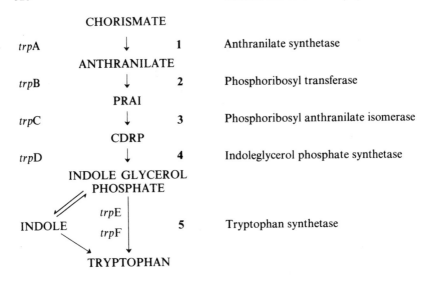

P. putida, trp genes (Gunsalus *et al.*, 1968)

Figure 8.32. Biosynthetic pathway for tryptophan (Gunsalus *et al.*, 1968; Calhoun *et al.*, 1973)

were derepressed for *trp*ABD and while the expression of these three genes was coordinate there was no correlation with the synthesis of the enzymes specified by *trp*C or *trp*EF. The mutation conferring derepression of the *trp*ABD cluster was not cotransducible with any of the *trp* genes. The *trp*ABD group of genes have the characteristics of an operon under repression control and if the regulatory mutation described by Maurer and Crawford (1971) specifies a repressor, then for this part of the pathway the regulation of tryptophan biosynthetic enzymes is by the classical repression of bio-synthetic enzymes by an unlinked regulator gene.

Earlier attempts to use analogues of the aromatic acids for the isolation of derepressed, or feedback inhibition-resistant mutants of *P. aeruginosa* were unsuccessful. The fluorophenylalanine-resistant mutants obtained by Waltho and Holloway (1966) had decreased incorporation of FPA into protein and frequently exhibited pleiotropic characters which were considered to be due to translational defects (Dunn & Holloway, 1972). However, Calhoun and Jensen (1972) found that 4-fluorotryptophan was a potent growth inhibitor at only 10 μg/ml, and 4FT-resistant mutants were isolated which were derepressed for the biosynthetic enzymes of the *trp*ABD group. In a more

Table 8.8. Tryptophan analogues and effects on *Pseudomonas* species

Compound	References	Effect
DL-4-Fluorotryptophan (4FT)	1	Inhibited growth *P. aeruginosa*. Depressed mutants obtained
DL-5-Fluorotryptophan (5FT)	2	Inhibited growth *P. putida*. Derepressed mutants obtained
	3	Inhibited induction by tryptophan of first two catabolic enzymes and tryptophan permease. Inhibited tryptophan transport in *P. acidovorans*
	5	Binds to catalytic site of tryptophan oxygenase
DL-5 Methyltryptophan	3	Inhibited growth *P. putida*
DL-α-Methyltryptophan	3	Inhibited growth *P. putida*
	4	Inducer of catabolic enzymes in *P. fluorescens* but only indirectly
	5	Binds to regulatory site of tryptophan oxygenase
DL-5-Hydroxymethyl-tryptophan	3	Inhibited growth *P. putida*
DL-7-Azatryptophan	3	Enhanced induction by tryptophan of catabolic enzymes in *P. acidovorans*
5-Fluoroindole	2	Inhibited growth *P. putida*. Derepressed mutants obtained
7-Methylindole	2	Inhibited growth *P. putida*. Derepressed mutants obtained

1. Calhoun & Jensen (1972).
2. Maurer & Crawford (1971).
3. Rosenfeld & Fiegelson (1969a,b).
4. Tremblay, Gottlieb & Knox (1967).
5. Forman & Fiegelson (1972).

extensive analysis Calhoun, Pierson and Jensen (1973a) showed that a 5FT-resistant mutant produced high levels of the three *trp*ABD enzymes and, unlike the *P. putida* derepressed mutants, excreted tryptophan into the medium. Exogenous tryptophan did not repress the level of the *trp*ABD enzymes in the wild-type strain, but starvation of some of the tryptophan auxotrophs for tryptophan led to derepression, so that the normal functioning of the pathway is such that this group of enzymes is in the repressed state. No changes in the activity of the *trp*C enzymes could be detected in various nutritional conditions and it was concluded that, as in *P. putida*, the phosphoribosylanthranilate isomerase was a constitutive enzyme. Tryptophan synthetase, specified by *trp*EF, was repressed about 10-fold by exogenous tryptophan and the two proteins were induced by indoleglycerol phosphate with increases in activities up to 100-fold. Some revertants were obtained from a *trp*E auxotroph that had acquired an additional mutation which enabled them to produce high levels of activity in the *trp*F protein in the presence of tryptophan. The mutants therefore appeared to be altered in both the induction and repression controls. The mutation determining this alteration was not characterized further but is a very interesting one in the light of the inducibility and repressibility of this biosynthetic enzyme.

In transduction experiments Jensen and his coworkers (1973) confirmed the separation of the *trp* genes into three separate groups, but suggest that the *trp*C is linked to a cysteine gene rather than one of the methionine genes as suggested by Fargie and Holloway (1965) for *P. aeruginosa* and Gunsalus and his associates (1968) for *P. putida*.

Extensive studies have been made of the properties of the enzyme proteins for tryptophan biosynthesis of *P. putida*. Enatsu and Crawford (1968) showed that the first 4 enzymic activities occurred on separate protein molecules. This contrasts with the position in *Escherichia coli* where the anthranilate synthetase and phosphoribosyl transferase activities are catalysed by a protein aggregate formed from the products of the first two genes and the enzymic activities responsible for steps 3 and 4 are catalysed by a single protein determined by one gene. The last step is catalysed by tryptophan synthetase which, in *P. putida* as in *E. coli*, contains two different types of sub-units determined in *P. putida* by *trp*EF under inducible control by indoleglycerol phosphate. More detailed study of anthranilate synthetase showed that there were two sub-units. ASI defined by *trp*A carries the substrate binding site for chorismate and the feedback-inhibition site for tryptophan; the smaller sub-unit, ASII carries the glutamine-binding site. Queener and Gunsalus (1970) showed that a similar sub-unit structure occurred for the anthranilate synthetases of *P. aeruginosa, P. acidovorans, P. testosteroni, P. stutzeri* and *P. cepacia*. The separation of the anthranilate synthetases of these six species into the ASI and ASII sub-units allowed assays to be carried out for *in vitro* complementation between sub-units from different species. The results were very striking and showed complete complementation between *P. putida* and *P. aeruginosa* sub-units. There was also complete complementation between the sub-units of *P. acidovorans* and *P. testosteroni* but when one sub-unit was derived from a strain from the first group (the p-a group) and the other from the second group (the c-t group) the hybrids formed had only 10–20 % of the activity of the native enzymes or hybrids formed with sub-units from the same group (Figure 8.33). The components prepared from *P. aeruginosa* and *P. putida* were of comparable sizes, with ASI about 64,000 and ASII about 18,000, while for *P. acidovorans* and *P. testosteroni*, the smaller ASII sub-unit was about the same size and the ASI sub-unit was slightly larger with a molecular weight of 71,000. These results add to the weight of evidence that *P. aeruginosa* and *P. putida* are closely related and that *P. acidovorans* and *P. testosteroni* are closely related to each other and much less closely related to the other two species. Judged from the ease of separation of the sub-units on columns, the *P. stutzeri* anthranilate synthetase complex behaved like the readily

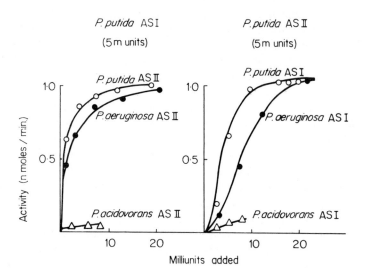

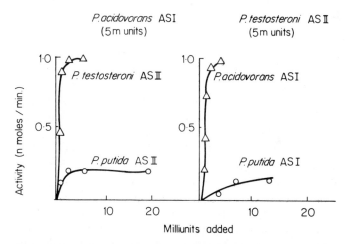

Figure 8.33. Sub-unit complementation of anthranilate synthetase. Activities of homologous and hybrid anthranilate synthetase complexes (activity expressed as nmoles/min). ASI and ASII sub-units were separated on DEAE–cellulose columns. In each case full activity was obtained when the ASI sub-unit from *P. putida* (or *P. aeruginosa*) was supplemented with ASII from *P. putida* (or *P. aeruginosa*) but only slight activity was found when the ASII sub-unit came from *P. acidovorans* or *P. tesosteroni* and *vice versa*. (Reproduced with permission from Queener and Gunsalus (1970) *Proceedings of the National Academy of Sciences, U.S.A.*, **67**, 1225)

dissociable p-a group and the *P. multivorans* enzyme like that of the c-t group which dissociates with difficulty. Mutants lacking anthranilate synthetase activity have all been defective in the larger ASI sub-unit, but the finding that ASII, as well as ASI, is overproduced in the *trp* regulatory mutant of Maurer and Crawford (1971) suggests that its gene may also be located within the *trp*ABD cluster.

Crawford and Yanofsky (1971) compared part of the sequence of the α sub-unit of the tryptophan synthetase of *P. putida* with that of *E. coli*. In the first 50 residues of the chain there were no additions or deletions, as compared with *E. coli*, with the exception of the *N*-terminal methionine which was absent from the *P. putida* polypeptide. Half of the residues were identical and for most of the differences the amino acids in equivalent positions were chemically similar. Half of the residues where differences were found, were specified by codons which differed by more than a single base. Two residues which in *E. coli* are known to mutate to enzymatically inactive missense proteins were conserved in *P. putida* indicating that they are probably also essential for catalytic activity in this organism. A further comparison of the 50 *N*-terminal residues of the typtophan synthetase α sub-units from a wider range of species indicated that the similarities to *E. coli* decreased in the order *Shigella dysenteriae* (*Salmonella typhimurium*, *Klebsiella aerogenes*), *Serratia marcescens*, *P. putida*.

The branch point for the biosynthesis of phenylalanine and tyrosine occurs at prephenate. In several organisms multienzyme complexes of aromatic pathway enzymes have been described which include in the complex the enzyme chorismate mutase which converts chorismate to prephenate (Lorence & Nester, 1967; Cotton & Gibson, 1968). Calhoun and colleagues (1973b) obtained a bifunctional protein complex from *P. aeruginosa* which had both chorismate mutase and prephenate dehydratase activity. This enzyme converted chorismate into a mixture of equivalent amounts of prephenate and phenylpyruvate, the immediate precursor of phenylalanine. The chorismate mutase activity was inhibited though not completely, by phenylalanine and prephenate, both of which appeared to compete with chorismate. Tyrosine and tryptophan had no effect and no evidence was found for repression control by tyrosine or phenylalanine. The prephenate dehydratase activity was very strongly inhibited by phenylalanine which suggested that it occupied the control point for phenylalanine biosynthesis, but raised the question of the provision of prephenate for tyrosine bio-synthesis. A second prephenate dehydratase II was isolated which had a molecular weight of about half that of the chorismate mutase–prephenate dehydratase I complex which was 134,000. The physiological significance of the second dehydratase which is not sensitive to inhibition by phenylalanine was not clear. Calhoun and his associates (1973b) suggest that the control of the two products of this part of the pathway is by a 'channel-shuttle system'. Tyrosine is the major inhibitor of the first enzyme, DAHP synthetase, and it is presumed that the basal level of this enzyme is set sufficiently high so that an adequate flow of precursors gets through even with a relatively high tyrosine concentration. When phenylalanine is low the chorismate mutase–prephenate dehydratase reactions would be channelled towards phenylalanine synthesis. When the phenylalanine concentration becomes high, with low tyrosine, then the second reaction of the chorismate mutase–prephenate dehydratase I complex would be inhibited and the prephenate would be available as a substrate for prephenate dehydrogenase for conversion to tyrosine, the 'shuttle reaction'.

In an analysis of phenylalanine-responding mutants of *P. aeruginosa*, Waltho (1972) established that six unlinked genes were concerned with the synthesis of phenylalanine, one of which corresponds to chorismate mutase. In this study Waltho (1973) found that chorismate mutase activity was inhibited by both phenylalanine and tyrosine, while prephenate dehydratase activity was inhibited by phenylalanine and stimulated by tyrosine.

Mutants were isolated by Calhoun and Jensen (1972) which had an altered prephenate dehydratase which was much less sensitive than that of the wild-type to inhibition by phenylalanine. These were obtained as mutants resistant to the analogue β-2-thienylalanine which inhibited growth of *P. aeruginosa* on minimal plates with fructose as the carbon source, but not when glucose was used. Phenylalanine relieved the inhibitory effect of β-2-thienylalanine on growth of the wild type in fructose minimal medium. It was suggested that the increased sensitivity of *P. aeruginosa* to growth inhibition by β-2-thienylalanine and to *p*-aminophenylalanine was due to a decrease in the level of precursors of the aromatic amino acids in the metabolic pool of cultures grown on fructose (Table 8.9).

Table 8.9. Phenylalanine and tyrosine analogues and effects on *Pseudomonas* species

Compound	References	Effect
DL-*p*-Fluorophenyl-alanine (FPA)	1, 2	Inhibits growth of *P. aeruginosa*. FPA-resistant mutants obtained
	3	Pleiotropic characters of FPA-resistant mutants
β-2-Thienylalanine	2	Inhibits growth of *P. aeruginosa* on fructose. Mutants isolated with prephenate dehydrase resistant to feedback inhibition by phenylalanine
p-Aminophenylalanine	2	Inhibits growth of *P. aeruginosa* on fructose

1. Waltho & Holloway (1966).   2. Calhoun & Jensen (1972).
3. Dunn & Holloway (1972).

Most strains of *P. aeruginosa* produce in addition to the yellowish fluorescent pigments, a characteristic blue–green pigment known as pyocyanin. A number of other phenazine pigments have also been recognized in *P. aeruginosa* (Chang & Blackwood, 1969). The precursor of the phenazine pigments has been considered for some time to be one of the intermediates of the aromatic amino acid biosynthetic pathway. Labelled shikimate can account for all the carbon atoms incorporated into the phenazine rings (Ingledew & Campbell, 1969).

Calhoun, Carson and Jensen (1972), using mutants blocked in various enzymes of the pathway, showed that the most probable intermediate at the branch point for pyocyanin biosynthesis is chorismate. Pyocyanin is an example of a secondary metabolite produced by a pathway branching from an essential biosynthetic sequence. The optimum condition for pyocyanin synthesis depend on the incubation of stationary bacteria, previously grown in complex medium, for up to 20 hours in a minimal salt medium in the absence of phosphate. Under these conditions the yield of pyocyanin can reach 80 to 90 μg/ml (Ingledew & Campbell, 1969; McGillivray, 1972). This system offers considerable possibilities for detailed genetic and biochemical studies on the regulation of pathways for secondary metabolites.

Many of the pseudomonads are able to use tryptophan as a growth substrate (Stanier *et al.*, 1966) and two distinct pathways are known. Tryptophan can be metabolized *via* anthranilate, catechol and the β-ketoadipate pathway discussed in Chapter 7, section 2, or through the quinolate pathway (Behrman, 1962). The initial enzymes for both pathways are the same and convert tryptophan to kynurenine which is then a substrate for one of the alternative pathways. Palleroni and Stanier (1964) examined the regulation of the early enzymes in *P. fluorescens* and this system provided the first example of sequential induction of groups of enzymes which were themselves coordinately regulated.

Rosenfeld and Fiegelson (1969a,b) showed that kynurenine was also the inducer of tryptophan oxygenase and formylkynurenine formamidase in a strain of *P. acidovorans* utilizing the quinoline pathway. The ratios of the two enzymes were similar in induced cultures but there was considerable deviation in the ratios found for induced cultures and the basal constitutive levels of these enzymes. This would seem to imply a lack of coordinate regulation of these enzymes in *P. acidovorans*. However, the activity of tryptophan oxygenase is under regulatory control by tryptophan itself which may complicate measurements of enzyme activity in cell extracts. Forman and Fiegelson (1972) showed that typtophan oxygenase possessed two types of binding site for tryptophan. Kinetic data indicated that three allosteric binding sites exist on the enzyme as well as the catalytic binding sites. The analogue α-methyltryptophan binds mainly at the allosteric sites and therefore has the effect of increasing the apparent activity of the enzyme at low tryptophan concentrations. 5-Fluorotryptophan, on the other hand binds specifically to the catalytic site. This analogue is not metabolizable by *P. acidovorans* and has a marked inhibitory effect on the induction of the first two enzymes and tryptophan permease. It also inhibits tryptophan permease activity in fully induced cells, and would thus appear to compete at the tryptophan binding site of the permease. Another analogue 7-azatryptophan had the unusual effect of enhancing induction by tryptophan of each of the two metabolic enzymes without itself being an inducer. It did not enhance the induction of tryptophan permease and its mode of action is unknown. It was particularly interesting that product induction by kynurenine extended not only to the two enzymes required to synthesize it from tryptophan, but also to the induction of tryptophan permease. However, the synthesis of permease was clearly not coordinated with that of the tryptophan oxygenase and formylkynurenine formamidase.

5. SUMMARY AND CONCLUSIONS

When we look at the accumulated information on the biosynthetic and catabolic pathways of *Pseudomonas* the most striking finding is the persistence of characteristic regulatory patterns throughout the species of this genus. The biosynthetic pathways for the amino acids present patterns of feedback inhibition, often with several effectors acting in concert, and we do not find the isofunctional enzymes for the branch points of biosynthetic pathways that are known for other genera. The little information presently available on the sequences of corresponding enzymes, suggests that homology between some of the biosynthetic enzyme proteins of *Pseudomonas* and *Escherichia–Salmonella* may be much greater than had been expected. This means that these organisms differ much more in the regulation of these pathways than in the enzymes themselves. The lack of clustering of the genes for the biosynthetic enzymes appears to go with patterns for the regulation of gene expression which again are very different from those found in *Escherichia–Salmonella*. Although end-product repression is clearly important for regulating the synthesis of the enzymes of several pathways, it does not appear to be a universal control of synthesis of all biosynthetic enzymes of *Pseudomonas*.

Feedback inhibition of biosynthetic enzymes is a sensitive and rapid mechanism of adjustment of the rate of operation of a particular pathway. It is well suited to controlling the rate of synthesis of one or more amino acids in response to the biosynthetic demands of the cell and changes in the metabolic pools. Repression is a slower response to higher concentrations of end-products which might be encountered in the external environment. Repression controls can enable some organisms to dispense almost entirely with the synthesis of all the enzymes of a biosynthetic pathway. It is known that derepressed mutants that produce biosynthetic enzymes constitutively are at a growth disadvantage. It might be asked whether pseudomonads are likely to encounter, in the natural environment, fluctuating levels of amino acids at such concentrations as would lead to a definite growth advantage for a general repression mechanism. The control patterns in which only one of the enzymes of a biosynthetic pathway, or a group of two or three enzymes, is under repression control may be equally effective even when occasionally high concentrations of the end-product are encountered. It would be interesting to discover whether the derepressed mutants for the *trp*ABD genes of the tryptophan pathway, isolated by Maurer and Crawford (1971) for *P. putida* and by Calhoun and his colleagues (1973a) for *P. aeruginosa*, are at a growth disadvantage compared with the wild-type parents. There are still many gaps in assigning enzyme functions and map positions to the genes for biosynthetic pathways although many auxotrophic mutants have been isolated. A more complete answer to the problem of the regulation of biosynthetic pathways requires further biochemical and genetic analysis.

In almost every case where studies have been carried out on catabolic pathways and their regulation, the results have shown that the pseudomonad being examined possessed an even greater metabolic potential than had been expected. Even within a single strain, three different pathways may exist for the catabolism of a single compound, catechol (see Chapter 7, section 3.h). In this case, closer examination revealed that this apparent metabolic redundancy permits cells to attack a range of substituted phenolic compounds with optimal physiological efficiency. In starting an investigation of a metabolic pathway in a pseudomonad it is well to be aware that it may be more complicated than it looks at first sight. The existence of alternative pathways for a single metabolite may explain in part some of the different results obtained for the same *Pseudomonas* species by different groups of workers. However, such differences can usually be related to the employment of different strains and this raises the problem of how much variation occurs between one strain of a species and another. We have frequently mentioned that one reaction is characteristic of *P. aeruginosa* and another of *P. putida*, but it must be admitted that in most cases single strains of a species have been compared. Sometimes the information is obtained from 'an unidentified pseudomonad'. Comparative studies between two or more species of *Pseudomonas* would be more valuable if they included comparisons between several strains of these species as well.

The clustering together of groups of genes for catabolic pathways, particularly where they occur on transmissible plasmids, appears to be a general characteristic of *Pseudomonas* which may be of advantage in allowing a population to exploit a very wide range of potential growth substrates. However, plasmids that are freely transmissible may also be freely lost and cultures maintained in the laboratory may lose the plasmid-borne genes for a large part of a metabolic pathway (see Chapters 5 and 6). If this occurs then the strain will have a negative phenotype for the character concerned. It is particularly interesting that Dunn and Gunsalus (1973) found that in one strain of *P. putida* the naphthalene genes were carried on a plasmid while in another strain they were on the chromosome. This could mean that for characters such as naphthalene utilization, some strains would be much more stable than others. The transmissibility of plasmids across species boundaries raises problems of another sort. One question which must then be asked is about the origin of the plasmid. Has it been long resident in the strains of this species or is it a fairly recent arrival from another species or even another genus? The implication of these questions in the evolution of pseudomonads is discussed in Chapter 9.

The multiplicity of pathways for the catabolism of the amino acids is interesting, and in most cases there are no obvious advantages for one pathway over another, although some have apparently evolved for the metabolism of the D-amino acids and others for the L-amino acids. With the exception of the studies of Wiame and his colleagues on the arginine pathways (see Chapter 8, section 4.i), little has yet been done on the ways in which the catabolic and biosynthetic pathways coexist. There is no evidence that the amino acid catabolic enzymes play any direct part in regulating the amino acid pools.

There are still many metabolic pathways of *Pseudomonas* yet to be explored and others where the biochemical pathways have been worked out but the details of the regulation and genetic structure are unknown. Very little is yet known about the biosynthesis of the macromolecular cell components and many questions are still to be answered on the biochemistry of such well-established phenomena as drug resistance, host controlled modification and multiple lysogeny.

6. BIBLIOGRAPHY

Behrman, E. J. (1962) *Nature, London*, **196**, 150.

Benson, A. M., Jarabak, R. & Talalay, P. (1971) *Journal of Biological Chemistry*, **246**, 7514.

Betz, J. L., Brown, J. E., Clarke, P. H. & Day, M. (1974) *Genetical Research*, **24**, in press.

Betz, J. L. & Clarke, P. H. (1972) *Journal of General Microbiology*, **73**, 161.

Betz, J. L. & Clarke, P. H. (1973) *Journal of General Microbiology*, **75**, 167.

Blumenthal, H. J. & Fish, D. C. (1963) *Biochemical and Biophysical Research Communications*, **11**, 239.

Brammar, W. J., Clarke, P. H. & Skinner, A. J. (1967) *Journal of General Microbiology*, **47**, 87.

Brammar, W. J., McFarlane, N. D. & Clarke, P. H. (1966) *Journal of General Microbiology*, **44**, 303.

Brill, W. J. & Magasanik, B. (1969) *Journal of Biological Chemistry*, **244**, 5392.

Brown, J. E., Brown, P. R. & Clarke, P. H. (1969) *Journal of General Microbiology*, **57**, 273.

Brown, J. E. & Clarke, P. H. (1970) *Journal of General Microbiology*, **64**, 329.

Brown, P. R. & Clarke, P. H. (1972) *Journal of General Microbiology*, **69**, 287.

Brown, P. R., Smyth, M. J., Clarke, P. H. & Rosemeyer, M. A. (1973) *European Journal of Biochemistry*, **34**, 177.

Buhlman, X., Vischer, W. A. & Bruhin, H. (1961) *Journal of Bacteriology*, **82**, 787.

Calhoun, D. H., Carson, M. & Jensen, J. A. (1972) *Journal of General Microbiology*, **72**, 581.

Calhoun, D. H. & Feary, T. W. (1969) *Journal of Bacteriology*, **97**, 210.

Calhoun, D. H. & Jensen, R. A. (1972) *Journal of Bacteriology*, **109**, 365.

Calhoun, D. H., Pierson, D. L. & Jensen, R. A. (1973a) *Molecular and General Genetics*, **121**, 117.

Calhoun, D. H., Pierson, D. L. & Jensen, R. A. (1973b) *Journal of Bacteriology*, **113**, 241.

Chang, P. C. & Blackwood, A. C. (1969) *Canadian Journal of Microbiology*, **15**, 439.

Chang, Y-F. & Adams, E. (1971) *Biochemical and Biophysical Research Communications*, **45**, 570.

Chapman, L. F. & Nester, E. W. (1969) *Journal of Bacteriology*, **97**, 1444.

Chapman, P. J. & Duggleby, R. G. (1967) *Biochemical Journal*, **103**, 7C.

Chasin, J. L. & Magasanik, B. (1968) *Journal of Biological Chemistry*, **243**, 5165.

Chou, I. N. & Gunsalus, I. C. (1971) *Bacteriological Proceedings*, 162.

Chou, C. S. & Rodwell, V. W. (1972) *Journal of Biological Chemistry*, **247**, 4486.

Clarke, P. H. (1972) *Journal of General Microbiology*, **71**, 241.

Clarke, P. H. & Tata, R. (1973) *Journal of General Microbiology*, **75**, 231.

Clarkson, C. E. & Meadow, P. M. (1971) *Journal of General Microbiology*, **66**, 161.

Cohen, G. N., Stanier, R. Y. & Le Bras, G. (1969) *Journal of Bacteriology*, **99**, 791.

Cooper, R. A. & Kornberg, H. L. (1964) *Biochemical Journal*, **91**, 82.

Coote, J. G. & Hassall, H. (1973a) *Biochemical Journal*, **132**, 409.

Coote, J. G. & Hassall, H. (1973b) *Biochemical Journal*, **132**, 423.

Cotton, R. G. H. & Gibson, F. (1968) *Biochimica et Biophysica Acta*, **160**, 188.

Crawford, I. P. & Gunsalus, I. C. (1966) *Proceedings of the National Academy of Sciences, U.S.A.*, **56**, 717.

Crawford, I. P. & Yanofsky, C. (1971) *Journal of Bacteriology*, **108**, 248.

Dagley, S. (1971) *Advances in Microbial Physiology*, **6**, 1.

Dagley, S. & Trudgill, P. W. (1965) *Biochemical Journal*, **95**, 48.

Dahms, A. S. & Anderson, R. L. (1969) *Biochemical and Biophysical Research Communications*, **36**, 809.

Dahms, A. S. & Anderson, R. L. (1972a) *Journal of Biological Chemistry*, **247**, 2222.

Dahms, A. S. & Anderson, R. L. (1972b) *Journal of Biological Chemistry*, **247**, 2228.

Dahms, A. S. & Anderson, R. L. (1972c) *Journal of Biological Chemistry*, **247**, 2233.

Dahms, A. S. & Anderson, R. L. (1972d) *Journal of Biological Chemistry*, **247**, 2238.

Datta, P. & Gest, H. (1964) *Nature, London*, **203**, 1259.

Delafield, F. P., Cooksey, K. E. & Doudoroff, M. (1965a) *Journal of Biological Chemistry*, **240**, 4023.

Delafield, F. P., Doudoroff, M., Palleroni, N. J., Lusty, C. J. & Contopoulos, R. (1965b) *Journal of Bacteriology*, **90**, 1455.

DeLey, J. & Doudoroff, M. (1957) *Journal of Biological Chemistry*, **227**, 745.

Dhawale, M. R., Creaser, E. H. & Loper, J. C. (1972) *Journal of General Microbiology*, **73**, 353.

Draper, P. (1967) *Journal of General Microbiology*, **46**, 111.

Dunn, N. W. & Gunsalus, I. C. (1973) *Journal of Bacteriology*, **114**, 974.

Dunn, N. W. & Holloway, B. W. (1972) *Genetical Research*, **18**, 185.

Emmer, M., de Crombrugghe, B., Pastan, I. & Perlman, R. (1970) *Proceedings of the National Academy of Sciences, U.S.A.*, **66**, 480.

Enatsu, T. & Crawford, I. C. (1968) *Journal of Bacteriology*, **95**, 107.

Entner, N. & Doudoroff, (1952) *Journal of Biological Chemistry*, **196**, 853.

Evans, W. C. & Smith, B. S. W. (1951) *Biochemical Journal*, **49**, x.

Evans, W. C., Smith, B. S. W., Moss, P. & Fernley, H. N. (1971) *Biochemical Journal*, **122**, 509.

Fan, C. L., Miller, D. L. & Rodwell, V. W. (1972) *Journal of Biological Chemistry*, **247**, 2283.

Fargie, B. & Holloway, B. W. (1965) *Genetical Research*, **6**, 284.

Farin, F. & Clarke, P. H. (1974) *Proceedings of the Society for General Microbiology*, **1**, 56.

Feary, T. W., Williams, B., Calhourn, D. H. & Walker, T. A. (1969) *Genetics*, **62**, 673.

Forman, H. J. & Fiegelson, P. (1972) *Journal of Biological Chemistry*, **247**, 256.

Foster, J. W. (1962) *Antonie van Leeuwenhoek*, **28**, 241.

Fuchs, A. (1965) *Antonie van Leeuwenhoek*, **31**, 323.

George, D. J. & Phillips, A. T. (1970) *Journal of Biological Chemistry*, **245**, 528.

Gholson, R. K., Baptist, J. N. & Coon, M. J. (1963) *Biochemistry*, **2**, 1155.

Gorini, L. W., Gundersen, W. & Burger, M. (1961) *Cold Spring Harbor Symposium Quantitative Biology*, **26**, 173.

Gryder, R. M. & Adams, E. (1969) *Journal of Bacteriology*, **97**, 292.

Gunsalus, I. C., Gunsalus, C. F., Chakrabarty, A. M., Sikes, S. & Crawford, I. P. (1968) *Genetics*, **60**, 419.

Haas, D., Kurer, V. & Leisinger, T. (1972) *European Journal of Biochemistry*, **31**, 290.

Halpern, Y. S. & Grossowicz, N. (1957) *Biochemical Journal*, **65**, 716.

Hamlin, B. T., Ng, F. M-W. & Dawes, E. A. (1967) *Microbial Physiology Continuous Culture, Proceedings International Symposium 3rd*, 211.

Harper, D. B. & Blakeley, E. R. (1971) *Canadian Journal of Microbiology*, **17**, 1015.

Hartman, P. E., Hartman, Z., Stahl, R. C. & Ames, B. N. (1971) *Advances in Genetics*, **16**, 1.

Hatfield, G. W. & Burns, R. O. (1970) *Proceedings of the National Academy of Sciences, U.S.A.*, **66**, 1027.

Hechtman, P. & Scriver, C. R. (1970) *Journal of Bacteriology*, **104**, 857.

Hegeman, G. D. & Rosenberg, S. L. (1970) *Annual Review of Microbiology*, **24**, 429.

Hermann, M., Thevent, N. J. Coudert-Maratier, M. M. & Vandecasteele, J. P. (1972) *European Journal of Biochemistry*, **30**, 100.

Hoet, P. P. & Stanier, R. Y. (1970a) *European Journal of Biochemistry*, **13**, 65.

Hoet, P. P. & Stanier, R. Y. (1970b) *European Journal of Biochemistry*, **13**, 71.

Holloway, B. W., Egan, J. B. & Monk, M. (1960) *Australian Journal of Experimental Biology and Medical Science*, **38**, 321.

Hopper, D. J., Chapman, P. J. & Dagley, S. (1968) *Biochemical Journal*, **110**, 798.

Horowitz, N. H. (1945) *Proceedings of the National Academy of Sciences, U.S.A.*, **31**, 153.

Horowitz, N. H. (1965) in *Evolving Genes and Proteins*, (Eds. V. Bryson and H. J. Vogel), Academic Press, New York, p. 15.

Horvath, I., Varga, J. M. & Szentirmai, A. (1964) *Journal of General Microbiology*, **34**, 241.

Hurlbert, R. E. & Jakoby, W. B. (1965) *Journal of Biological Chemistry*, **240**, 2772.

Hynes, M. J. & Pateman, J. A. (1970) *Journal of General Microbiology*, **63**, 317.

Ingledew, W. M. & Campbell, J. J. R. (1969) *Canadian Journal of Microbiology*, **15**, 535.

Inouye, H., Nakawaga, T. & Morihara, K. (1963) *Biochimica et Biophysica Acta*, **34**, 117.

Isaac, J. H. & Holloway, B. W. (1972) *Journal of General Microbiology*, **73**, 427.

Jacoby, G. A. (1964) *Biochemical Journal*, **92**, 1.

Jacoby, G. A. & Gorini, L. (1967) *Journal of Molecular Biology*, **24**, 41.

Jakoby, W. B. & Fredericks, J. (1959) *Journal of Biological Chemistry*, **234**, 2145.

Jayaraman, K., Müller-Hill, B. & Rickenberg, H. V. (1966) *Journal of Molecular Biology*, **18**, 339.

Jeffcoat, R., Hassall, H. & Dagley, S. (1969a) *Biochemical Journal*, **115**, 969.

Jeffcoat, R., Hassall, H. & Dagley, S. (1969b) *Biochemical Journal*, **115**, 977.

Jensen, R. A., Calhoun, D. H. & Stenmark, S. L. (1973) *Biochimica et Biophysica Acta*, **293**, 256.

Jensen, R. A., Nasser, D. S. & Nester, E. W. (1967) *Journal of Bacteriology*, **94**, 1582.

Kay, W. W. & Gronlund, A. F. (1969a) *Journal of Bacteriology*, **7**, 273.

Kay, W. W. & Gronlund, A. F. (1969b) *Journal of Bacteriology*, **100**, 276.

Kay, W. W. & Gronlund, A. F. (1969c) *Biochimica et Biophysica Acta*, **193**, 444.

Kay, W. W. & Gronlund, A. F. (1969d) *Journal of Bacteriology*, **98**, 116.

Kay, W. W. & Gronlund, A. F. (1971) *Journal of Bacteriology*, **105**, 1039.

Kelly, M. & Clarke, P. H. (1962) *Journal of General Microbiology*, **27**, 305.

Kimura, T. (1959) *Journal of Biochemistry, Tokyo*, **46**, 1271.

Kohn, L. D. & Jakoby, W. B. (1968a) *Journal of Biological Chemistry*, **243**, 2465.

Kohn, L. D. & Jakoby, W. B. (1968b) *Journal of Biological Chemistry*, **243**, 2472.

Kohn, L. D. & Jakoby, W. B. (1968c) *Journal of Biological Chemistry*, **243**, 2486.

Kohn, L. D. & Jakoby, W. B. (1968d) *Journal of Biological Chemistry*, **243**, 2494.

Kohn, L. D., Packman, P. M., Allen, R. H. & Jakoby, W. B. (1968) *Journal of Biological Chemistry*, **243**, 2479.

Kovach, J. S., Phang, J. M., Ference, M. & Goldberger, R. T. (1969) *Proceedings of the National Academy of Sciences, U.S.A.*. **63**, 481.

Kusunose, M., Kusunose, E. & Coon, M. J. (1964a) *Journal of Biological Chemistry*, **239**, 1374.

Kusunose, M., Kusunose, E. & Coon, M. J. (1964b) *Journal of Biological Chemistry*, **239**, 2135.

Lack, L. (1959) *Biochimica et Biophysica Acta*, **34**, 117.

Lack, L. (1961) *Journal of Biological Chemistry*, **236**, 2835.

Laishley, E. J. & Bernlohr, R. W. (1968) *Biochimica et Biophysica Acta*, **167**, 547.

Leidigh, B. J. & Wheelis, M. L. (1973) *Molecular and General Genetics*, **120**, 201.

Leisinger, T., Hass, D. & Hegarty, M. V. (1972) *Biochimica et Biophysica Acta*, **262**, 214.

Lessie, T. G. & Neidhardt, F. C. (1967a) *Journal of Bacteriology*, **93**, 1337.

Lessie, T. G. & Neidhardt, F. C. (1967b) *Journal of Bacteriology*, **93**, 1800.

Lessie, T. & Vander Wyk, J. C. (1972) *Journal of Bacteriology*, **110**, 1107.

Lessie, J. G. & Whiteley, H. R. (1969) *Journal of Bacteriology*, **100**, 878.

Lorence, J. & Nester, E. W. (1967) *Biochemistry*, **6**, 1541.

Lusty, C. J. & Doudoroff, M. (1966) *Proceedings of the National Academy of Sciences, U.S.A.*, **56**, 960.

Maas, W. K. (1972) *Molecular and General Genetics*, **119**, 1.

MacDonald, D. L., Stanier, R. Y. & Ingraham, J. L. (1954) *Journal of Biological Chemistry*, **210**, 809.

McFarlane, N. D., Brammar, W. J. & Clarke, P. H. (1965) *Biochemical Journal*, **95**, 24C.

McGillivray, A. D. (1972) 'Pyocyanine formation by *Pseudomonas aeruginosa*', *Ph.D. Thesis*, University of Glasgow.

McGilvray, D. & Morris, J. G. (1969) *Biochemical Journal*, **112**, 657.

McKenna, E. J. (1972) in *Degradation of Organic Molecules in the Biosphere*, National Academy of Sciences, Washington, U.S.A.

McKenna, E. J. & Kallio, R. E. (1965) *Annual Review of Microbiology*, **19**, 183.

Marinus, M. G. & Loutit, J. S. (1969a) *Genetics*. **63**, 547.

Marinus, M. G. & Loutit, J. S. (1969b) *Genetics*, **63**, 557.

Markovitz, A., Klein, H. P. & Fischer, E. H. (1956) *Biochimica et Biophysica Acta*, **19**, 267.

Marshall, V. P. & Sokatch, J. R. (1972) *Journal of Bacteriology*, **110**, 1073.

Martin, R. G., Berberich, M. A., Ames, B. C., Davis, W. D., Goldberger, R. F. & Yourno, J. D. (1971) *Methods in Enzymology*, (Eds. H. Tabor & C. W. Tabor), Academic Press, New York and London, p. 17B.

Maurer, R. & Crawford, I. P. (1971) *Journal of Bacteriology*, **106**, 331.

Meagher, R. B. & Ornston, L. N. (1973) *Biochemistry*, **12**, 3523.

Mee, B. J. (1969) 'Histidine biosynthesis in *Pseudomonas*', *Ph.D. Thesis*, University of Melbourne.

Mee, B. J. & Lee, B. T. O. (1967) *Genetics*, **55**, 709.

Mee, B. J. & Lee, B. T. O. (1969) *Genetics*, **62**, 687.

Meiss, H. K., Brill, W. J. & Magasanik, B. (1969) *Journal of Biological Chemistry*, **244**, 5382.

Miller, D. L. & Rodwell, V. W. (1971) *Journal of Biological Chemistry*, **246**, 2758.

Minson, A. C. & Creaser, E. H. (1969) *Biochemical Journal*, **114**, 49.

Morihara, K. (1963) *Biochimica et Biophysica Acta*, **73**, 113.

Morihara, K. (1964) *Journal of Bacteriology*, **88**, 745.

Mossel, D. A. A. & van Zadelhoff, C. (1971) *Journal of General Microbiology*, **69**, xiv.

Neidhardt, F. C. & Magasanik, B. (1957) *Journal of Bacteriology*, **73**, 253.

Ornston, L. N. (1966a) *Journal of Biological Chemistry*, **241**, 3787.

Ornston, L. N. (1966b) *Journal of Biological Chemistry*, **241**, 3795.

Ornston, L. N. & Stanier, R. Y. (1966) *Journal of Biological Chemistry*, **241**, 3776.

Palleroni, N. J. & Doudoroff, M. (1957) *Journal of Bacteriology*, **74**, 180.

Palleroni, N. J. & Stanier, R. Y. (1964) *Journal of General Microbiology*, **35**, 319.

Parke, D., Meagher, R. B. & Ornston, L. N. (1973) *Biochemistry*, **12**, 3537.

Patel, R. N., Meaher, R. B. & Ornston, L. N. (1973) *Biochemistry*, **12**, 3531.

Paulus, H. & Gray, E. (1964) *Journal of Biological Chemistry*, **239**, 4008.

Pemberton, J. & Holloway, B. W. (1972) *Genetical Research*, **19**, 251.

Peterson, J. A., Basu, D. & Coon, M. J. (1966) *Journal of Biological Chemistry*, **241**, 5162.

Peterson, J. A., Kusenose, M., Kusenose, E. & Coon, M. J. (1967) *Journal of Biological Chemistry*, **242**, 4334.

Pollock, M. R. (1962) in *The Bacteria, A Treatise on Structure and Function*, (Eds. I. C. Gunsalus & R. Y. Stanier), Vol. IV, Academic Press, New York, p. 121.

Quay, S. C., Friedman, S. B. & Eisenberg, R. C. (1972) *Journal of Bacteriology*, **112**, 291.

Queener, S. F. & Gunsalus, I. C. (1970) *Proceedings of the National Academy of Sciences, U.S.A.*, **67**, 1225.

Razin, S., Gery, I. & Bachrach, U. (1959) *Biochemical Journal*, **71**, 551.

Robert-Gero, M., Poiret, M. & Cohen, G. N. (1970) *Biochimica et Biophysica Acta*, **206**, 17.

Robyt, J. F. & Ackerman, R. J. (1971) *Archives of Biochemistry and Biophysics*, **145**, 105.

Rosenfeld, H. & Fiegelson, P. (1969a) *Journal of Bacteriology*, **97**, 697.

Rosenfeld, H. &. Fiegelson, P. (1969b) *Journal of Bacteriology*, **97**, 705.
Rothman-Dienes, L. & Martin, R. G. (1971) *Journal of Bacteriology*, **106**, 227.
Rubenstein, K. E., Striebel, E., Massey, S., Lapi, L. & Cohen, S. (1972) *Journal of Bacteriology*, **112**, 1213.
Shilo, M. (1957) *Journal of General Microbiology*, **16**, 472.
Shilo, M. & Stanier, R. Y. (1957) *Journal of General Microbiology*, **16**, 482.
Shuster, C. W. & Doudoroff, M. (1967) *Archiv für Mikrobiologie*, **59**, 279.
Sierra, G. (1957) *Antonie van Leeuwenhoek*, **23**, 15.
Silverstone, A. E., Magasanik, B., Reznikoff, W. S., Miller, J. H. & Beckwith, J. R. (1969) *Nature, London*, **221**, 1012.
Singer, C. E. & Smith, C. R. (1972) *Journal of Biological Chemistry*, **247**, 2989.
Sistrom, W. R. & Stanier, R. Y. (1954) *Journal of Biological Chemistry*, **210**, 821.
Skinner, A. J. & Clarke, P. H. (1968) *Journal of General Microbiology*, **50**, 183.
Smyth, P. F. & Clarke, P. H. (1972) *Journal of General Microbiology*, **73**, ix.
Stalon, V., Ramos, F., Pierard, A. & Wiame, J. M. (1967) *Biochimica et Biophysica Acta*, **139**, 91.
Stalon, V., Ramos, F., Pierard, A. & Wiame, J-M. (1972) *European Journal of Biochemistry*, **29**, 25.
Stanier, R. Y., Palleroni, N. J. & Doudoroff, M. (1966) *Journal of General Microbiology*, **43**, 159.
Tristram, H. & Thurston, C. F. (1966) *Nature*, **212**, 74.
Trust, T. J. & Millis, N. F. (1970) *Journal of General Microbiology*, **61**, 245.
Trust, T. J. & Millis, N. F. (1971) *Journal of Bacteriology*, **105**, 1216.
Twarog, R. (1972) *Journal of Bacteriology*, **111**, 37.
Udaka, S. (1966) *Journal of Bacteriology*, **91**, 617.
Vandecasteele, J-P. & Hermann, M. (1972) *European Journal of Biochemistry*, **31**, 80.
van der Linden & Thijsse (1965) *Advances in Enzymology*, **27**, 469.
Voellmy, R. & Leisinger, T. (1972) *Journal of General Microbiology*, **73**, xiii.
Waltho, J. (1972) *Journal of Bacteriology*, **112**, 1070.
Waltho, J. (1973) *Biochimica et Biophysica Acta*, **320**, 232.
Waltho, J. A. & Holloway, B. W. (1966) *Journal of Bacteriology*, **92**, 35.
Weimberg, R. (1961) *Journal of Biological Chemistry*, **236**, 629.
Weimberg, R. & Doudoroff, M. (1955) *Journal of Biological Chemistry*, **217**, 606.
Wheelis, M. L., Palleroni, N. J. & Stanier, R. Y. (1967) *Archiv. für Mikrobiologie*, **59**, 302.
Wheelis, M. & Stanier, R. Y. (1970) *Genetics*, **66**, 245.
Wu, T. T., Lin, E. C. C. & Tanaka, S. (1968) *Journal of Bacteriology*, **96**, 447.
Zucker, M. & Hankin, L. (1970) *Journal of Bacteriology*, **104**, 13.

CHAPTER 9

Evolutionary Prospects for *Pseudomonas* Species

M. H. RICHMOND and P. H. CLARKE

Many microbiologists think of *Pseudomonas* species as exceptional in metabolic diversity and ability to adapt to a wide range of environmental conditions. Certainly the material described earlier in this book illustrates their biochemical versatility, and the complexity of the genetic organization argues that this group of organisms must be at least as well endowed in this respect as other bacteria. But what of the evolutionary potential of *Pseudomonas*? Do we know anything about possible developments that may occur in this group of bacteria? This is a subject which has only recently been investigated to any purpose and experiments and observations are still few and far between. In practice the detailed changes have been studied most intensively at a biochemical level, but some examples which alter the response of the organisms to selection pressure have been analysed in genetic terms. In this chapter we will describe some of these studies and speculate a little about the future.

1. GENETIC ORGANIZATION AND EVOLUTIONARY POTENTIAL

Although there is some suggestion that *P. aeruginosa* may carry more genetic information than other Gram-negative bacteria (Leth Bak, Christiansen & Stenderup, 1970), this is not well established and the group, though very flexible genetically, probably does not differ significantly in this respect from *Escherichia coli* and other members of the Enterobacteriaceae. Moreover, the overall organization of the DNA in *Pseudomonas* spp. seems so similar in principle to that of *Escherichia coli* that it seems likely that many of the genetic arrangements with evolutionary potential already examined in the Enterobacteriaceae will be found in pseudomonads as well.

As is the case with other bacterial species, plasmids are by far the most important genetic elements with evolutionary potential in pseudomonads

(Jones & Sneath, 1970; Richmond & Wiedeman, 1974). These extrachromosomal elements have already been detected in many strains and species of *Pseudomonas*, and it seems more and more likely that it is only exceptional isolates that lack one at least of these adventitious pieces of DNA. Already the molecular diversity of the reactions catalysed by plasmids in pseudomonads is known to be great (Chapters 5 and 7) and one gains the impression that a plasmid location for a metabolic gene is no disadvantage to the cell that carries it. It may even be an advantage to the population of which the cell forms part. Furthermore, since many plasmids are transmissible, the location on a plasmid of metabolic genes with great selective advantage allows one to speculate that they may have arrived only relatively recently from some other species by transfer, and are maintained in the new host only by the advantage they confer to it.

In practice, plasmids aid two important types of evolutionary development: first they may allow genes to be mobilized from the bacterial chromosome, a step which can allow gene duplication to be maintained (Broda, Beckwith & Scaife, 1964; Asheshov, 1969). If so, this is potentially important since it permits a 'spare' gene copy to evolve without the risk of lethal consequences to the cell. Secondly, since many plasmids are self-transmissible they provide a vehicle for genes to pass between bacterial strains and species. Even when they are not self-transmissible a plasmid location enhances their chance of transfer either by transduction or mobilization by another plasmid. This ability, taken in conjunction with chromosomal gene mobilization, ensures that it is possible for genetic information from the *Pseudomonas* chromosome to be passed to a range of potential recipients (see Chapter 6). If transfer is by transduction, the range of accessible strains is limited to the host range of the phage involved, but if transfer is by conjugation the number of possibilities is much greater. For example some of the RP plasmids of *P. aeruginosa* can promote the transfer of genes—and among them chromosomal genes—to nine members of the Pseudomonadales (Stanisich & Holloway, 1971; 1972; Olsen & Shipley, 1973).

So far there does not seem to be any example where the mobilization of chromosomal genes has provided an obvious selective advantage to any strain of *Pseudomonas* under natural conditions although this can be set up easily enough in the laboratory. Both the RP and FP plasmids can mobilize chromosomal markers and these can repair auxotrophic lesions in other species after growth of the two strains together (Stanisich & Holloway, 1971; 1972). It seems only a matter of time therefore before some example of this sort of evolutionary step is detected in a natural population. As with many transfer events however the rarity of the event makes it difficult to be watching at the time that transfer occurs and in the last analysis the evidence is likely to be strictly circumstantial.

As yet the only detailed study on the role of plasmids in the evolution of *Pseudomonas* cultures concerns the acquisition of antibiotic resistance—notably resistance to carbenicillin—by strains of *Pseudomonas aeruginosa* infecting burned patients. This sequence of events occurred in the Burns' Unit in the Birmingham Accident Hospital and it has been analysed in increasing depth in a number of reports (Lowbury *et al.*, 1969; Sykes & Richmond, 1970; Lowbury, Babb & Roe, 1972; Ingram, Richmond & Sykes, 1973).

Up to March 1969 few carbenicillin resistant strains of *P. aeruginosa* had been detected in the Burns Unit in Birmingham, and certainly none of them owed their resistance to β-lactamase production. About the middle of that month, however, carbenicillin resistant strains of *P. aeruginosa* which owed their resistance to an R factor mediated β-lactamase appeared simultaneously in a number of sero- and phage types of *P. aeruginosa* (Table 9.1)

Table 9.1. Dates of isolation, serotypes and phage types of *Pseudomonas aeruginosa* strains carrying R factors isolated in the Burns' Unit of the Accident Hospital, Birmingham

Serotype	Phage type	Date of isolation
NT	44/F8/109/119X/1214	March 3, 1969
8	7/21/68/119X	March 4, 1969

Up to the 3rd March 1969 no R factor carrying *Pseudomonas aeruginosa* strains had been detected in the Burns' Unit.

(Lowbury *et al.*, 1969; Sykes & Richmond, 1970). So resistant were these organisms that the value of carbenicillin for therapy was severely undermined and alternative treatment had to be instituted. Carbenicillin was withdrawn in the Unit and the resistant pseudomonads disappeared: only sensitive strains could be isolated from the burns. After about six months without resistant strains being isolated and without carbenicillin being used, the antibiotic was administered once more and the resistant strains—many of them with different phage and serotypes than those found before—appeared once more. The resistance was once again R factor mediated. Withdrawal of carbenicillin resulted in a further disappearance of resistance; and it seems quite clear that the rise and fall of the incidence of R factors in the pseudomonads in this Unit reflected the fluctuating selection pressure from an alternating policy for carbenicillin use.

Molecular studies on the nature of the R factors in the two phases of carbenicillin resistance showed that the plasmids involved were extremely similar (Table 9.2). They all had a molecular weight of about 40×10^6, a

Table 9.2. Characteristics of the plasmids isolated from the two phases of carbenicillin resistance in *Pseudomonas aeruginosa* and from the intervening period in ampicillin resistant *Klebsiella aerogenes*

Plasmid	Species of isolation	Molecular weight	G + C of plasmid DNA (%)	Marker pattern	Phase of isolation
RP1	*P. aeruginosa*	40×10^6	60	Ap Km Ne Tc	I
RP2	*P. aeruginosa*	40×10^6	60	Ap Km Ne Tc	I
RK1	*K. aerogenes*	40×10^6	60	Ap Km Ne Tc	Intermediate
RP9	*P. aeruginosa*	40×10^6	60	Ap Km Ne Tc	II

Abbreviations: Ap, resistance to ampicillin and carbenicillin; Km, kanamycin resistance; Ne, neomycin resistance; Tc, tetracycline resistance.

G + C content of the plasmid DNA of about 60 %, a contour length under the electron microscope of about 18 μ and a marker pattern AP Ne Km Tc. Furthermore the plasmids all had a similar base sequence to their DNA in that they hybridized with one another virtually completely (Igram, Richmond & Sykes, 1973). Such a close similarity in properties was not detected in other groups of R factors isolated in other circumstances, and it seemed probable that the same R factor had infected strains of *P. aeruginosa* to produce an evolutionary change during two periods separated by about six months.

Since no carbenicillin resistant pseudomonads were found in the Unit while this antibiotic was withheld, the question arose as to where the reservoir of this particular R factor might be. After a prolonged search, strains of *Klebsiella aerogenes* were found with an R factor indistinguishable from that found in *P. aeruginosa* and there was good evidence, in one patient at least, that the *K. aerogenes* carrying the relevant R factor appeared in his burns before the plasmid was detected in *P. aeruginosa*. In this case the circumstantial evidence for the transfer of this particular R factor (RP1) from *K. aerogenes* to *P. aeruginosa* was strong. Moreover the pattern of events suggests that the plasmid in question was maintained in *Klebsiella* strains in the Unit—possibly in the gut of patients—by selection pressure afforded by ampicillin use (Ingram, Richmond & Sykes, 1973). The β-lactamase gene gives resistance to ampicillin in enteric species but this point is irrelevant for *P. aeruginosa* as this antibiotic is inactive against these strains. Ampicillin therefore selects R factors in enteric bacteria while carbenicillin applies pressure to both enterics and *P. aeruginosa* (Lowbury, Babb & Roe, 1972).

The ability of certain plasmids—notably the RP plasmids—to transfer antibiotic determinants successfully between *P. aeruginosa* and a wide range of other Gram-negative genera (Sykes & Richmond, 1970; Datta *et al.*, 1971;

Olsen & Shipley, 1973) raises the question as to whether all genes might be expected to be transferable in this way or whether the phenomenon is restricted to the nature of the gene products concerned. For example, might one expect the genes coding for *Pseudomonas* ribosomal proteins, or DNA or RNA polymerases or complex metabolic pathways to be able to repair appropriate lesions in *Escherichia coli*, or *vice versa*? At this stage we can only guess. Certainly the genes for type IIIa β-lactamase synthesis, for neomycin phosphorylation and probably also for other aminoglycoside modifying enzymes can express themselves equally well in members of these two species, and equally certainly the penicillin resistance due to the β-lactamase gene can be used to select for RP1 or RP4 transfer in a wide range of genera—for example in *Salmonella, Proteus, Enterobacter, Acinetobacter, Azotobacter, Vibrio, Rhodospirillum, Rhodopseudomonas*, and *Rhizobium*, (Sykes & Richmond, 1970; Datta *et al.*, 1971; Olsen & Shipley, 1973). But these genes specify the synthesis of enzymes which do not, apparently, form part of any integrated series of enzyme reactions, and consequently the restrictions that need to be placed on their location in the cell may be less than would be the case with ribosomal proteins, for example. It is certain that the ribosome is a complex organelle where the component proteins probably have to interact with great precision (Nomura, 1970). In this case it seems that there might be restrictions to the successful transfer across intergeneric boundaries, not because the DNA in question did not survive in the recipient and express itself but because the protein product of the gene would not fit accurately enough into its functional role in the cell. In between these extremes, one might expect certain metabolic genes, which do not seem to have such a closely integrated function as some of the examples already mentioned, to be expressed satisfactorily in a wide range of hosts.

2. ENZYMIC VARIETY AND EVOLUTION

When we look more closely at the enzymes themselves we have to accept that a very large variety of enzymes exist as separate entities, and we can ask several sorts of question about their origins. If we look at the very recent history of a single enzyme we might be able to say that the parent strain acquired the gene for the enzyme in a fully evolved state from an outside source. We have already discussed how this might take place by the agency of a transmissible plasmid or by transduction from the same or a related species. However, such transfer only amounts to the sharing of existing information over a wider range of microorganisms. Can we explain the presence of enzyme activities which did not exist previously in any organism?

If we ask this question we could go back a long way and consider the primitive catalysts and how they developed into the type of enzyme we now recognize. This is not open to experimental approach in any obvious way but some clues can be found from comparative studies of the sequences of known enzymes. The classification of enzymes by physiological function in the cell gives a way of choosing enzymes for comparison from a wide variety of sources. Broadly speaking, there are very few sorts of enzymes in spite of the enormous array of different specificities and different metabolic pathways. This was recognized by the Enzyme Commission of the International Union of Biochemistry in 1961 who drew up a list of six main classes of enzymes grouped together by the types of reaction catalysed. The position is therefore something akin to the number of stories that can be told. It is said that there are only a few main plots and that these can be recognized in all the novels, plays and stories that are written. Can one say that among the enzymes there are only a few main patterns of molecular structures and from these have arisen all the variety of specific enzymes that can be isolated?

Recent studies on protein sequences suggest that the early enzymes may have got bigger by a process of partial duplication and that this could have continued, to produce an enzyme with both a catalytic site and a regulatory site (Engel, 1973). Duplication of the gene itself could allow the two gene copies to acquire different mutations and to evolve into enzymes with separate and distinct functions. By this time comparative studies of proteins have become an important part of evolutionary theories and it is known that some proteins, like cytochrome c, are remarkably similar in closely related animal species and that more differences appear when comparisons are made between species which are biologically widely separated (Smith, 1970). For this very important molecule it is also known that part of the sequence is invariant in cytochromes c isolated from a wide range of plant and animal sources. The number of enzyme sequences known is constantly increasing and it is now clear that enzymes of similar function may have similar amino acid sequences around the active sites irrespective of their origins (Hartley, 1974). If two enzymes carrying out similar functions are related, it would be expected that there would be regions of considerable homology around the active centre, and that there would be some divergences directly related to the differences in their specificities, and others which need have no obvious basis. If the two enzymes had evolved separately, then much larger differences would be expected between them. There are very few complete sequences of *Pseudomonas* proteins but Ambler and Wynn (1973) report surprisingly large differences between the sequences of cytochromes c-551 isolated from three species of *Pseudomonas*. If this is a general characteristic then enzymes carrying out the same biological functions in different *Pseudomonas* species might show similarly large differences in sequences but it should still be possible to determine which were the homologous proteins and which were not. The sequences of cytochromes c-551 from eight different isolates of *P. aeruginosa* were identical showing the constancy of the protein in this species (Ambler, 1974).

Comparison of *Pseudomonas* enzymes that carry out similar reactions in different pathways within the same strain, could shed some light on whether two enzymes have a common origin and could have arisen by gene duplication and divergence. A start has been made on this, and Ornston and colleagues have shown that there is considerable similarity between two pairs of corresponding enzymes in the two branches of the β-ketoadipate pathway in *P. putida*. The two lactonizing enzymes, muconate lactonizing enzyme and carboxymuconate lactonizing enzyme (Figure 7.10) are similar in molecular weight and *N*-terminal sequences. The next two enzymes, muconolactone isomerase and carboxymuconolactone decarboxylase, also resemble each other, and lead to the production of β-ketoadipate from the catechol and protocatechuate branches of the pathway respectively (see Chapter 8 and Meagher & Ornston, 1973; Parke, Meagher & Ornston, 1973; Patel, Meagher & Ornston, 1973). There are many other candidates for comparisons of this sort and it will be of very great interest to see if the complexity of the pathways for the metabolism of aromatic compounds by pseudomonads can be traced to a common archetype.

One particularly interesting suggestion has been made by Jeffcoat and Dagley (1973). They discuss the problem which arises in attempting to work out whether or not catabolic pathway enzymes could have evolved by retrograde evolution as first postulated by Horowitz (1945; 1965). One of the main objections has always been that the previous enzyme of a metabolic sequence often catalyses a totally different type of reaction. Jeffcoat and Dagley (1973) suggest that the important characteristic is not the reaction catalysed but the mechanism of action of the enzymes. In the degradation of D-glucarate by *Escherichia coli* and *Klebsiella aerogenes* the action of a hydrolyase is followed by an aldolase (Figure 9.1a). They consider that the mechanism of both enzymes is the same, requiring Mg^{2+} ions and initiated by an electron shift with release of a proton. In *P. acidovorans* there is no aldolase and the second step is a decarboxylation (Figure 9.1b). The glucarate hydrolyases of *K. aerogenes* and *P. acidovorans* have been purified and both have slight aldolase activity although only *K. aerogenes* employs the aldolase route. Jeffcoat and Dagley (1973) suggests that the weak aldolase activity of glucarate hydrolyase makes it a suitable condidate for gene duplication and that mutational divergence could have produced a suitable aldolase for the pathway. It would be interesting to test other enzymes which follow one another in a metabolic sequence for weak activities of this sort and to compare the properties of the enzyme proteins.

The types of mutational change that can occur in a protein and render it inactive as an enzyme are now well known. The elegant studies of Yanofsky and his colleagues showed for example, that the substitution of glutamic acid or arginine for glycine in the α-subunit of tryptophan synthetase was sufficient to abolish enzyme activity although a protein was produced with

348

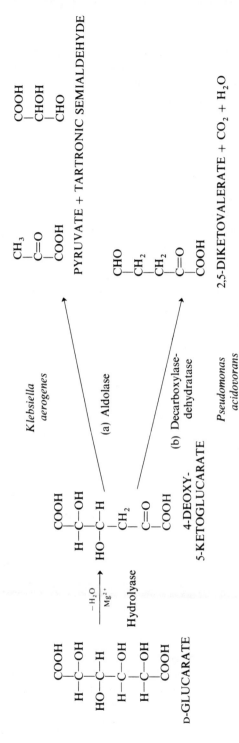

Figure 9.1. Metabolic pathways for D-glucarate in *Pseudomonas acidovorans* and *Klebsiella aerogenes* (after Jeffcoat & Dagley, 1973)

enough resemblance to the active enzyme to give cross-reaction with the antiserum (Yanofsky, 1964). Enzyme activity may be abolished by other mutations including frameshift, nonsense or deletion, where no recognizable fragments of the original protein need be produced. Could we expect any of these mutations to lead to an enzyme with new activity or, in other words, is positive mutation the same sort of process as negative mutation or is it more complex?

Koch (1972) has a rather pessimistic view of the value of single site mutations for 'improving' an enzyme. He suggests that after a certain stage in evolution it would be essential for gene duplication to occur and for one gene copy to acquire multiple mutations to get significant changes. He stresses the importance of gene duplication followed by a period during which the redundant gene lies dormant acquiring mutations. One mechanism he suggests for this is for an intervening period of untranslatable intermediates which could result from the presence of a nonsense mutation. The original gene could continue to carry out its previous function while the second gene, having acquired the requisite amount of mutational change, could then become active again by a supression of the mutation which had rendered it untranslatable. It is difficult to see now this concept could be explored by experiment.

If one takes the view that evolution of enzymes is still continuing then it should be possible to set up an experimental system and observe evolution in action and find out what sort of mutations are occurring (Clarke, 1974). The *Pseudomonas* species are obvious subjects for such a study since, not only are they known to have a vast range of biochemical activities, but they have long been a problem to taxonomists because of their variability. Tests for nutritional characters give clear cut results under standard conditions, but if cultures are incubated for longer periods then positive reactions may be obtained from strains originally considered to be negative for a particular substrate. Indeed, Stanier (1953) remarked of the pseudomonads that, 'The taxonomists' misfortune is the evolutionists' opportunity'.

Many of the studies made on the biochemical pathways of pseudomonads have been done with strains isolated from enrichment culture. After the strain has been isolated, the metabolic pathway is worked out and the characteristics of the enzymes are noted. However, it might be asked whether the enzymes were there before they were looked for, or whether the act of looking for them allowed them to be selected. Some of the recorded sources of cultures in the collection examined by Stanier and colleagues (1966) indicate that this may not be altogether an idle question. A culture described as isolated 'from clay suspended in kerosene for three weeks', may have acquired rather better enzymes for metabolizing kerosene than it possessed at the start of this isolation procedure. With enrichment culture, the act of isolation may also be a process of hastening evolution. It is not obvious how to set up controls to test this hypothesis.

What kind of precursor would be expected for an enzyme with new activity? Even if it existed, there is no way to watch a protein with no enzyme activity evolve into a useful enzyme, and indeed there is no reason to suppose that such a blank protein would exist although an 'all-purpose enzyme precursor' is an attractive proposition. A report of a new β-galactosidase in *Escherichia coli* (Campbell, Lengyel & Langridge, 1973) makes it clear that this new enzyme could not have been derived from the well-known z gene of the *lac* operon but from some other gene of unknown function. What is more promising than looking for a haphazard emergence of new enzyme activity, is to take an enzyme which has well-characterized properties and see if it can be made to evolve a different substrate specificity or to carry out a modified catalytic reaction. A catabolic enzyme offers less problems than a biosynthetic one since changes in activity need not interfere with the central metabolism of the cell. It is more than likely that a catabolic enzyme would be under induction and repression controls and some way will be needed for lifting this restriction in order to study the evolution of the enzyme itself. We will return later to the problem of enzyme regulation.

An enzyme which has proved ideally suited for experimental evolution is the aliphatic amidase which enables *P. aeruginosa* to grow on acetamide and propionamide (see Chapter 8). The specificity of the wild-type enzyme is fairly narrow and the rates of hydrolysis of formamide, acetamide, propionamide and butyramide under standard conditions are in the ratio $0.1:1.0:3.0:0.2$. The aliphatic amides provide a very suitable range of potential substrates since evolutionary changes in the enzyme allow *P. aeruginosa* to utilize amides of longer chain length or with substituents either in the side-chain or on the amido nitrogen group. The first altered enzyme, B amidase, was obtained by a single mutational step from a constitutive mutant which itself produces wild-type A amidase (Brown, Brown & Clarke, 1969). The A and B amidases differ in electrophoretic mobility and specific activities under standard conditions and while the apparent K_m of A amidase for butyramide is 500 mM, that of B amidase is 70 mM. From mutant B6, producing B amidase, were obtained a series of mutants producing V-type amidases which hydrolyse valeramide at rates sufficient for growth to occur. Further mutational steps give mutant enzymes which enable phenylacetamide to be utilized (Figure 9.2). These successive changes in the enzyme protein alter the relative affinities for the amides, the specific activities and in some cases produce marked alterations in thermal stability. Of five Ph amidases' produced by phenylacetamide-utilizing mutants belonging to different phenotypic classes, two are relatively stable to heating at 60 °C for 10 minutes, while the other three lose activity although at different rates. These mutants give cross-reactions with antisera to the wild-type amidase

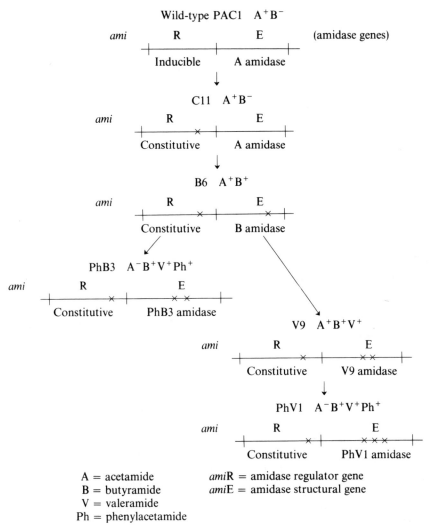

A = acetamide   *ami*R = amidase regulator gene
B = butyramide   *ami*E = amidase structural gene
V = valeramide
Ph = phenylacetamide

Figure 9.2. Derivation of some of the *Pseudomonas aeroginosa* mutants which are able to utilize butyramide, valeramide or phenylacetamide

and are thought to differ between themselves with respect to one or two amino acids (Betz & Clarke, 1972).

One interesting new enzyme obtained in a single step from a constitutive mutant producing the wild-type enzyme is AI3 amidase, which enables growth to occur on acetanilide (*N*-phenylacetamide). This mutant enzyme differs from A amidase only in the substitution of an isoleucine for a threonine residue, and has a high specific activity (Brown & Clarke, 1972). It is quite

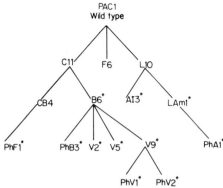

Figure 9.3. Family tree of *Pseudomonas aeruginosa* amidase mutants. Those producing mutant enzymes are marked with an asterisk. Details in Table 9.3

clear, in this instance, that a single site mutation has been able to confer new properties on an enzyme protein and that a positive mutation has allowed the utilization of a substrate not available to the parent strain.

Figure 9.3 shows the family tree for the evolution of the mutant amidases and Table 9.3 shows the range of substrate specificities. The shift in substrate specificity which allows mutant PhB3 to utilize phenylacetamide has resulted

Table 9.3. *Pseudomonas aeruginosa* amidase mutants shown in Figure 9.3

Strain no.	Series no.	Phenotype	Genotype
PAC1	Wild type	Ind A^+B^-	*amiR*$^+$*amiE*$^+$
PAC111	C11	Con A^+B^-	*amiR*11*amiE*$^+$
PAC153	F6	F-ind A^+B^-	*amiR*43*amiE*$^+$
PAC142	L10	Con A^+B^+ But-r Crp-r	*amiR*33*amiE*$^+$*crp*-7
PAC128	CB4	Con A^+B^+ But-r	*amiR*11,37*amiE*$^+$
PAC351	B6	Con A^+B^+	*amiR*11*amiE*16
PAC366	AI3	Con $A^+B^-AI^+$Crp-r	*amiR*33*amiE*56*crp*-7
PAC326	LAm1	(Ind)A^-(Crp-r)	*amiR*33*amiE*34*crp*-7
PAC392	PhF1	Con $A^{(+)}B^+V^+Ph^+$	*amiR*11,37*amiE*82*crp*-7
PAC377	PhB3	Con $A^-B^+V^+Ph^+$	*amiR*11,*amiE*16,67
PAC353	V2	Con $A^-B^+V^+$	*amiR*11*amiE*16,23
PAC356	V5	Con $A^-B^+V^+$	*amiR*11*amiE*16,26
PAC360	V9	Con $A^+B^+V^+$	*amiR*11*amiE*16,30
PAC391	PhA1	Con $A^{(+)}B^+V^+Ph^+$	*amiR*33*amiE*34,81*crp*-7
PAC388	PhV1	Con $A^-B^+V^+Ph^+$	*amiR*11*amiE*16,30,78
PAC389	PhV2	Con $A^{(+)}B^+V^+Ph^+$	*amiR*11*amiE*16,30,79

A = acetamide; B = butyramide; V = valeramide; Ph = phenylacetamide; Ind = inducible; Con = constitutive; But-r = Resistant to butyramide repression; Crp-r = Resistant to catabolite repression; + = Good growth; (+) = Trace growth; − = No growth; *amiR* = amidase regulator gene; *amiE* = amidase structural gene.

in a loss of activity at the other end of the range, since it is now unable to grow on acetamide. If the gene for PhB3 amidase could be introduced into the wild-type strain we would have an organism producing two amidases of different specificities which could be described as 'acetamidase' and 'phenylacetamidase'. It is interesting that some strains of *P. putida* are able to grow on both acetamide and phenylacetamide (Betz & Clarke, 1973).

P. putida, unlike *P. aeruginosa*, can metabolize phenylacetate so that while *P. aeruginosa* mutant PhB3 utilizes phenylacetamide solely as a nitrogen source, *P. putida* utilizes it as both nitrogen and carbon source. The ability to make use of the aromatic moiety would make it advantageous for *P. putida* to have evolved a phenylacetamidase, while it would be of little advantage to *P. aeurginosa* since the limitation for growth in the natural environment is the availability of carbon compounds and not a shortage of nitrogen. The results of this exercise in experimental evolution show that very few mutational steps can produce large changes in substrate specificities leading to the acquisition of new phenotypes. The evolution of phenylacetamidases in *P. aeruginosa* in the laboratory may have mimicked that which occurred in nature in another *Pseudomonas* species.

We have left out of the discussion so far any consideration of the regulation of amidase synthesis. As inducers, the activities of the amides are acetamide > propionamide > formamide. Butyramide and some of the higher amides are not inducers, but compete at the inducer binding site and repress synthesis in inducible and a many constitutive strains. A mutation to increased inducibility by formamide leads to better growth on succinate + formamide plates (Brammar, Clarke & Skinner, 1967). Mutations to constitutivity can lead to growth on butyramide, but only if they are accompanied by resistance to butyramide repression (Brown & Clarke, 1970). The response of *P. aeruginosa* to butyramide is curious and may be fortuitous. Although the activity of the wild-type amidase towards this amide is low, it can be utilized by those constitutive strains which are resistant to butyramide repression and butyrate is readily utilized as a carbon source. The acetamidase of *P. putida* is very similar in substrate specificity to that of *P. aeruginosa* but butyramide can be utilized by *P. putida* A87 since in this strain butyramide is able to induce amidase synthesis (Clarke, 1972). As yet no butyramide-inducible mutants of *P. aeruginosa* have been isolated.

Changes in the regulation of an enzyme can therefore lead to a new phenotype with the utilization of a substrate not available to the parent strain, and the most commonly occurring mutation of this class is a mutation to constitutivity. All the amidase altered enzyme mutants discussed earlier were constitutive and this is inherently extravagant from the point of view of cell economy. If the mutants with the new enzymes were to be able to compete in the natural environment it would probably be essential for them to acquire another mutation allowing them to be induced by their new substrate.

Indeed, a major factor in maintaining the stability of the biochemical pheno-type of any species is probably the dependence of catabolic enzymes on both substrate and inducer specificities, and the improbability of both mutations occurring at the same time.

A fascinating problem is the evolution of the regulatory systems themselves. In Chapters 7 and 8 we discussed the ways in which the regulatory patterns for both catabolic and biosynthetic pathways were characteristic for the various bacterial groups and that the regulator genes probably appeared later than those for the enzymes they control. Nothing is yet known about the origins of regulator genes. Since substrate and inducer specificities overlap, it is not unreasonable to speculate that the regulator gene for an inducible enzyme may have arisen from a duplication of the gene for the enzyme. No regular proteins have yet been isolated in *Pseudomonas* and it has not yet been possible to compare any of the sequences of regulator proteins and enzymes in *Escherichia coli*.

The acquisition of new metabolic capacity by a constitutive mutation for an enzyme possessing weak activity for the novel growth substrate, is well-known among the enterobacteria. Wood and Mortlock and their colleagues have provided many examples of this for the utilization of 'unnatural' pentoses and pentitols by *Escherichia coli* and *Klebsiella aerogenes* (Mortlock & Wood, 1964; Camyre & Mortlock, 1965; Oliver & Mortlock, 1971). A constitutive strain can produce very large amounts of a normally inducible enzyme. It is common for constitutive enzyme to comprise 2 to 5 % of the total cell protein and some mutants, including *P. aeruginosa* amidase mutants, may synthesize a single enzyme at a rate which enables it to reach 10 % or more of the total cell protein. Still higher values have been reported for some enzymes produced by mutants generated in continuous culture but this is usually due to multiple gene copies which are readily lost when the selection pressure is removed (Horiuchi, Tomizawa & Novick, 1962).

We have seen that some *Pseudomonas* enzymes have fairly broad specifi-cities and it might be possible to select constitutive mutants by using one of the poorer substrates in the way that butyramide is used as a selective agent for constitutive mutants of *P. aeruginosa* amidase. This would not only be a useful trick for mutant selection for genetic studies but could be a way of selecting mutants producing high yields of an enzyme required for industrial use. Another of the regulatory systems controlling the amount of enzyme synthesized, is catabolite repression and the selection of catabolite repression-resistant mutants on selective media again offers a way of increasing enzyme yield.

From the point of view of getting a particular enzyme, or group of enzymes, made at the maximum rate it would not matter which of the many types of mutation which lead to decreased catabolite repression was selected. One rather special type of selection for increased enzyme yield is to devise a method for getting 'super promotor' mutants. This was achieved for *P. aeruginosa* amidase by an indirect method. A large number of mutants with an amidase-defective phenotype were selected and it was predicted that among these would be some which would have promotor site mutations. It was also predicted that some of the promotor mutants would be able to revert to a positive phenotype by a second mutation in the promotor region which would make them simultaneously resistant to catabolite repression. Selection of catabolite repression-resistant resistants gave a mutant strain which had an exceptionally high rate of amidase synthesis (a possible 'super-promotor mutant'). This method was possible because the characteristics of the enzyme and its regulator system were fairly well understood, but this general approach should be possible with many other enzymes. By screening a large number of mutants with a negative phenotype it should be possible to get promotor site mutations which result in higher rates of enzyme synthesis.

If we consider the changing environment of this planet we can think of biological evolution in terms of adaptations to new organic molecules. After adapting themselves for millennia to the degradation of compounds which arose by biological activity in the natural environment, the world of today offers the microbial populations compounds which come straight from the laboratory and the chemical plant. We have discussed already one group of such compounds, the semi-synthetic antibiotics such as ampicillin and penicillin. Some of the enzymes which attack the naturally occurring penicillins were able to be changed to attack these modified penicillins. But can we predict which of the compounds produced by the organic chemist would be biodegradable and in fact how much degradability do we want? Attention has been focussed in recent years on the persistance in the soil of certain pesticides and the accumulation of others in the food chain. We can now see by hindsight why some compounds are more readily degraded than others. Dagley (1972) points out that where a compound can follow a pathway already existing for a natural product the new pesticide molecule is going to be degraded at a reasonable rate. He suggests that 2,4-D, (2,4-dichloro-phenoxyacetic acid) is just the sort of herbicide which would have been designed on the basis of our current knowledge of the aromatic pathways. If we consider that the straight-chain aliphatic hydrocarbons and fatty acids are much more readily degraded than the branched-chain isomers, then it is obvious that the latter are less suitable as the basis for detergent molecules, but it took several years of problems caused by the non-biodegradability of the branched-chain detergents before this was accepted.

One group of herbicides comprises the chlorinated aniline-based compounds which include phenylcarbamates, phenylureas and acylanilides. Some of these are known to be degraded at least partially by soil micro-organisms. *P. striata* hydrolyses phenylcarbamates and acylanilides (Kearney

& Kaufman, 1965). However, an acetanilide utilizing mutant of *P. aeruginosa* was obtained by selective mutation of the naturally occurring aliphatic amidase and it is possible that more effective enzymes for attacking these novel organic molecules could be obtained by laboratory evolution.

Some enzymes which metabolize chemicals which have been introduced fairly recently into the environment appear to be easily lost on subculture. This could be due to the enzymes being located on a plasmid which is lost from the cell when alternative nutrients are available. A number of plasmids carrying genes for catabolic enzymes were discussed in earlier chapters. One of which is relevant to a consideration of the degradation of newly introduced organic molecules is the plasmid carrying genes for the breakdown of the detergent molecule, benzene sulphonate (Johnston & Cain, 1973). Careful isolation of cultures from water and soil where effluents might have introduced such compounds, would probably reveal many more degradative systems of this sort.

In summary we can say that there is no reason to fear that the evolution of *Pseudomonas* is at an end. We have suggested ways in which it may continue in the natural environment and ways in which it might be encouraged in the laboratory. In nature the plasmids seem to present the most potent source of evolutionary change. Not only may they transfer both extra-chromosomal and chromosomal genes within species of *Pseudomonas*, certain of them can also transfer on a much wider scale. Their particular potential—as is the case with all processes of gene transfer—is to pass blocks of genetic information already refined by selection pressure elsewhere from strain to strain, species to species and sometimes genus to genus. Evolutionary change may therefore be expected to occur by some substantial steps as well as by the more conventional selection of point mutations, many of which may be deleterious unless protected by gene duplication (Koch, 1972). The successful transfer of blocks of genes—by whatever means—will be limited by a number of factors: is there a vector that will bridge the gap in question? Will the incoming DNA survive in the recipient or will it be detected and destroyed by the restriction systems of the recipient (see Boyer, 1971, for a review). Will the gene products once transferred and expressed be able to fulfil their potential physiological role in the recipient? Could we expect to be able to transfer the gene for altered enzymes such as those for the mutant amidases through a wide variety of Gram-negative genera using a plasmid such as RP1 as the vector? So far our knowledge of these points is scrappy. RP1 and RP4 are perhaps the most effective vectors we have in pseudo-monads. The FP plasmids can mobilize chromosomal material but transfer it only within a narrow range of *Pseudomonas aeruginosa* strains, and most of the transducting phages that have been examined so far have more re-stricted range than the FP particles. This very neglected area of research has enormous importance. Not only does it concern those interested in transferring genes between strains to aid their fundamental studies, it also aids those who need to transfer genes for industrial processes. But the underlying theme that we have tried to stress in this book is that it allows an examination of the fundamental processes of evolution in microbial species.

3. BIBLIOGRAPHY

Ambler, R. P. (1974) *Biochemical Journal*, **137**, 3.
Ambler, R. P. & Wynn, M. (1973) *Biochemical Journal*, **131**, 485.
Asheshov, E. H. (1969) *Journal of General Microbiology*, **59**, 289.
Betz, J. L. & Clarke, P. H. (1972) *Journal of General Microbiology*, **73**, 161.
Betz, J. L. & Clarke, P. H. (1973) *Journal of General Microbiology*, **75**, 167.
Boyer, H. B. (1971) *Annual Review of Microbiology*, **25**, 153.
Brammar, W. J., Clarke, P. H. & Skinner, A. J. (1967) *Journal of General Microbiology*, **47**, 87.
Broda, P., Beckwith, J. R. & Scaife, J. (1964) *Genetical Research, Cambridge*, **5**, 489.
Brown, J. E., Brown, P. R. & Clarke, P. H. (1969) *Journal of General Microbiology*, **57**, 273.
Brown, J. E. & Clarke, P. H. (1970) *Journal of General Microbiology*, **64**, 329.
Brown, P. R. & Clarke, P. H. (1972) *Journal of General Microbiology*, **69**, 287.
Campbell, J. H., Lengyel, J. A. & Langridge, J. (1973) *Proceedings of the National Academy of Sciences, U.S.A.*, **70**, 1841.
Camyre, K. P. & Mortlock, R. P. (1965) *Journal of Bacteriology*, **90**, 1157.
Clarke, P. H. (1972) *Journal of General Microbiology*, **71**, 241.
Clarke, P. H. (1974) *Symposium of the Society of General Microbiology*, **24**, 183.
Dagley, S. (1972) in *Degradation of Synthetic Organic Molecules in the Biosphere*, National Academy of Sciences, Washington, U.S.A., p. 338.
Datta, N., Hedges, R. W., Shaw, E. J., Sykes, R. B. & Richmond, M. H. (1971) *Journal of Bacteriology*, **108**, 1244.
Engel, P. C. (1973) *Nature, London*, **241**, 118.
Hartley, B. S. (1974) *Symposium of the Society for General Microbiology, Microbial Evolution*, **24**, 151.
Horiuchi, T., Tomizawa, J. & Novick, A. (1962) *Biochimica et Biophysica Acta*, **55**, 152.
Horowitz, N. H. (1945) *Proceedings of the National Academy of Sciences, U.S.A.*, **31**, 153.
Horowitz, N. H. (1965) in *Evolving Genes and Proteins*, (Eds. V. Bryson & H. J. Vogel), Academic Press Inc., N.Y.
Ingram, L. C., Richmond, M. H. & Sykes, R. B. (1973) *Antimicrobial Agents and Chemotherapy*, **3**, 279.
Jeffcoat, R. & Dagley, S. (1973) *Nature, New Biology*, **241**, 186.
Johnston, J. B. & Cain, R. B. (1973) *Federation of European Biochemical Societies, Dublin, Abstract*, 137.
Jones, D. & Sneath, P. H. A. (1970) *Bacteriological Reviews*, **34**, 40.
Kearney, P. C. & Kaufman, D. D. (1965) *Science*, **147**, 740.
Koch, A. L. (1972) *Genetics*, **72**, 297.
Leth Bak, A., Christiansen, C. & Stenderup, A. (1970) *Journal of General Microbiology*, **64**, 377.
Lowbury, E. J. L., Babb, J. R. & Roe, E. (1972) *Lancet*, **ii**, 941.

Lowbury, E. J. L., Kidson, A., Lilly, H. A., Ayliffe, G. A. J. & Jones, R. J. (1969) *Lancet*, **ii**, 448.

Meagher, R. B. & Ornston, L. N. (1973) *Biochemistry*, **12**, 3523.

Mortlock, R. P. & Wood, W. A. (1964) *Journal of Bacteriology*, **88**, 838.

Nomura, M. (1970) *Bacteriological Reviews*, **34**, 228.

Oliver, E. J. & Mortlock, R. P. (1971) *Journal of Bacteriology*, **108**, 287.

Olsen, R. H. & Shipley, P. (1973) *Journal of Bacteriology*, **113**, 772.

Parke, D., Meagher, R. B. & Ornston, L. N. (1973) *Biochemistry*, **12**, 3537.

Patel, R. N., Meagher, R. B. & Ornston, L. N. (1973) *Biochemistry*, **12**, 3531.

Richmond, M. H. & Wiedeman, B. (1974) *Symposium of the Society of General Microbiology*, **24**, 59.

Smith, E. L. (1970) in *The Enzymes*, Vol. 1. (Ed. P. D. Boyer), 3rd ed., Academic Press, London and New York.

Stanier, R. Y. (1953) in *Adaptations of Micro-organisms*, Symposium of the Society of General Microbiology, (Eds. E. F. Gale & R. Davies), Vol. 3.

Stanier, R. Y., Palleroni, N. J. & Doudoroff, M. (1966) *Journal of General Microbiology*, **43**, 159.

Stanisich, V. & Holloway, B. W. (1971) *Genetical Research, Cambridge*, **17**, 169.

Stanisich, V. & Holloway, B. W. (1971) *Genetical Research, Cambridge*, **19**, 91.

Sykes, R. B. & Richmond, M. H. (1970) *Nature, London*, **226**, 952.

Yanofsky, C. (1964) in *The Bacteria*, Vol. 5, (Eds. I. C. Gunsalus & R. Y. Stanier), p. 373.

Index